Least Cost Analysis of Social Landscapes

Least Cost Analysis of Social Landscapes

ARCHAEOLOGICAL CASE STUDIES

edited by

Devin A. White and Sarah L. Surface-Evans

THE UNIVERSITY OF UTAH PRESS
Salt Lake City

The Defiance House Man colophon is a registered trademark of the University of Utah Press. It is based upon a four-foot-tall, Ancient Puebloan pictograph (late PIII) near Glen Canyon, Utah.

16 15 14 13 12 1 2 3 4 5

LIBRARY OF CONGRESS CATALOGING-IN-PUBLICATION DATA

Least cost analysis of social landscapes : archaeological case studies / edited by Devin A. White and Sarah L. Surface-Evans.
p. cm.
Includes bibliographical references and index.
ISBN 978-1-60781-171-8 (cloth : alk. paper)
1. Landscape archaeology—Costs—Case studies. 2. Social archaeology—Costs—Case studies. 3. Cultural landscapes—Case studies. 4. Cost accounting—Case studies. 5. Cost effectiveness—Case studies. 6. Archaeology—Economic aspects—Case studies. 7. Archaeology—Data processing—Case studies. 8. Geographic information systems—Case studies. I. White, Devin Alan. II. Surface-Evans, Sarah L.
CC77.L35L43 2012
930.1—dc23

2011032971

Printed and bound by Sheridan Books, Inc., Ann Arbor, Michigan.

In memory
of our ancestors
and in commitment
to future generations

Contents

Figures

Color Plates

Following page 187

Tables

CHAPTER 1

An Introduction to the Least Cost Analysis of Social Landscapes

Sarah L. Surface-Evans and Devin A. White

There is no shortage of books discussing the role and use of Geographic Information Science (GIS) in archaeological inquiry in recent years (see Allen et al. 1990; Lock 2000; Lock and Stancic 1995; Mehrer and Wescott 2006; Snead et al. 2010; Wescott and Brandon 2000; Wheatley and Gillings 2002). Certainly there are a growing number of archaeologists applying GIS technologies to their research problems and questions. Advances in the types of geospatial data available to the archaeologist, analytical techniques to extract information from them, the growth of GIS-centric interdisciplinary collaboration, and the decreasing cost of computing power allow us to ask ever more sophisticated questions and develop increasingly elaborate models on numerous aspects of past human behavior. Least cost analysis (LCA) is one such avenue of inquiry.

This edited volume presents a series of case studies concerning the intersection of archaeology and LCA modeling at the practical, methodological, and theoretical levels. While least cost studies are not new to the social sciences in general, LCA is a relatively new area of inquiry in archaeology. The contributions in this book showcase the richness and diversity of its application to archaeological questions, especially within the past five years. The case studies also highlight the challenges that come with injecting geospatial technology into the archaeological research process.

This volume is designed to be a guidebook for the archaeologist interested in using LCA to answer behavioral questions. While it would be impossible to provide something for everyone, the goal of this volume is to present a wide cross-section of practical examples for both novices and experts, situated within ongoing research projects. The contributions encompass a broad spectrum of cultural complexity, time period, geographic extent, technical skill, and theoretical focus.

The reader will quickly see that, apart from following a few basic ground rules and adopting some similar solutions to frequently encountered technical issues, the authors in this book perform LCA in very different ways. This demonstrates that the correct application of LCA greatly depends on the archaeological question that is being addressed. While there is currently no unified technique—and no argument is being made that there should be—there are common modeling issues that researchers must consider. In particular, successful applications of LCA should contain overt links between behavioral questions, archaeological data, and analytical technique. It is imperative that any models of human behavior be based on documented human conditions, constraints, and actions (e.g., Binford 2001, 1980); without this, a model does nothing to help our understanding of the past. Equally important is understanding the limits of a technology and the answers it can provide. LCA is a means to an end, not an end in itself. It is a tool, much like a trowel, which helps archaeologists do their job. As such, it has strengths and weaknesses that must be understood before it can be used properly. Constructive criticism and critical discussion, centered on both methodological and theoretical issues related to the use of LCA in archaeology, are also necessary to advance this burgeoning area of inquiry.

This book offers the reader a sample of some of the various avenues of research with LCA that are currently ongoing in the archaeological community. The contributions demonstrate that even simple LCA applications can be used to explore sophisticated theoretical research questions that are not usually thought to be within the domain of GIS. The remarkable diversity of archaeological cultures and social questions explored by the book's contributors is noteworthy and suggests the influential nature of LCA with respect to investigating human behavior.

The chapters are arranged in a technological and theoretical progression from beginner applications (complete with the kinds of challenges and pitfalls alluded to by the contributors to Part 4, who provide constructive criticism) to more complex applications that require advanced programming skills and years of experience. Ideally, novice readers can start at the beginning of the book and work their way through to the end as they became more technologically and theoretically proficient, but readers at all skill levels will be able to learn something from every chapter.

1.1. What Is Least Cost?

Archaeologists and other social scientists have applied the concept of least cost to modeling human behavior for a variety of social and economic circumstances (see Anderson and Gillam 2000; Clarke 1977; Hunt 1992; Kelly 1992; Vita-Finzi and Higgs 1970). At the most general level, LCA assumes that humans will tend to economize many aspects of their behavior, encompassing everything from speech to movement, a phenomenon referred to by Zipf (1949:7) as the Principle of Least Effort.

One area where this behavioral assumption can be seen is in how people interact with the landscape. Humans will attempt to limit the costs of traveling in a landscape, however cost is defined, by using all available knowledge to formulate an appropriate path. The tendency to reduce cost results in patterned behaviors of interaction with the landscape: as cost increases, the likelihood of interaction decreases. In other words, people are more apt to travel to and interact with areas with greater ease of access. This phenomenon includes both social and physical interaction with the landscape. While there are many reasons that humans do not always choose to limit costs (see Branting this volume), this assumption is a heuristic device for developing baseline models of patterned movement within a landscape that can be used as part of a comparative framework to further explore complex archaeological questions.

The utility of the LCA approach is that it allows archaeologists to formulate hypothetical networks of travel and interaction for cultures long past. In other words, LCA is a means of reconstructing extinct connections between peoples and places, connections that are at the heart of many complex social, political, and economic questions of interest to archaeologists. The connections themselves might have physical expression in the form of prehistoric roads or trails, which may or may not still exist. Alternatively, they can also be entirely symbolic in nature, such as ideological connections.

1.2. How Least Cost Analysis Works

LCA addresses a relatively straightforward minimization problem: How does a traveler get from Point A to Point B (within a landscape, or in GIS terms, on the same surface) in a manner that keeps accumulated cost as low as possible? If the cost of traveling some discrete distance (hereafter referred to as a *cell*) is exactly the same for any direction, then the best possible route is a straight line (Figure 1.1).

The solution becomes more difficult when the cost varies from one cell to the next. In Figure 1.1, the total travel cost is twenty units. When lower costs are assigned to certain cells, the most cost-efficient path between A and B changes (Figure 1.2).

In this case, the most efficient path costs seventeen units but does not travel in a straight line. The situation described here highlights two important aspects of least cost analysis: directionality of movement and distance accumulation.

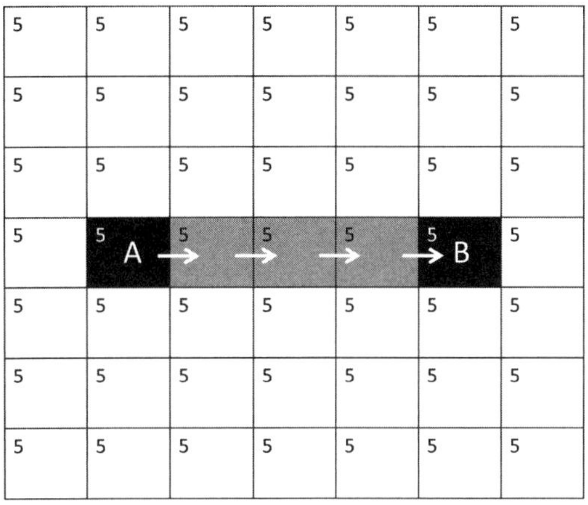

FIGURE 1.1. Movement across a landscape with uniform cost. In this example, each possible move on the given landscape incurs the same arbitrary cost of five units. The most cost-efficient path between A and B is a straight line.

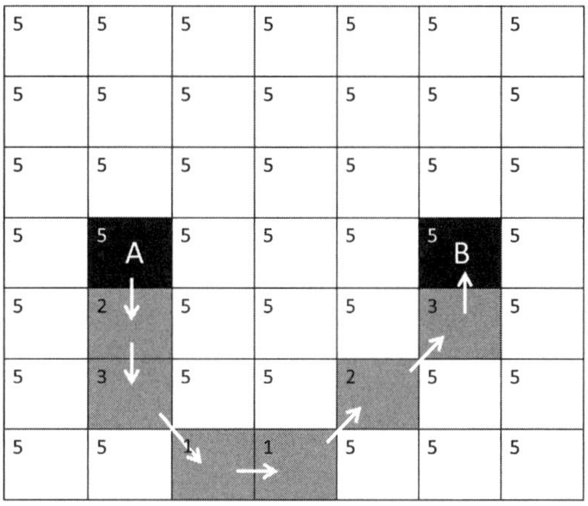

FIGURE 1.2. Movement across a landscape with non-uniform cost. In this example, the cost of movement is lower for some cells, which results in a least cost route between A and B that is not a straight line.

1.2.1. Directionality

So far, we have considered travel into and out of every cell to be equally costly—a concept referred to as *isotropy*. This is not necessarily realistic, depending on what the cost in each cell represents. For example, if the surface represents a set of terrain heights (e.g., a digital elevation model), travel from one cell to the next will be uphill, downhill, or flat. Additionally, travel *into* a cell may be downhill, which implies that travel *out of* the cell is uphill. Each mode of travel will likely incur a different cost depending on the direction of travel. For example, walking downhill is generally easier than walking uphill. The concept of differential cost based on travel direction is referred to as *anisotropy* and is explored by multiple authors in this book (Branting, Kantner, Livingood, Phillips and Leckman, Ullah and Bergin, White). The most tangible outcome of considering anisotropy for building pathways is that the least costly route between two locations, A and B, may be different depending on whether one is moving to B from A or to A from B. In other words, travel direction can substantially drive path morphology.

1.2.2. Distance

Physical distance accumulation, measured in either two or three dimensions, often plays a secondary role in LCA—if it is taken into account at all. However, the cost of travel distance, on top of any other form of cost, is an important consideration when one is working with a surface that is a representation of the real world in some form. In the second example, the least costly route between A and B may still have been a straight line if minimizing accumulated distance was also important to the traveler—especially if the primary cost being minimized did not directly involve physical distance. Twenty units of cost spread over four moves may have been a better choice than seventeen units of cost spread over seven. In reality, the choice of a particular path all comes back to the motivations of the traveler, which can be difficult to approximate archaeologically.

It is also important to note that while the above example relies on the concept of travel across a gridded (raster) surface, there are other ways to connect points on a landscape to one another and find paths between them. The chapters by Branting and White explore some of these alternatives, paying special attention to graph-based networks.

1.2.3. Model Algorithms

In the simple examples of route choice shown in Figures 1.1. and 1.2, the reader can essentially eyeball a good solution to the least cost path problem. However, archaeologists often deal with large landscapes where (1) the costs assigned to each cell vary significantly, (2) barriers to travel may exist, and (3) the task of modeling human movement is very complex. These realities push us toward approaches that involve computer algorithms. While many variations exist, one of two algorithms is generally used to solve least cost problems: Dijkstra and A* (pronounced "A-star"). Both Dijkstra and A* enjoy widespread use within the computer science community and the video game industry.

The Dijkstra algorithm, originally proposed by Dutch computer scientist Edsger Dijkstra (1959), is designed to find the lowest cost routes between a given location and every other location on the surface. Unless directed otherwise, the algorithm will perform an exhaustive search of the entire surface. So if the goal is simply to find the least costly path between two specific locations, the algorithm can be stopped after that condition has been met, which generally speeds up the process. It is a brute-force "greedy" algorithm, in that it is always searching for the least costly way to reach a given location regardless of other possible mitigating factors, such as accumulated travel distance.

A*, proposed by Hart, Nilsson, and Raphael (1968) as an extension to Dijkstra's algorithm, uses a distance-plus-cost heuristic function to determine how it will search the surface and choose the next least costly move. As A* searches, it follows a path of the lowest *known* cost but also keeps a list of alternate paths it could have taken. If it moves in a direction along the current path that is more costly than one of the alternatives in the list, it switches to that one and continues the search. This process is repeated until the goal location is reached. The search space itself is smaller than that used by Dijkstra's algorithm, owing to (1) the focus on travel between two locations and (2) the addition of the heuristic function, which constantly provides estimates of how much cost is involved in traversing the distance between a given intermediate point and the goal. Without the heuristic function, the entire surface would have to be searched, resulting in a process identical to that used by Dijkstra's algorithm.

For the purposes of this book, the main difference between Dijkstra and A* is that Dijkstra may be slower but is guaranteed to return the least costly path every time. A* is generally faster but gains its speed advantage by sacrificing accuracy—the amount of which depends on the specifications made by the researcher. Dijkstra is embedded in all standard GIS software packages because of its ease of implementation, so most archaeologists (including all but one author in this book) conducting LCA have used it. Use of A* within archaeology has been extremely limited, but it shows great potential (see Livingood this volume for an example).

1.3. Modeling Considerations in Least Cost Analysis

As mentioned previously, there is no single or correct way to perform LCA. While the specific approach for LCA modeling and the variables used are largely based on the particular question being considered, there are several general considerations that anyone using least cost modeling must reflect on before proceeding with analysis:

- What variables are necessary to include in the model?
- What data are needed to create the modeling universe?
- How will costs be measured?
- Which software tool should be used to perform the analysis?

1.3.1. Variables

Archaeologists can consider numerous variables that may assist or constrain human movement in a landscape. There are three major classes of variables: environmental, cultural, and physiological. Features of the terrain, such as topography, hydrology, and vegetation, are significant *environmental* factors. For example, rivers may impede or aid travel, depending on whether the culture considered had access to watercraft. Steep bluffs and canyons may hinder movement between areas, while ridges can facilitate travel. Additionally, *cultural* features of the landscape may be important for understanding how people move on a landscape. Trails and traces may ease movement, while territorial borders, distance from/between points of interest, and village locations may place constraints on travel. *Physiological* attributes of human biology may also be considered in the modeling of costs, such as the expenditure of caloric energy (see the chapters by Rademaker et al. and White).

The particular variables incorporated into a LCA model are largely determined by one's research question. While all models must always be recognized as approximations of "reality," the more complexity we can include in a model, the better it can potentially capture a "real-looking" representation of the past. Theoretically, the more variables that can be incorporated into the model, the more realistic the model will be. However, as the contributors focused on theoretical concerns (Branting, Kantner, Anderson) point out, it is easy to overmodel an archaeological problem. Therefore, care must be taken in the choice of variables, how they are constructed, and how they are combined. It is often better to start simple and work toward more complex representations as needed.

1.3.2. Data

Two types of data are necessary for building LCA models in a GIS platform: cultural and environmental. Settlement and land use data from archaeological investigations form the basis of the "cultural landscape" as start and end points for pathways of interaction. While these data are often represented as sites, specific locales, or more general regions on the landscape, a nonsite approach can also be employed. For example, Phillips and Leckman (Chapter 4) use the distribution of ceramic and lithic scatters on the landscape. Whether a site or distributional approach is used, data concerning the cultural landscape must be developed by the archaeologist. A researcher must choose the sites or locales of interest, as well as how best to represent them in digital form, based on his or her social and cultural research questions. As the reader will see, the contributors to this book approach this aspect of model building at a variety of scales—from site level to regional level.

The remaining data needed for creating a modeling universe are mainly the environmental features of the landscape. These can include slope, terrain roughness, hydrological features, vegetative habitats, line-of-sight analyses/viewsheds, distance-to-water maps, surface types, or any other notable environmental feature of the landscape. Again, the environmental features that one chooses to include in a model are largely dependent on the types of questions one asks. For example, if one wants to consider how rivers may have helped or hindered regional transportation networks, a topic taken up by Livingood (Chapter 10), then river hydrology is an important dataset for the LCA model.

Fortunately, the availability of environmental datasets has increased significantly in recent years. There are numerous online clearinghouses in the United States at the federal and state levels where environmental data can be easily accessed and downloaded (e.g., USGS National Map Seamless Server and NRCS Geospatial Data Gateway). National-level datasets are increasingly available for other parts of the world as well, generally through national GIS data clearinghouses (e.g., Natural Resources Canada and the Instituto Nacional de Estadística y Geografía in Mexico). This does not mean, however, that these data are free of error or that they are always compatible with an archaeologist's research goals and interests. Environmental datasets frequently need to be converted and "massaged" to be useful for research purposes. In some cases, available environmental data are not at appropriate scales for archaeological research questions or are simply not available for the region of study. Occasionally, it is necessary to collect environmental data and create necessary data layers from scratch, as was the case for the precipitation data discussed by Nolan and Cook (Chapter 5).

1.3.3. Cost

At a most basic level, there are three different measures of cost available to archaeologists conducting LCA. The most obvious and most commonly considered is cumulative distance, in other words, the total distance that is traveled from one locale to another. Duration or travel time is another measure for cost. Often time and distance measures are roughly equivalent, but not always. A third potential measure of cost is in terms of energetic expenditures, or calories. Energetic costs are rarely equivalent to distance or time measures. For example, the shortest path is probably not the easiest or least costly in terms of energy in a rugged mountainous landscape. Yet another category of measures is more abstract, containing such items as visibility between sites, social distance, and spheres of influence. It is likely that each of these measures of costs may play a role in how people choose to travel, whether they are consciously aware of it or not. Researchers conducting LCA must determine which measure, or combination of measures, makes the most sense for the particular culture and landscape they are examining. As the reader will see, the contributors to this book define cost differently, arriving at their calculations of cost through a wide array of theoretical and technical avenues.

To illustrate the significance of choosing an appropriate measure of cost, let's consider an example from contemporary urban life. In this example, you want to travel from one side of a modest-sized city to another. If traveling by car, you might elect to take an interstate bypass—which is not the most direct path in terms of distance but takes less time than driving through town in stop-and-go traffic. If you are traveling by bike, however, you would probably elect to take the shorter path through town (not to mention that you are prohibited by law from traveling on the interstate on a bike). In this example, your decisions are determined by the technology you have available to you as well as the constraints of an existing transportation network and associated social concerns (law breaking). Past peoples made similar decisions regarding how to travel through a landscape, using the tools at their disposal and their knowledge of the constraints of the landscape that may facilitate or hinder travel.

1.3.4. Software

Least cost analysis is, by its very nature, a computational approach to answering archaeological questions. Although small network graphs and simple cost problems can be solved manually, to do anything more complex or at a larger scale requires access to software that can (1) integrate the various data layers one wishes to use (as well as create new ones, if needed), (2) combine them to produce cost estimates, and (3) process the estimates to create the desired path outputs. These may seem like straightforward tasks, but they generally require robust GIS software such as ArcGIS (commercial) or GRASS (open source). For most archaeological applications, these "off the shelf" software packages are more than sufficient, as long as they are used with awareness of the software requirements and limitations. Most of the contributions to this book use prepackaged software and give the reader a broad survey of the tools available in such software, as well as information on how to use them effectively. However, these readily available tools were not built with archaeologists in mind, so it is not uncommon for researchers to partner with computer scientists or to strike out on their own to create custom software that better meets their needs. Three of the chapters address this subject and showcase several different ways in which LCA can be conducted, in part or in whole, outside the world of prepackaged GIS software (Livingood, Ullah and Bergin, White).

1.4. Avenues of Least Cost Inquiry in Archaeology

The case studies explored in this book examine different regions, cultures, levels of social complexity, temporal periods, scales of analysis, and social questions, yet they are unified by their investigations into social uses of space and human movement from a least cost perspective. They employ equally diverse techniques to understand human-landscape interactions.

This book is organized thematically by both technique and level of technical skill. Part 1 eases the reader into LCA with relatively traditional applications within a standard GIS platform. While the techniques are fairly straightforward, the archaeological questions explored by the chapters in this part are varied and complex. Rissetto considers how landscape and distance may have structured Paleolithic lithic procurement in northern Spain. Rademaker et al. use LCA to consider how Paleoindian peoples came to settle and explore the southern highland region of Peru. Phillips and Leckman examine potsherd distributions and use LCA to consider how these artifact distributions may represent patterns of movement and interaction in the arid environments of the Jornada Mogollon region of the American Southwest. Nolan and Cook model the hypothesis that regional topography and climate patterns influenced spheres of interaction among Fort Ancient peoples of the U.S. midcontinent.

Part 2 explores "out of the box" approaches that still use standard GIS software and technology. In other words, the chapters in this section couple LCA with other types of spatial analysis. The chapters by Hudson and Richards-Rissetto use a combination of LCA and space syntax analysis to explore social dimensions of space and land use in the U.S. Southwest and Mayan Honduras, respectively. Surface-Evans develops a method for calculating catchment zones based on least cost considerations for Archaic hunter-gatherers of the U.S. midcontinent.

Part 3 highlights approaches that go beyond what is currently available in traditional GIS software. Ullah and Bergin create a hybrid methodology that integrates agent-based modeling with landscape process and least cost modeling to examine agropastoral landscapes in the Mediterranean. Livingood meticulously develops modeling algorithms to consider river-based travel for Mississippian chiefdoms of the southern Appalachian region. White uses a fusion of remote sensing, GIS, human biodynamics, and graph theory to conduct a cost-based exploration of the possible functions of prehistoric trail networks in the Western Papaguería of the American Southwest.

Last, but certainly not least, Part 4 shines a critical spotlight on LCA—both the successes and limitations, as illustrated by the case studies in this book. Authors in this section (Branting, Kantner, Anderson) also address the broader theoretical concerns of LCA and ways to refine our analytical techniques and be more aware of our modeling assumptions.

References

Allen, K. M. S., S. W. Green, and E. B. Zubrow (editors)
1990 *Interpreting Space: GIS and Archaeology*. Taylor and Francis, London.

Anderson, D. G., and C. J. Gillam
2000 Paleoindian Colonization of the Americas: Implications from an Examination of Physiography, Demography, and Artifacts. *American Antiquity* 65(1):43–66.

Binford, L. R.
1980 Willow Smoke and Dogs' Tails: Hunter-Gatherer Settlement Systems and Archaeological Site Formation. *American Antiquity* 45(1):4–20.
2001 *Constructing Frames of Reference: An Analytical Method for Archaeological Theory Building Using Hunter-Gatherer and Environmental Data*. University of California Press, Berkeley.

Clarke, D. L. (editor)
1977 *Spatial Archaeology*. Academic Press, New York.

Dijkstra, E. W.
1959 A Note on Two Problems in Connexion with Graphs. *Numerische Mathematik* 1:269–271.

Hart, P. E., N. J. Nilsson, and B. Raphael
1968 A Formal Basis for the Heuristic Determination of Minimum Cost Paths. *IEEE Transactions on Systems Science and Cybernetics* 4(2):100–107.

Hunt, E.
1992 Upgrading Site-Catchment Analysis with the Use of GIS: Investigating the Settlement Patterns of Horticulturalists. *World Archaeology* 24(2):283–309.

Kelly, R. L.
1992 Mobility/Sedentism: Concepts, Archaeological Measures, and Effects. *Annual Review of Anthropology* 21:43–66.

Lock, Gary (editor)
2000 *Beyond the Map: Archaeology and Spatial Technologies*. NATO Science Series A: Life Sciences, Vol. 321. IOS Press, Amsterdam.

Lock, Gary, and Voran Stancic (editors)
1995 *Archaeology and Geographical Information Systems: A European Perspective*. Taylor and Francis, London.

Mehrer, Mark W., and Konnie L. Wescott (editors)
2006 *GIS and Archaeological Site Location Modeling*. CRC Press, Boca Raton, Florida.

Snead, J. E., C. L. Erickson, and J. A. Darling
2010 *Landscapes of Movement: Trails, Paths, and Roads in Anthropological Perspective*. University of Pennsylvania Museum of Archaeology and Anthropology, Philadelphia.

Vita-Finzi, C., and E. S. Higgs
1970 Prehistoric Economy in the Mount Carmel Area of Palestine: Site Catchment Analysis. *Proceedings of the Prehistoric Society* 36:1–37.

Wescott, K., and J. Brandon (editors)
2000 *Practical Applications of GIS for Archaeologists: A Predictive Modeling Kit*. Taylor and Francis, London.

Wheatley, D., and M. Gillings
2002 *Spatial Technology and Archaeology: The Archaeological Applications of GIS*. Taylor and Francis, London.

Zipf, George K.
1949 *Human Behavior and the Principle of Least Effort: An Introduction to Human Ecology*. Addison–Wesley, Cambridge, Massachusetts.

PART 1

Traditional Applications of Existing Methods

CHAPTER 2

Using Least Cost Path Analysis to Reinterpret Late Upper Paleolithic Hunter-Gatherer Procurement Zones in Northern Spain

John D. Rissetto

Where and how far did late Upper Paleolithic (21,000–10,000 BP) hunter-gatherer groups travel to procure necessary resources in the high-relief topography of north-central Spain? The generally accepted resource procurement models established for this region argue that these hunter-gatherers collected the majority of their resources (e.g., food, fuel, lithics) within a relatively small geographic radius (~25 km) (Butzer 1986; González Sainz 1991; Ibáñez and González Urquijo 1998; Marín Arroyo 2004, 2008; Sarabia 1999; Straus 1986; Utrilla 1981). These radii are almost always represented in the scientific literature as a single ring or series of concentric rings centering on an individual site and encompassing specific resource locations. In almost all cases, the distance between resource loci and occupied site is calculated using a straight-line, or "as the crow flies," measurement. While these local resource procurement models are supported by faunal evidence rich in nonmigratory game (e.g., red deer, ibex, riverine resources) (Marín Arroyo 2004, 2008) and clustered settlement patterns centered along high-relief river valleys (Butzer 1986; Straus 1986), the models only assume that sources of utilized lithic materials, specifically chert, are also located in this same approximately 25-km circular radius.

Lithic sourcing research involving two Lower Magdalenian (17,000–12,500 BP) and three Upper Magdalenian (12,500–11,000 BP) assemblages located within the Asón Valley of eastern Cantabria tested this local lithic source assumption posited by the small geographic radius model. To test the assumption, I used a multitiered analytical methodology that included macroscopic, petrographic, and trace element analyses that compared Magdalenian chert artifacts with natural chert samples from sources located across north-central Spain. The comparison results indicated that the majority of utilized chert artifacts recovered from these assemblages originate from geological sources located outside the 25-km procurement radius proposed by the small resource procurement radius model (e.g., Butzer 1986; Straus 1986; Utrilla 1981). Thus I argue that this model must be modified to include a wider geographic mobility pattern in which late Upper Paleolithic groups traveled to procure necessary resources, specifically chert. In this chapter, least cost path (LCP) analysis is used to examine and interpret a new chert sourcing–based resource mobility and procurement pattern model for Magdalenian groups that occupied these sites in the Asón Valley. Specifically, LCP analysis is used in place of traditional straight-line measurements to determine the most energetically efficient distance, travel route, and travel cost among the five Magdalenian assemblages located in the Asón Valley and their utilized chert sources, which are located primarily outside the previously assumed 25-km boundary.

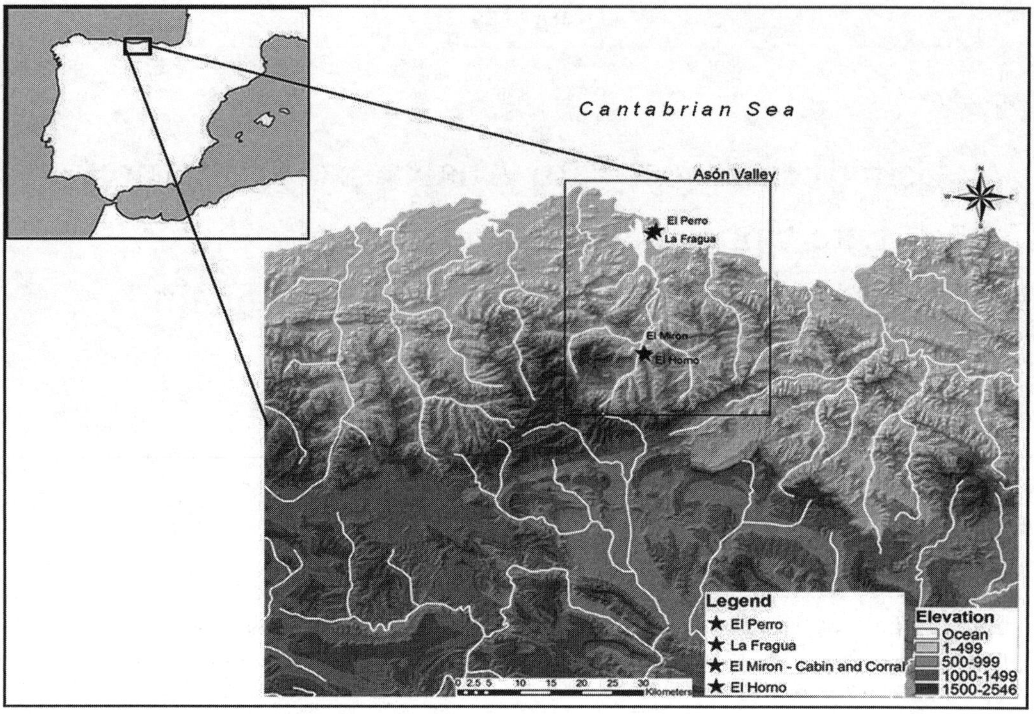

FIGURE 2.1. Eastern Cantabria and the Asón Valley, Spain.

Unlike straight-line measurements, LCP analysis provides an empirical, adaptable, and replicable method by which to gauge the effects that natural (e.g., rivers, lakes, mountains) and/or social (e.g., group territories, sacred landscapes, resource availability) variables had on determining the geographic movement of these groups to and from resource loci. Using LCP, individual researchers can also modify the effects of these variables according to their own interpretation of the natural or social surroundings at specific chronological or climatological points in time. In addition, the distance and travel routes determined by LCP analysis can be used as predictive models for identifying new culturally relevant occupational areas, such as settlement location, hunting context (e.g., jumps, drives, surrounds), and additional resource procurement areas (García Moreno 2010). Ultimately, the goal of this chapter is to present a straightforward application of how least cost path analysis can be used to help evaluate hunter-gatherer mobility, resource procurement, and land use patterns through lithic sourcing data.

2.1. Research Setting

The modern administrative province of Cantabria consists mainly of a narrow strip of land situated along latitude 43° north between the Bay of Biscay/Cantabrian Sea to the north and the Cantabrian Cordillera to the south (Figure 2.1). The province stretches approximately 140 km east-west and 50 km north-south. Cantabria has a mix of ecological and topographic environments ranging from low-relief coastal plains to high-relief mountain ranges. It is assumed that these zones provided unique seasonally predictable subsistence resources procurable during the Magdalenian period (Butzer 1986; Freeman 1973, 1981; Marín Arroyo 2008; Straus et al. 2002).

The high-relief topography begins south in the Cantabria Cordillera and extends northward, creating numerous steep, narrow river valleys that end in the Cantabrian Sea. The river valley ridges range in height from 500 to 2,500 m above sea level and run from <25 to 50 km to the coast. It is argued that the high-relief topography dividing the southern and middle portions of these valleys would have been a serious impediment to routinc, east-west mobility patterns of both humans and animals during the Late Pleistocene (Straus et al. 2002). The position of the steep northern coastline has changed little — receding only 4 to 12 km to the south — since the lowest exposure of the narrow continental shelf during the height of the Last Glacial Maximum (González Morales and Straus 2000; Straus 1992). During this period, the larger coastal zone would have facilitated easier east-west movement by humans and animals restricted by the deep river valleys (Straus 1992).

The Asón River is located in the eastern section of Cantabria Province and has a drainage basin of approximately 800 km^2. The topographically well-defined Asón River basin system includes three major tributaries, the Gándara, Calera, and Carranza, and two minor tributaries, the Aras and Silencio. Today the Asón is approximately 40 km long and, including its tributaries, flows south to north through karstic terrain from its headwaters at the base of the Cantabrian Cordillera/Sierra del Escudo de Cabuerniga to the Cantabrian coastline. Even with the addition of 4–12 km for the Late Glacial coastline, it is assumed to have been a very manageable two- to three-day walk from one end of the river to the other (Straus et al. 2002). The Asón drainage is divided into two main regions, the Lower and Upper Asón (Straus et al. 2002). The Lower Asón includes the estuary and confluence sections, and the Upper Asón contains the intermontane and cordilleran sections. During the Magdalenian period, the Lower Asón would have provided exploitable seasonal or year-round resources from coastal, estuary, and riparian ecozones (Marín Arroyo 2004), while the Upper Asón would have provided seasonally or annually available resources from both forested and alpine ecozones (Marín Arroyo 2008). This research analyzes archaeological assemblages from both the estuary and intermontane sections of the Lower and Upper Asón regions.

2.2. Chert-Sourcing Research

In order for archaeologists to define hunter-gatherer mobility patterns, they must examine the geographic distribution of these groups (Goodyear 1989). A direct empirical method to do so is through lithic raw material provenience studies. A lithic provenience study incorporates independent, qualitative, and quantitative techniques (e.g., individual attribute, petrographic, and trace element analyses) designed to characterize and compare utilized lithic artifacts to natural lithic samples taken directly from identified geologic formations. Through these comparisons it becomes possible to establish which type and amount of utilized lithic artifacts originate from which natural geologic formations located locally or nonlocally to the occupied archaeological site. Fundamental to provenience studies is a thorough knowledge of the regional lithologic environment and the characteristics of materials present across that landscape (Féblot-Augustins 2007; Straus 1980). Materials that are dissimilar to the identified regional materials raise questions regarding the social or cultural mechanisms responsible for their presence, such as individual contacts, trade and exchange, and/or local group migrations (Féblot-Augustins 2007).

In this chapter, hunter-gatherer mobility is defined as a settlement strategy that solves the problems of the spatial and temporal differences between resources and individuals by redistributing people across a known environment (Kelly 1983). It is argued that hunter-gatherer groups with different mobility patterns would likely organize their technology and resource procurement strategies differently (Binford 1977; Kelly 1988). This creates a link between mobility patterns and technological organization because "mobility simultaneously dictates tool needs and access to raw material" (Kelly 1988:718). Therefore, by documenting the strategies of a group's pattern of raw material procurement, we can make inferences about that group's mobility strategies.

The theoretical basis for the mobility and land use model for northern Spain late Upper Paleolithic groups originated in part from ethnoarchaeological studies of modern hunter-gatherers that closely connect subsistence practices with settlement patterns (Binford 1978, 1979, 1980). According to Binford (1980), hunter-gatherer groups can be categorized along a continuum from "forager" to "collector" mobility strategies. Briefly, the forager strategy requires the establishment of a high number of residential sites that house the entire group as it moves across the landscape in search of available resources. Groups that use the collector strategy are logistically organized "by supplying themselves with specific resources through specially organized task groups" (Binford 1980:10). In other words, for forager groups, consumers are brought to resources whereas for collector groups, resources are brought to consumers.

In northern Spain, late Upper Paleolithic hunter-gatherers are interpreted as maintaining a collector-style acquisition strategy (Butzer 1986; Straus 1986; Utrilla 1981). These groups are assumed to have focused on maintaining a few consistently occupied residential base camps situated in resource-rich zones and numerous briefly occupied logistical camps located in proximity to specific resources. Residential base camps served as centralized loci from which groups could initiate logistical forays to procure resources located outside these river-valley settings. Archaeological support for this residential/resource model exists across the Cantabrian region. The region contains fairly low numbers of residential sites with high densities of diverse cultural assemblages, and large numbers of logistical sites with low densities of specialized types of cultural artifacts (e.g., Altuna and Merino 1984; Barandiarán et al. 1985; Clark and Straus 1983; González Echegaray and Barandiarán 1981; Straus and Clark 1986; Straus et al. 1986).

The region's high topographic relief and a settlement/subsistence spectrum focused on the exploitation of non- or minimally migratory game support these models, but what is not explicit in these models is the role lithic procurement played in defining or influencing the mobility and land use strategies of these same groups. Because lithic sourcing information is not directly incorporated in the establishment of these models, it is only assumed that the locations of utilized lithic materials, namely chert, were similar to subsistence resources and were abundantly located in the same occupied valleys/territories.

2.3. Asón Valley Magdalenian Assemblages

The examined chert artifacts originated from both Lower and Upper Magdalenian assemblages located throughout the Asón Valley. The Lower Magdalenian artifacts came from two assemblages located in the same cave site of El Mirón (Cabin level 17 and Corral level 116) (Straus and González Morales 1998, 2007). El Mirón is located in the montane Upper Asón Valley approximately 0.5 km from the Upper Magdalenian site of El Horno. Both assemblages are considered unique occupations of the cave because of (1) different radiocarbon ages and (2) separate physical locations within the cave site (Straus et al. 2008). The Cabin assemblage is situated near the front of the cave, and the Corral assemblage is found at the rear of the cave's vestibule. The two assemblages are separated by 9 m. The Upper Magdalenian artifacts came from three sites: El Perro (Levels 2B/2C–3) and La Fragua (Level 4), located on the modern Cantabrian coast in the Lower Asón Valley (González Morales 1990, 1999; González Morales and Diez Casado 1992), and El Horno, located in the montane interior of the Upper Asón Valley, adjacent to El Mirón cave (Levels 1–3) (Fano 2005, 2008).

While the Asón Valley contains sources of both chert and nonchert material, there are three main reasons why the sourcing of chert artifacts is the focus of this research. First, based on previous survey research across the region, the majority of primary and secondary source areas for nonchert raw materials (e.g., mudstone, quartzite, quartz, limestone) have been identified both within and outside the Asón Valley (Rissetto 2005). These sources are primarily located in the river bed loads <5 km from the examined archaeological sites. Second, Magdalenian hunter-gatherers in north-central and eastern Spain and elsewhere in western Europe almost exclusively (>90 percent) used chert to produce their utilitarian and non-utilitarian stone artifacts (Freeman 1991; González Sainz 1991; Straus 1980, 1991; Straus et al. 2002). Third, the geographic/geologic locations for many of the most widely utilized chert types by late Upper Paleolithic groups that occupied north-central Spain are currently unknown.

The chert artifacts and chert source samples used for the source analysis fit specific criteria. Chert artifacts had to be either retouched tools as defined by Sonneville-Bordes and Perrot (1954) or pieces of unretouched debris ≥2 cm in size. Samples were selected from natural chert sources that best represented the source's physical (i.e., macroscopic) variability, such as color, texture,

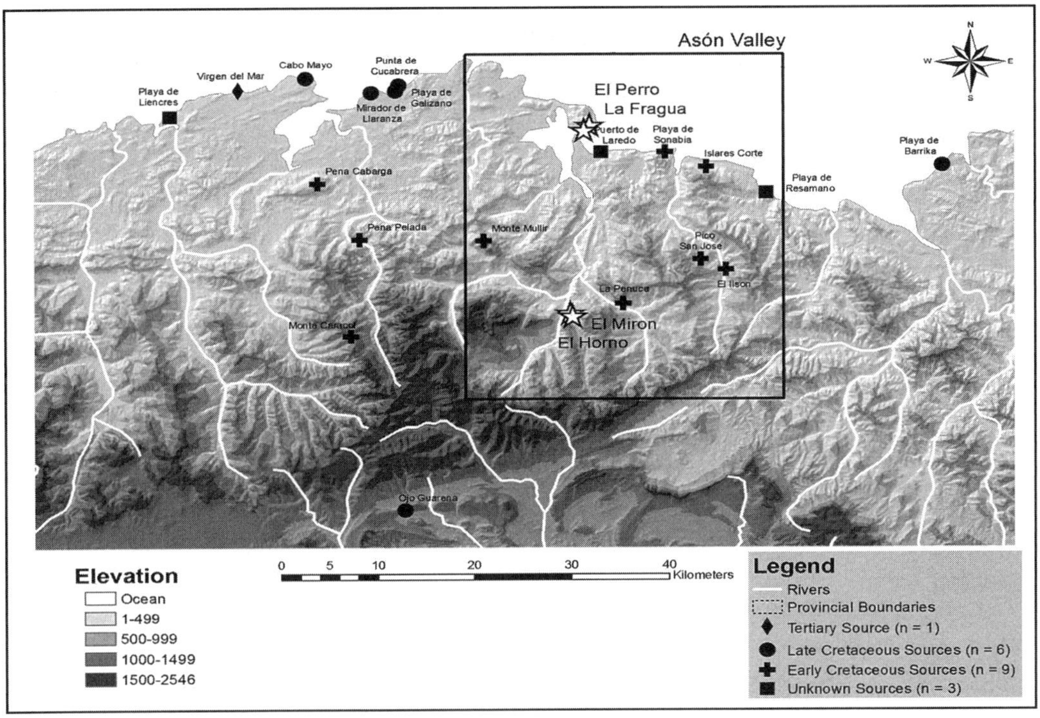

FIGURE 2.2. Identified, characterized, and sampled chert source areas by geologic time period.

inclusions, and fracture pattern. A total of nineteen chert source areas were identified and sampled during the pedestrian survey portion of this research across central and eastern Cantabria (Figure 2.2). However, samples from only seventeen source areas were used in the comparative portion of the analysis. Samples collected from the Islares Corte and El Ilson source areas were not used because of issues associated with raw material size and quality. For analytical purposes, all individual sources regardless of geologic age are considered independent loci.

2.4. Chert Sourcing Analysis Results

The sourced chert artifact data from the El Mirón Cabin and Corral Lower Magdalenian assemblages indicates that these hunter-gatherer groups likely maintained similar overall patterns of chert exploitation and geographic mobility. In general, these data show that both assemblages consist predominantly of chert artifacts produced from chert originating from source areas located outside the defined Asón Valley territory (Table 2.1). This result is demonstrated through the cumulative percentages of chert from the four shared outside–Asón Valley chert source areas (Puerto de Laredo, Playa de Liencres, Playa de Barrika, and Mirador de Llaranza). Combined, these materials represent 59 percent of the Cabin-sourced assemblage and 95 percent of the Corral-sourced assemblage.

There are both consistencies and differences regarding the procurement of chert resources among all three Upper Magdalenian assemblages. The most prominent consistency among the assemblages is the large presence of chert from the Playa de Barrika source area, located outside and to the east of the Asón Valley (Table 2.2). Chert from this source area represents almost half—El Horno, 58.3 percent; El Perro, 50.7 percent; and La Fragua, 46.6 percent—of the chert identified within each sourced stone assemblage. A noticeable difference among the Upper Magdalenian assemblages is that the second-highest percentage of sourced chert in each assemblage originates from a unique source area. In the montane assemblage of El Horno, the second most prevalent chert raw material type is comparable to materials from the Llaranza source area (21.7 percent). This outside–Asón Valley material was

TABLE 2.1. Percentage of Sourced Lower Magdalenian Chert Artifacts by Assemblage.

Utilized Source Area	Within or Outside Asón Valley	El Mirón Corral Artifact Totals	% of Corral-Sourced Assemblage	El Mirón Cabin Artifact Totals	% of Cabin-Sourced Assemblage
Puerto de Laredo	**Within**	**23**	**3.10**	**19**	**2.30**
Playa de Sonabia	Within	—	—	106	13.10
Playa de Resamano	Within	—	—	137	16.90
Monte Mullir	Within	—	—	14	1.70
Mirador de Llaranza	**Outside**	**321**	**42.80**	**304**	**37.60**
Playa de Barrika	**Outside**	**239**	**39.10**	**215**	**26.70**
Playa de Liencres	**Outside**	**74**	**9.90**	**14**	**1.70**
Punta de Cucabrera	Outside	35	4.10	—	—
Playa de Galizano	Outside	3	1.00	—	—
Totals		**695**	**100.00**	**809**	**100.00**

Note: Boldface indicates similar sources in two or more assemblages.

TABLE 2.2. Percentage of Sourced Upper Magdalenian Chert Artifacts by Assemblage.

Utilized Source Areas	Within or Outside Asón Valley	El Horno Artifact Totals	% of El Horno–Sourced Assemblage	La Fragua Artifact Totals	% of La Fragua–Sourced Assemblage	El Perro Artifact Totals	% of El Perro–Sourced Assemblage
Puerto de Laredo	**Within**	**25**	**6.20**	**16**	**5.40**	**58**	**7.60**
Playa de Resamano	**Within**	**45**	**11.10**	**30**	**10.10**	**194**	**25.40**
Monte Mullir	**Within**	**9**	**2.20**	—	—	**4**	**0.40**
Playa de Sonabia	Within	2	0.50	—	—	—	—
Playa de Barrika	**Outside**	**236**	**58.30**	**138**	**46.60**	**388**	**50.70**
Mirador de Llaranza	**Outside**	**88**	**21.70**	**22**	**7.40**	**75**	**9.90**
Playa de Galizano	**Outside**	—	—	**24**	**8.10**	**44**	**5.80**
Peña Cabarga	Outside	—	—	61	20.70	—	—
Playa de Liencres	Outside	—	—	5	1.70	—	—
Ojo Guareña	Outside	—	—	—	—	2	0.20
Totals		**405**	**100.00**	**296**	**100.00**	**765**	**100.00**

Note: Boldface indicates similar sources in two or more assemblages.

also used equally if not more frequently during the Lower Magdalenian occupations in both El Mirón assemblages. The second most abundant chert source type within the coastal La Fragua assemblage is comparable to material from the Peña de Cabarga source area (20.6 percent), located outside and west of the Asón Valley. A procurement pattern unique to the coastal site of El Perro is the increased exploitation of chert from the Playa de Resamano source area (25.4 percent), located east of the Asón drainage. In addition, the presence of a small quantity of chert originating from a source area located south of the Asón Valley, Ojo Guareña, may represent the increased mobility patterns of the groups that occupied El Perro or evidence for trade or exchange between groups at El Perro and others located south of the valley. This hypothesis is strengthened by the presence of Upper Magdalenian occupation and cave art sites in and around the Ojo Guareña cave system (Corchón 2006; Delibes de Castro and Martín 2006). These southern sites may contain chert originating from chert sources located along the Cantabrian coast.

The results of the macroscopic and petrographic analyses indicate that the large majority of chert recovered within both the Lower and Upper Magdalenian sites is most comparable to chert from source areas located outside the Asón Valley, regardless of site location. In addition, the consistent use of chert from similar source areas throughout the Magdalenian may indicate a consistent mobility and land use pattern as well as the exploitation of a similar geographic radius. However, the difference in the quantities of chert procured from source areas located east, west, or south of the Asón Valley may represent procurement and mobility patterns focused on the incorporation of a greater amount of economic resources (Freeman 1991; González Sainz 1991; Straus 1980) or social networks (Conkey 1980; Gould 1978; Gould and Saggers 1985; Schwendler 2004) located outside the Asón Valley. Using the new methods and techniques associated with LCP, we can further examine these differences between Magdalenian assemblages by identifying the geographic distance and travel routes that groups may have taken to and from utilized chert sources and their occupied sites.

2.5. Least Cost Path Analysis

Through the chert sourcing research, specific points of resource procurement were identified across the geographic landscape of Magdalenian hunter-gatherers in the Asón Valley. Rather than determine the distance between points of chert extraction and human settlements using as-the-crow-flies, or straight-line, measurement, I generated an LCP using a Geographic Information System (GIS). The purpose of using the LCP analysis is to calculate the most energetically efficient path between two points — chert sources and archaeological sites — based on the value of slope or incline across a digitized landscape (Conolly and Lake 2006; Longley et al. 2005; Wheatley and Gillings 2002). For this project, slope was derived from a digital elevation model (DEM) of northern Spain so that a raster-based cost surface with a 90-m resolution could be created. The cost surface represents a continuous plane incorporating geographic and topographic features, where slope has been assigned as the weighted numerical value for each pixel. The GIS sums the slope values for each pixel situated between two points and then assigns the LCP to the route with the lowest accumulated slope values. These values are ultimately represented as travel cost, expressed in terms of the total accumulated slope amounts between two points. The travel cost value is thus correlated with, but not derived from, the amount of time and energy spent moving from one place to another.

During the field research portion of this project, the locational information for each identified chert source and archaeological site was collected with a Garmin eTrex Summit global positioning system (GPS). The locational data for each source and site were recorded using the Universal Transverse Mercator (UTM) coordinate system with the North American Datum of 1983 (NAD83). The UTM data were then downloaded into an Excel spreadsheet, uploaded into the ArcGIS 9.2 program, and turned into a shapefile. I purchased 90-m-resolution digital elevation model (DEM) data of the Asón Valley and vicinity from an online GIS data website. Unfortunately, I could not find higher-resolution data within my budget. While it is possible that higher-resolution data may change the distance and path cost from the sites and sources, these differences are assumed to be minor. In addition, higher-resolution data would be unlikely to change the travel routes between the sites and sources due to the research area's high topographical relief, which naturally restricts movement. The purchased DEM data were then converted to a slope. Once the slope was created, a cost-distance surface and cost-direction surface were produced using the slope surface. Finally, the "spatial analyst" function in the ArcGIS 9.2 toolbar produced the least cost paths from the individual sites to their respective utilized chert sources. For this analysis, natural barriers such as rivers and mountains were not weighted differently than the natural ground surface because of the unknown nature and locations of these types of features in the Asón Valley during the Late Pleistocene. In addition, only least cost paths that linked a single site to the multiple utilized sources were produced. Paths were not created that ran from or between multiple sites and/or multiple sources.

In this research, an LCP was created to identify the costs of travel for Upper and Lower Magdalenian sites and the location of chert source areas identified within the assemblages. The LCP represents the most energetically efficient path between two points. This is based on an assumption grounded in the Principle of Least Effort (Zipf 1949), which is summarized in Surface-Evans and White (Chapter 1, this volume). Briefly, this principle states that humans will choose the solution to a given problem based on which one involves the least amount of effort. Here this principle refers to the physical path humans chose to transport chert materials from a given source area to the occupied site. Because these heavy materials were carried sufficient distances, it is likely that groups would have sought paths that took the least amount of effort to traverse.

There are several conceptual issues associated with using the LCP models for anthropological interpretation. First, the least effort principle does not take into account that humans do not always act efficiently. A lack of efficiency could stem from individual, social, cultural, or environmental reasons. Thus, it is also highly likely that other, less energetically efficient paths were used by Magdalenian period hunter-gatherers who occupied the Asón Valley to collect and transport chert raw materials. Second, it is extremely difficult, if not impossible, to determine from current archaeological information whether a certain modeled path was indeed used during the Magdalenian. However, since the LCP travel costs and directions are based on the measurement of topographic slope, which has changed little since the end of the late Upper Paleolithic period, least cost path analysis provides an empirical and heuristic model for evaluating past human movement within a landscape. If it is valid to assume that hunter-gatherer groups maintained economically efficient strategies for traversing long distances, it is likely that LCP models represent frequently utilized paths for moving lithic resources from site to source and back again. Regardless, LCP models certainly constitute a significant improvement over traditional straight-line distance measures of physical distances between two points (Sarabia 1999; Straus et al. 1986; Tarriño 2006). Simple straight-line distance models do not consider how landscape constraints affect human movement. Therefore, in its consideration of how the cost of traversing a landscape potentially influences the movement between habitation sites and chert extraction locales, this research represents a step forward.

2.6. Least Cost Path Results

2.6.1. Lower Magdalenian

2.6.1.1. *El Mirón Corral*

The LCP analysis for the El Mirón Corral assemblage indicates that the Lower Magdalenian groups focused on the use of chert materials from source areas located outside the Asón Valley along the present-day Cantabrian coast. The LCP-based travel routes indicate that if these groups procured chert materials directly from these source areas, they would have traveled down the Asón River Valley until they reached the coast and then headed either east or west across the coastal plain to the designated source area (Figure 2.3). None of the LCP included diagonal or straight-line paths through or over the mountainous terrain that separates the source areas from the site. The ranking of travel distances from the Corral assemblage to utilized chert sources is similar to the travel costs calculated by the LCP (Table 2.3). As described in Table 2.3, the source areas with the lowest or highest amount of travel cost, Laredo and Liencres, respectively, also represent the sources with the shortest and longest distances from the site. In addition, if the Corral groups followed the LCP, they would have covered a cumulative additional 139.09 km when compared with the simple straight-line path. This significant increase in distance traveled by the LCP highlights the fact that cost-based models of movement do not always correspond to the shortest path between two points.

The least cost calculated distances show that chert materials were collected within the three procurement zones. The first zone includes the distance of 0–40 km from the site and represents the smallest percentage, 3.0 percent of the sourced assemblage and 1.5 percent of the total assemblage (sourced and unsourced), of the chert materials identified within the assemblage. The second zone includes the distance of 40–70 km from the site and represents the largest percentage, 87.1 percent of the sourced assemblage and 42.8 percent of the total assemblage, of sourced chert materials identified within the assemblage. The third zone includes the distance of 70 km and up and represents a moderate percentage, 9.9 percent of the sourced assemblage and 4.9 percent of the total assemblage, of the overall assemblage.

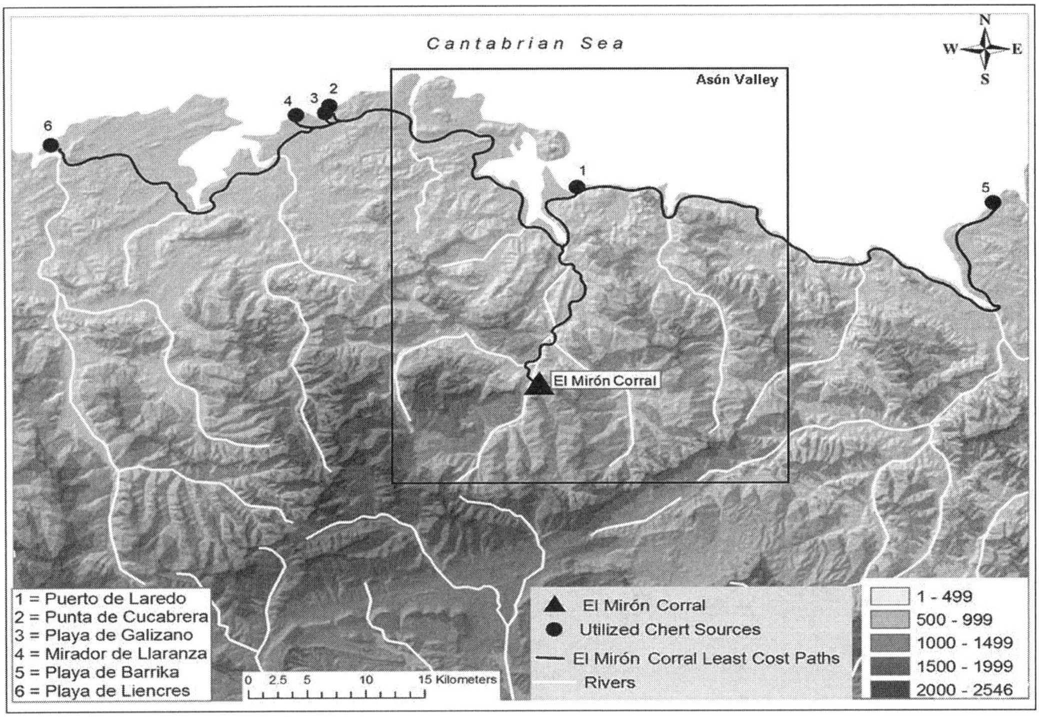

FIGURE 2.3. Least cost path total cost, distance, and direction from chert sources to the El Mirón Corral assemblage.

TABLE 2.3. El Mirón Corral: Comparison of Least Cost Path and Straight-Line Distances.

Source Number	Source Name	% of Sourced Assemblage	% of Total Assemblage	LCP Distance from Source to Site (km)	Travel Cost	Straight-Line Distance (km)	Difference between LCP and Straight-Line Distance (km)
1	Laredo	3.00	1.50	26.61	97,173	19.3	7.31
2	Cucabrera	4.10	2.30	54.53	112,850	31.95	22.58
3	Galizano	1.00	0.20	54.91	113,487	32.3	22.61
4	Llaranza	42.90	21.10	57.33	116,444	33.32	24.01
5	Barrika	39.10	19.20	68.93	125,336	42.24	26.69
6	Liencres	9.90	4.90	83.21	129,137	47.32	35.89
Totals		**100.00**	**49.10**	**345.52**		**206.43**	**139.09**

Overall, the largest percentage of sourced materials identified within the Corral assemblage originates from the Barrika and Llaranza source areas, which are located outside the Asón Valley in procurement zone 2.

2.6.1.2. *El Mirón Cabin*

The LCP analysis for the El Mirón Cabin assemblage demonstrates that the Lower Magdalenian groups focused on the procurement of chert materials from source areas located in both a montane and a coastal physiographic context (Figure 2.4). Because the Cabin and Corral assemblages originate from the same site location, the two assemblages have identical travel distances, routes, and costs for similar source areas. However, the Cabin assemblage, unlike the Corral assemblage, includes chert from the montane source area of Monte Mullir. While none of the other paths travels a straight line to the source area, the LCP from

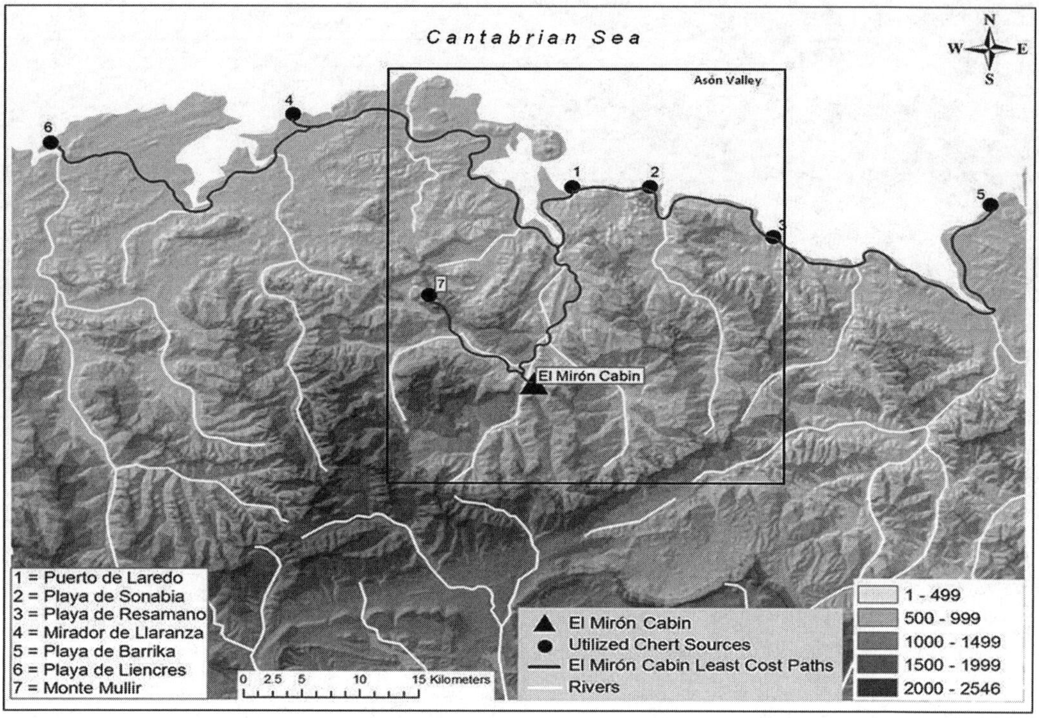

FIGURE 2.4. Least cost path total cost, distance, and direction from chert sources to the El Mirón Cabin assemblage.

TABLE 2.4. El Mirón Cabin: Comparison of Least Cost Path and Straight-Line Distances.

Source Number	Source Name	% of Sourced Assemblage	% of Total Assemblage	LCP Distance from Source to Site (km)	Travel Cost	Straight-Line Distance (km)	Difference between LCP and Straight-Line Distance (km)
1	Laredo	2.30	1.60	26.61	98,783	19.3	7.31
2	Sonabia	13.10	8.90	37.95	100,100	21.31	16.64
3	Resamano	16.90	11.50	53.68	110,423	24.6	29.08
4	Llaranza	26.70	18.00	57.33	114,547	33.32	24.01
5	Barrika	37.60	25.40	68.93	125,336	42.24	26.69
6	Liencres	1.70	1.20	83.21	128,735	47.32	35.89
7	Mullir	1.70	1.20	13.84	142,266	12.46	1.38
Totals		**100.00**	**67.60**	**341.55**		**200.55**	**141.00**

El Mirón to Mullir is very close. This is evident in the only slight difference between the LCP and the straight-line distance from site to source.

The LCP results indicate the same valley-floor-to-coast travel routes to utilized sources for the Lower Magdalenian groups that created the Cabin assemblage as for those that created the Corral assemblage. And as for the Corral assemblage, if groups followed the LCP, they would have covered an additional 141.00 km when compared with the straight-line path. With the exception of the Monte Mullir source area, travel cost corresponds directly to travel distance from the Cabin assemblage to the utilized source areas (Table 2.4). While Mullir is geographically closer to the site than any other source area, it contains the highest calculated travel cost, due to either its extreme terrain or its sociocultural significance.

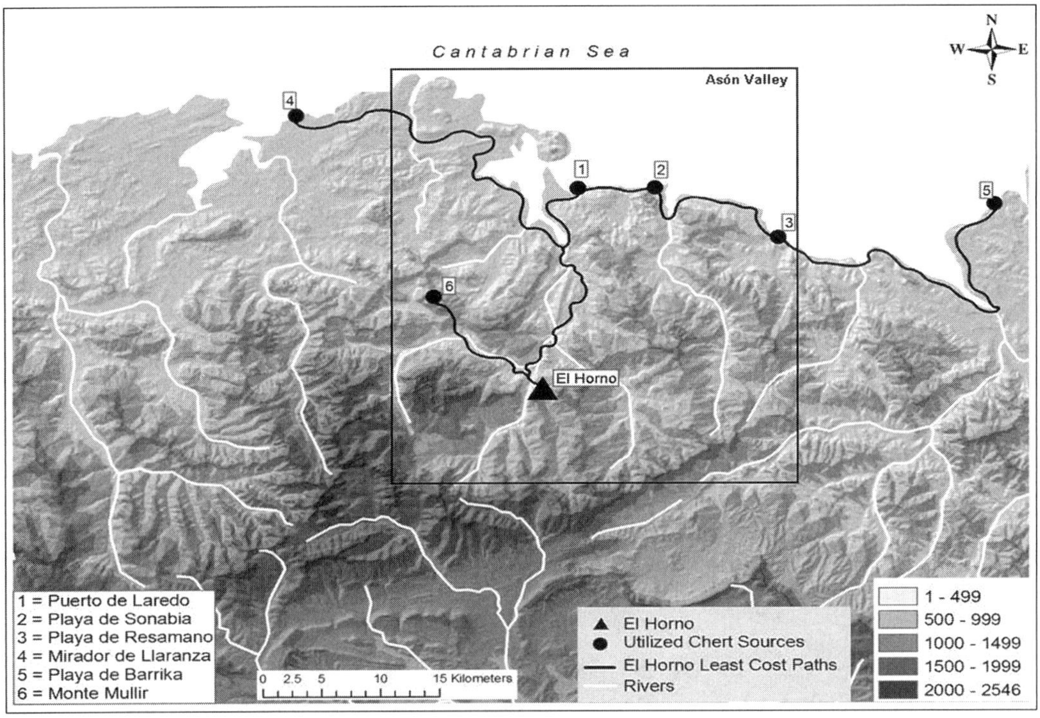

FIGURE 2.5. Least cost path total cost, distance, and direction from chert sources to the El Horno assemblage.

The Cabin assemblage also contained chert from source areas located in three procurement zones (0–40 km, 40–70 km, 70+ km). As with the Corral assemblage, the majority of the chert recovered within the Cabin assemblage corresponded to source areas located in procurement zone 2. The second zone (40–70 km) represents the largest percentage, 81.2 percent of the sourced assemblage and 54.8 percent of the total assemblage, of sourced chert materials identified within the assemblage.

2.6.2. Upper Magdalenian

2.6.2.1. *El Horno*

The LCP analysis for the Upper Magdalenian assemblage in El Horno demonstrates that these groups focused on the procurement of chert from both montane and coastal source areas (Figure 2.5). As for the Lower Magdalenian groups that occupied the montane site of El Mirón, the most energetically efficient route from El Horno to utilized chert source areas follows the valley floor to the coast. The LCP analysis also indicates a similar almost straight-line path from El Horno to Monte Mullir. Judging by the location of the sourced materials, it is evident that the large majority of chert material originated from source areas to the east of the Asón Valley. However, there appears to have been a targeted procurement strategy in place for the collection of materials from the Llaranza source area, located west of the Asón.

With the exception of the Monte Mullir, the ranking of distances traveled between El Horno to utilized chert sources are similar to the travel costs calculated by the LCP (Table 2.5). If the El Horno groups followed the LCP, they would have traveled 106.79 km farther than if they took the straight-line path. This extra distance may have influenced social or logistical decisions about which groups traveled to procure chert from these sources and when. The majority of the sourced chert artifacts are comparable to materials from sources located in procurement zone 2. Chert from procurement zone 2 represents the largest percentage, 91.1 percent of the sourced assemblage and 47.4 percent of the total assemblage, of sourced chert materials identified within the assemblage. The only other zone in which chert was procured is zone 1 (0–40 km).

TABLE 2.5. El Horno: Comparison of Least Cost Path and Straight-Line Distances.

Source Number	Source Name	% of Sourced Assemblage	% of Total Assemblage	LCP Distance from Source to Site (km)	Travel Cost	Straight-Line Distance (km)	Difference between LCP and Straight-Line Distance (km)
1	Laredo	6.20	3.20	27.01	101,287	19.4	7.61
2	Sonabia	0.50	0.30	38.35	102,604	21.46	16.89
3	Resamano	11.10	5.80	54.08	112,927	24.7	29.38
4	Llaranza	21.70	11.30	57.73	117,051	33.51	24.22
5	Barrika	58.30	30.30	69.33	127,841	42.09	27.24
6	Mullir	2.20	1.20	14.24	144,770	12.79	1.45
Totals		**100.00**	**52.10**	**260.74**		**153.95**	**106.79**

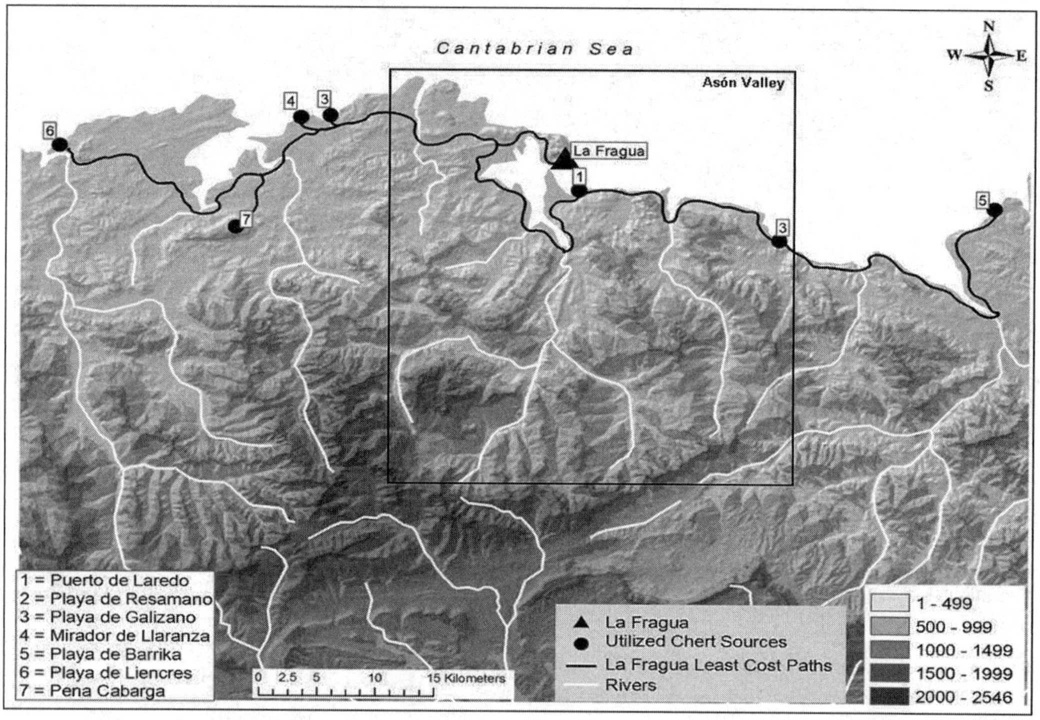

FIGURE 2.6. Least cost path total cost, distance, and direction from chert sources to the La Fragua assemblage.

Material from these sources equals 8.9 percent of the sourced and 4.7 percent of the total lithic assemblage. A Chi-square comparison identifies a statistically significant difference between the number of artifacts from zone 1 (n = 369) and zone 2 (n = 36) (1, n = 405) = 3380.7, $p < 0.001$.

2.6.2.2. *La Fragua*

The LCP analysis for the La Fragua assemblage indicates that Upper Magdalenian groups focused on the procurement of chert materials from source areas located mainly along the coastal plain and in the higher-elevation mountains along the coast (Figure 2.6). The analysis shows that the most energetically efficient paths to the utilized chert source areas stretch along the Cantabrian coast. It also indicates that the most efficient path to the montane source area of Peña Cabarga would be to follow the coast and then move to the source area from the north. This is in stark contrast to approaching the source area from the southern interior, where groups would

TABLE 2.6. La Fragua: Comparison of Least Cost Path and Straight-Line Distances.

Source Number	Source Name	% of Sourced Assemblage	% of Total Assemblage	LCP Distance from Source to Site (km)	Travel Cost	Straight-Line Distance (km)	Difference between LCP and Straight-Line Distance (km)
1	Laredo	5.40	3.10	3.98	24,816	3.78	0.2
2	Resamano	10.10	5.80	24.22	33,676	19.78	4.44
3	Galizano	8.10	4.70	31.27	37,702	20.42	10.85
4	Llaranza	7.40	4.30	35.13	40,659	22.95	12.18
5	Barrika	46.70	26.70	39.56	48,589	37.05	2.51
6	Liencres	1.70	1.00	58.79	53,075	43.44	15.35
7	Cabarga	20.60	11.80	51.08	103,951	29.03	22.05
Totals		**100.00**	**57.40**	**244.03**		**176.45**	**67.58**

have had to cross rivers and high-relief topography.

The majority of the La Fragua chert assemblage originates from source areas located to the east of the Asón Valley (62.2 percent) (Table 2.6). However, chert materials also appear to have been routinely procured from source areas located west of the site. With the exception of Peña Cabarga, the ranking of straight-line distances traveled between La Fragua to utilized chert sources is similar to the travel costs calculated by the LCP. Even though Peña Cabarga is only a short distance from the coast, it is in a high-relief environment. This high relief creates a stark difference in the amount of travel cost, almost double, incurred by groups procuring materials from this source when compared with the others. However, because both La Fragua and the majority of the utilized source areas are located along the coast, there is a considerably smaller difference in distance between the LCP and the straight-line paths (67.58 km).

The La Fragua assemblage is distinct from the other four assemblages in that the majority of chert assemblage originates from source areas located within procurement zone 1 (0–40 km). Materials from this first zone include 77.7 percent of the sourced assemblage and 44.6 percent of the total assemblage (sourced and unsourced) of the chert materials identified within the assemblage. The second zone (40–70 km) represents only a moderate percentage of the chert assemblage, 24.3 percent of the sourced assemblage and 12.8 percent of the total assemblage. There are no artifacts comparable to source area materials identified in the third procurement zone (>70 km).

2.6.2.3. *El Perro*

The LCP analysis demonstrates that the Upper Magdalenian groups that occupied El Perro procured chert materials from source areas located in montane, coastal, and meseta physiographic contexts (Figure 2.7). As was the case for the occupants of the montane sites (El Mirón, El Horno), the LCP shows a valley floor–based travel route for coastal groups that moved to procure chert from sources in the mountains or interior. Thus it is assumed that groups would have traveled along the river floodplains until they came to a low-elevation pass in order to reach the Ojo Guareña source area in the southern meseta region.

With the exception of Monte Mullir, the ranking of straight-line distances traveled between El Perro to utilized chert sources is similar to the travel costs calculated by the LCP (Table 2.7). The source area with the highest travel cost is Ojo Guareña, which has a travel cost three times the amount of the next closest source area, Monte Mullir. This increased cost is due mostly to the various changes in slope between the source and El Perro. As for La Fragua, the coastal location of the site allows for only a moderate difference between the LCP and straight-line distances (57.26 km). Even though El Perro has samples from the two source areas with the highest travel cost, the difference between the LCP and straight-line distances are the lowest among all five assemblages.

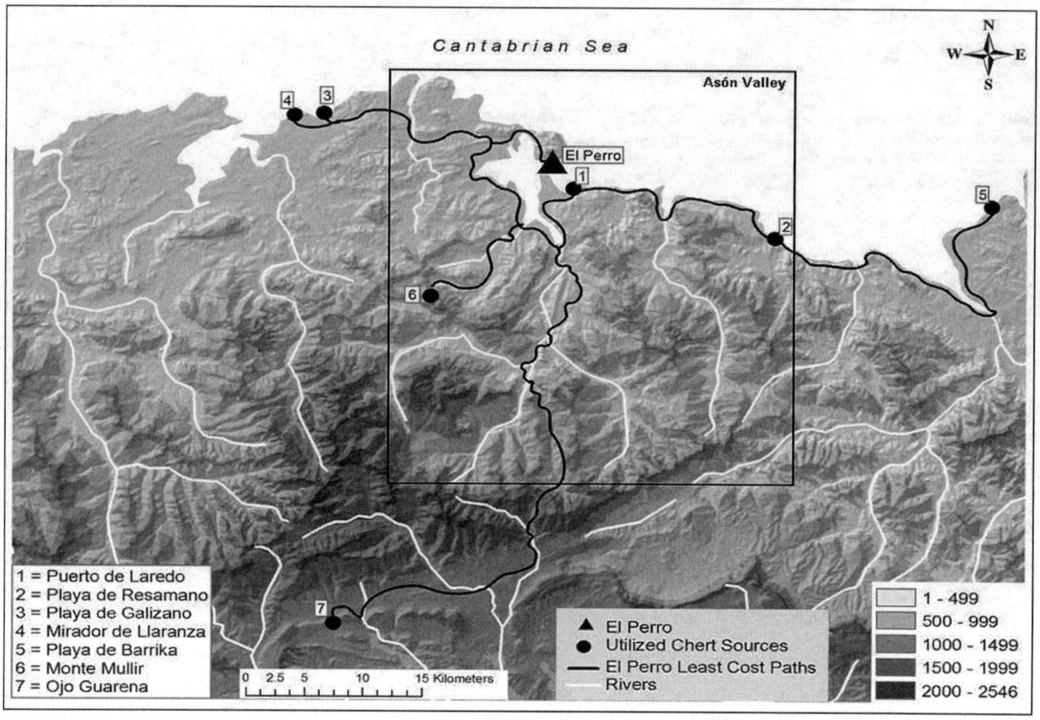

FIGURE 2.7. Least cost path total cost, distance, and direction from chert sources to the El Perro assemblage.

TABLE 2.7. El Perro: Comparison of Least Cost Path and Straight-Line Distances.

Source Number	Source Name	% of Sourced Assemblage	% of Total Assemblage	LCP Distance from Source to Site (km)	Travel Cost	Straight-Line Distance (km)	Difference between LCP and Straight-Line Distance (km)
1	Laredo	7.60	4.20	4.54	14,471	3.24	1.3
2	Resamano	25.40	14.20	25.07	24,446	20.21	4.86
3	Galizano	5.80	3.20	32.12	28,391	20.01	12.11
4	Llaranza	9.90	5.50	35.98	31,349	22.35	13.63
5	Barrika	50.70	28.40	40.41	41,302	37.61	2.8
6	Mt. Mullir	0.40	0.30	21.06	105,618	16.79	4.27
7	Ojo Guareña	0.20	0.10	66.73	316,698	48.44	18.29
Totals		**100.00**	**56.00**	**225.91**		**168.65**	**57.26**

The LCP-calculated distances indicate that chert materials were procured within the first two procurement zones. The first procurement zone (0–40 km) represents approximately half the chert materials identified within the assemblage, 49.1 percent of the sourced assemblage and 27.5 percent of the total assemblage (sourced and unsourced). The other half of the assemblage originates from source areas found in procurement zone 2 (40–70 km), 50.9 percent of the sourced assemblage and 28.5 percent of the total assemblage. There are no artifacts identified in the assemblage originating from source areas in the third zone (>70 km).

2.7. Discussion and Interpretation

2.7.1. Lower Magdalenian Interpretation

According to the chert sourcing analyses, the Lower Magdalenian groups that occupied the Corral and Cabin areas of the El Mirón site trav-

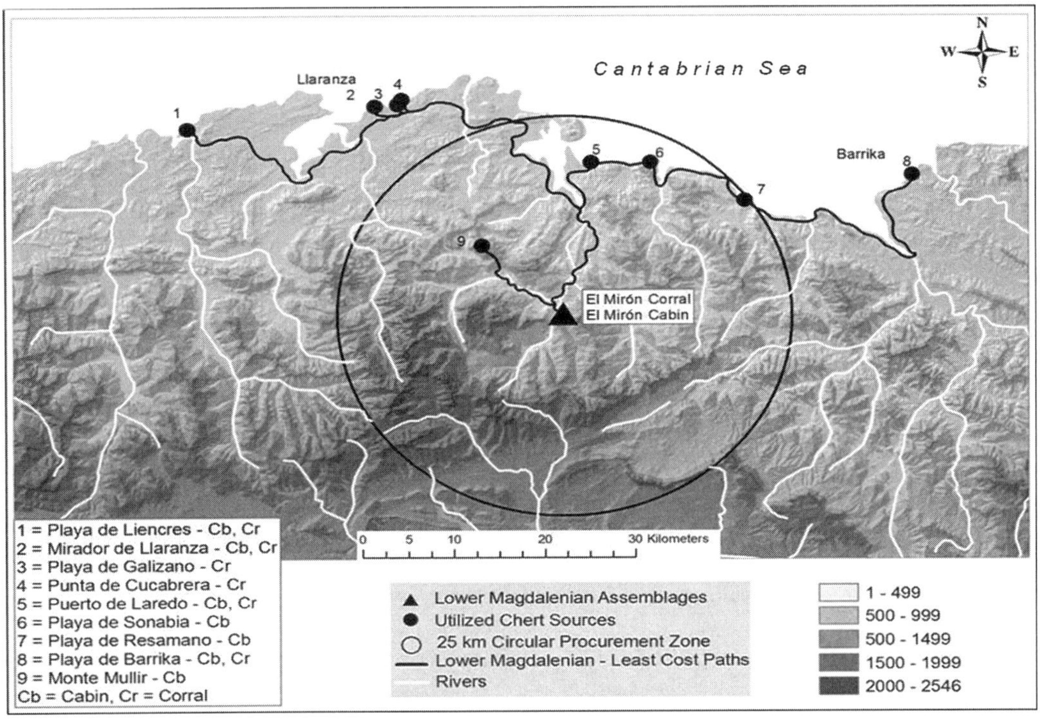

FIGURE 2.8. LCP between Lower Magdalenian archaeological assemblages (Cb = Cabin, Cr = Corral) and utilized source areas.

eled outside the topographic boundaries of the Asón Valley to procure the majority of their chert resources (Figure 2.8). As indicated by the travel distance and routes determined by the LCP analysis, these groups traveled on average 55 km down the Asón Valley floodplain and then in an easterly or westerly direction along the Cantabrian coastline to reach specific chert sources. This form of dual-directional procurement strategy was likely due in part to the technological need to procure high-quality chert from sources such as Playa de Barrika (east) and Mirador de Llaranza (west) for the production of blade and bladelet artifacts.

The LCP analysis also indicated that the source areas with the highest travel cost were those among the farthest from El Mirón site. The source areas with travel costs from 110,000 to 125,000 represent the largest majority of chert materials identified within both assemblages. Interestingly, the source area that was the shortest distance from the site, Monte Mullir, also had the highest travel cost. This anomaly is due to the fact that the Mullir source is located at the highest elevation of Monte Mullir. Since the large majority of both sourced chert assemblages originates from source areas located outside the Asón Valley, the small (~25 km) procurement and mobility hypothesis is not supported by the data presented in this research. Clearly, a reinterpretation of the mobility and lithic procurement patterns for the Lower Magdalenian hunter-gatherers in the Asón Valley is necessary. Based on this conclusion, it appears that these groups may have focused on the procurement of chert from geographically diverse source areas that are located approximately 50 km from the occupied site.

An explanation for this extended procurement strategy is possibly based in the need by Magdalenian hunter-gatherers to procure mainly chert materials of the highest quality in order to produce their bladelet-dominated lithic technology (Freeman 1991; González Sainz 1986; Straus 1980). Bladelet artifacts are small (<2 cm long), narrow (<0.5 cm wide), parallel-sided lithic artifacts produced from a specifically prepared and shaped core. These artifacts were slotted into grooved bone, antler, or wood shafts and used for various cutting, slicing, and projectile purposes. Because of their size and specific shape, bladelets were primarily made from the highest-quality chert material available. This interpretation is supported by the assemblage data, which show that the highest quantities of artifacts originate from sources with the consistently highest-quality chert material, Playa de Barrika and Mirador de Llaranza. Because groups needed materials from these sources, it is likely that they would have used the most energetically efficient path to and from the site to the source. However, this interpretation assumes that these groups were going straight to the source and returning directly to El Mirón without stopping at other sites along the way. Future sourcing research may help address this issue. In addition, through increased LCP analysis, new predictive models may be created to help identify other routes to and from sites and sources.

2.7.2. Upper Magdalenian Interpretations

Chert recovered from the three Upper Magdalenian assemblages was procured from source areas located along the modern Cantabrian coast, from high-elevation montane settings, or from the southern meseta region (Figure 2.9). In all three assemblages, the largest percentage of chert originates from the Playa de Barrika source area, located east of the Asón Valley. The second most numerous chert type varies among the assemblages; in the Horno assemblage, it is the Llaranza source area. The constant use of Llaranza chert within montane sites may indicate similarities in site use, seasonal or annual mobility patterns, subsistence foci, or social or cultural activities.

In the La Fragua site, the second most numerous chert type originates from the Peña Cabarga source area. This source area is located west of the Asón Valley and is very close to the Llaranza source area. It is possible that chert from the Cabarga source was selected over the more ubiquitous Llaranza material because of its proximity to other specific resource procurement areas such as mountain goat hunting areas or social or ritual sites such as cave art sites.

In the El Perro assemblage, chert materials from the Playa de Resamano were the second most numerous type identified within the site. The presence of this chert type, along with the high quantity of Barrika chert, may identify a pattern of resource (lithic or subsistence) procurement strategies focused on the collection of materials from areas east of the Asón Valley. However, because of the consistent presence of materials from other source areas located both west and south of the site, it is also possible that these groups incorporated a geographically larger mobility and procurement pattern than that used by groups that occupied the other two Upper Magdalenian sites. This is most evident in the presence of materials from the Ojo Guareña source area. The presence of this chert type could have various meanings, from the commonplace to the elaborate. Since the distance from El Perro to Ojo Guareña is within a 66-km radius of the site, which is less than the LCP distance from El Mirón to Liencres, it is possible that materials from this source and other sources adjacent to this area (Elorza and García-Garmilla 1997; Elorza and Bustillo 1989) were used more readily for the production of chert artifacts than originally assumed. It is also possible that because this material is recovered in such small quantities, it may represent trade items received through exchange transactions or networks with groups located in the meseta region.

In summary, the results of the travel-distances analysis between the utilized sources and both Lower and Upper Magdalenian assemblages show that the LCP distances are often double those created with straight-line measurements. This is because least cost paths do not necessarily take the shortest route but model the least costly route in terms of energy or time. However, it must be kept in mind that even though these routes may not have been the only ones taken between these sites and sources, it is likely that they were used more frequently than others because of their energetic efficiency to sources of necessary resources. Nevertheless, groups will often forgo the most efficient route between two places because of other social, symbolic, or ritual aspects intrinsic to either the route itself (e.g., religious pilgrimage routes) or the destination (Gould 1978; Gould and Saggers 1985; McBryde 1997; Taçon 1991).

The issue of whether a portion of the sourced chert artifacts were acquired via trade or exchange networks is difficult to address in this archaeological context because few natural chert sources have been identified across north-central

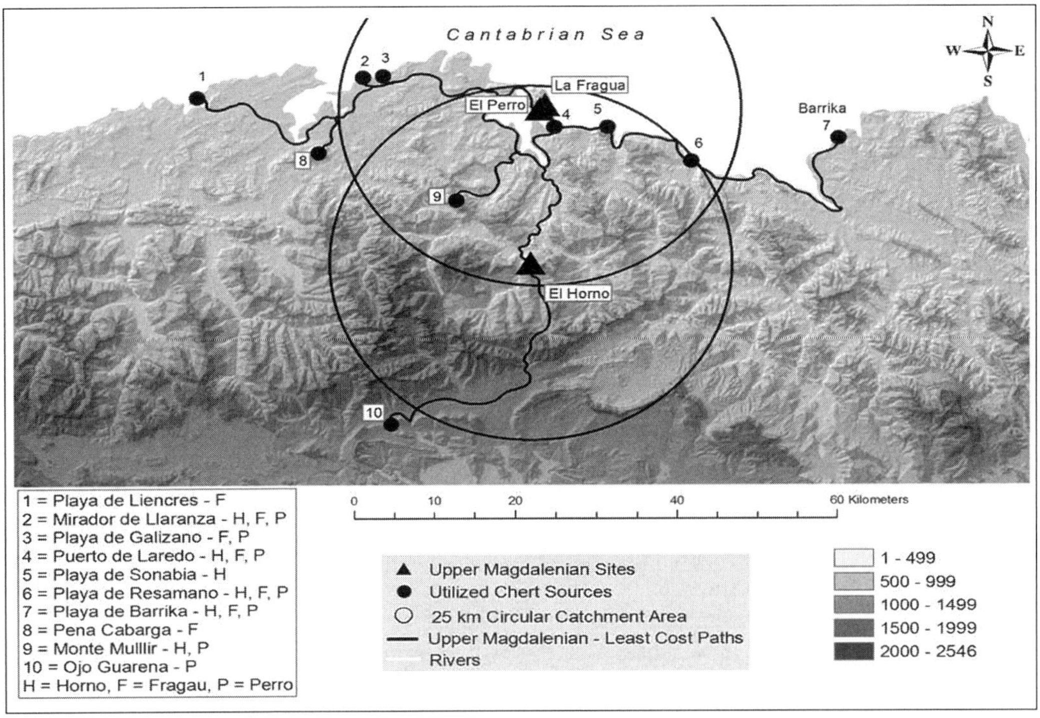

FIGURE 2.9. LCP between Upper Magdalenian archaeological assemblages (H = El Horno, F = La Fragua, P = El Perro) and utilized source areas.

Spain (Ibáñez and González Urquijo 1998; Rissetto 2009; Sarabia 1999; Straus et al. 1986; Tarriño 2006). Because so little is known about the visual, microscopic, and geochemical signatures of chert sources present in the region, it is extremely difficult to determine which types of artifacts were traded or exchanged between neighboring or distant groups. As more chert sources are identified and characterized across the region, it will be possible for us to more accurately define which sources groups are utilizing more frequently than others. The comparison of the amounts and types of artifacts being produced from chert from certain sources will help us differentiate between chert use patterns associated with direct procurement and trade/exchange.

2.8. Conclusion

Least cost path analysis is a successful alternative to straight-line measurement for determining the distance and travel route between two locations. In this research, LCP analysis helped determine the distance and travel routes possibly taken from Magdalenian period sites in the Asón Valley to utilized chert sources. LCP models were conducted between known habitation sites and chert source areas identified from the analysis of the stone tool assemblages at the sites.

The LCP analysis indicated that the travel distance from these sites to their utilized sources extended well beyond the previously assumed 25-km procurement radius. The analysis also provided the most energetically efficient travel routes between these locations, which were different from the straight-line routes in almost all instances. The LCP analysis showed that for most source areas there is a direct relationship between the energetic costs of travel and the frequency of costly chert in occupied assemblages. In other words, source areas with the highest travel cost are often represented by some of the highest percentages of material presence in the assemblages.

In the future, additional elements can be added to this research procedure to further elucidate the connection between LCP and Late Pleistocene hunter-gatherer resource procurement. First, higher-resolution data can be used to see if in fact new paths are created between sites and sources. Second, instead of isotropic tests, in which the slope is the same in all directions, anisotropic tests, which that take into account upslope and downslope, can be run to see if there are different paths to and from sites and sources (see Kantner 2004). Third, human biodynamics (energy expenditure) can be incorporated as a variable in the least cost path calculation (see Rademaker et al. this volume). Fourth, travel distance can be included as a cost accumulator to see if the path between a site and a source still goes along the base of the valley and then along the coast (e.g., El Mirón, El Horno) when distance is balanced against cost (see Livingood this volume). Fifth, barrier costs can be assigned to natural features (e.g., rivers, high peaks) or social territories (e.g., cultural taboos, conflict) that would have prevented groups from traveling in certain directions. These are just a few possible modifications that can be made to the existing dataset for further use of LCP analysis in hunter-gatherer research and analysis.

Through the continued use of least cost paths, we will obtain new empirical data to help us examine how the geographic movement on the landscape influenced resource procurement, group mobility patterns, and social organization and interaction. Even though the LCP routes may not have been the exact ones taken by these groups, the technique offers us an empirical and replicable way to identify the most energetically efficient route between two points of interest, while taking into account natural barriers that may have influenced group movement. These hypothetical routes can also be the starting point for future surveys aimed at identifying previously unknown archaeological sites or resource loci. In this chapter, least cost path analysis has shed light on our understanding of strategies used by Magdalenian hunter-gatherers in north-central Spain to procure necessary resources, move across the landscape, and possibly organize themselves within local and/or regional social networks.

References

Altuna, Jesús, and J. M. Merino
1984 *El yacimiento prehistórico de la cueva de Ekain.* Sociedad de Estudios Vascos, San Sebastián, Spain.

Barandiarán, Ignacio, Leslie Freeman, Joaquin González Echegaray, and Richard Klein
1985 *Excavaciones en la cueva del Juyo.* Monografías 14. Centro de Investigación y Museo de Altamira, Santander, Spain.

Binford, Lewis R.
1977 Forty-Seven Trips: A Case Study in the Character of Archaeological Formation Processes. In *Stone Tools as Cultural Markers: Change, Evolution, and Complexity*, edited by R. V. S. Wright, pp. 24–36. Australian Institute of Aboriginal Studies, Canberra.
1978 *Nunamiut Ethnoarchaeology.* Academic Press, New York.
1979 Organization and Formation Processes: Looking at Curated Technologies. *Journal of Anthropological Research* 35:255–273.
1980 Willow Smoke and Dogs' Tails: Hunter-Gatherer Settlement Systems and Archaeological Site Formation. *American Antiquity* 45:4–20.

Butzer, Karl
1986 Paleolithic Adaptations and Settlement in Cantabrian Spain. *Advances in World Archaeology* 5:201–252.

Clark, Geoffrey A., and Lawrence G. Straus
1983 Late Pleistocene Hunter-Gatherer Adaptations in Cantabrian Spain. In *Hunter-Gatherer Economy in Prehistory: A European Perspective*, edited by G. N. Bailey, pp. 131–148. Cambridge University Press, Cambridge.

Conkey, Margaret W.
1980 The Identification of Prehistoric Hunter-Gatherer Aggregation Sites: The Case of Altamira. *Current Anthropology* 21:609–630.

Conolly, James, and Mark Lake
2006 *Geographical Information Systems in Archaeology.* Cambridge University Press, Cambridge.

Corchón Rodríguez, María Soledad
2006 Las cuevas de la Griega y Palomera (Ojo Guareña) y la cuestión de la cronología del arte paleolítico en la meseta. In *El Paleolítico Superior en la meseta norte española*, edited by Germán Delibes de Castro and Fernano Diez Martín, pp. 75–112. Universidad de Valladolid, Valladolid, Spain.

Delibes de Castro, Germán, and Fernano Diez Martín
2006 Una meseta desolada? Estado actual de la investigación sobre el Paleolítico Superior en

las regions interiors de la península ibérica. In *El Paleolítico Superior en la meseta norte española*, edited by G. Delibes de Castro and F. Diez Martín, pp. 11–40. Universidad de Valladolid, Valladolid, Spain.

Elorza, J. J., and M. A. Bustillo

1989 Early and Late Diagenetic Chert in Carbonate Turbidites of the Senonian Flysch, Northeast Bilbao, Spain. In *Siliceous Deposits of the Tethys and Pacific Regions*, edited by J. R. Hein and J. Obradović, pp. 93–105. Springer, New York.

Elorza, J., and F. García-Garmilla

1997 Chert Types in the Cueva Carbonate Platform (Upper Cretaceous, Northern Spain). In *Siliceous Rocks and Culture*, edited by A. Ramos-Millan and M. A. Bustillo, pp. 57–73. Monográfica Arte y Arqueológica, University of Granada, Granada, Spain.

Fano, Miguel Angel Martínez

2005 La cueva de El Horno (Ramales, Cantabria): Un asentamiento de montaña del final del Magdaleniense. In *La cuenca Mediterránea durante el Paleolítico Superior (38.000–10.000 años)*, edited by J. L. Sanchidrián, A. Ma. Márquez, and J. Ma. Fullola, pp. 468–478. Actas del IV Simposio de Prehistoria Cueva de Nerja Reunión de la VIII Comisión del Paleolítico Superior U.I.S.P.P. Fundación Cueva de Nerja, Málaga, Spain.

2008 La prehistoria del valle del Asón: Excavaciones en la cueva de El Horno (Ramales de la Victoria), 2000–2001. In *Actuaciones arqueológicas en Cantabria, 2000–2003*, edited by R. Ontañón, pp. 73–77. Gobierno de Cantabria, Santander, Spain.

Féblot-Augustins, Jehanne

2007 Raw Materials and Early Mediterranean Neolithic Expansion Process in the Upper Rhône Basin: The Case of the Gardon Cave (Ambérieu-en-Bugey, Ain, France). *Arkeotek Journal* 1(3). Electronic document, http://www.thearkeotekjournal.org/.

Freeman, Leslie

1973 The Significance of Mammalian Faunas from Paleolithic Occupations in Cantabrian Spain. *American Antiquity* 38:3–44.

1981 The Fat of the Land: Notes on Paleolithic Diet in Iberia. In *Omnivorous Primates*, edited by R. Harding and G. Teleki, pp. 104–165. Columbia University Press, New York.

1991 What Mean These Stones? Remarks on Raw Material Use in the Spanish Paleolithic. In *Raw Material Economies among Prehistoric Hunter-Gatherers*, edited by A. Montet-White and S. Holen, pp. 73–126. Publications in Anthropology 19. University of Kansas, Lawrence.

García Moreno, Alejandro

2010 Patrones de asentamiento y ocupación del territorio en el Cantábrico oriental al final del Pleistoceno: Una aproximación mediante SIG. Unpublished Ph.D. dissertation, Universidad de Cantabria, Santander, Spain.

Goodyear, Albert C.

1989 A Hypothesis for the Use of Cryptocrystalline Raw Materials among Paleoindian Groups of North America. In *Eastern Paleoindian Lithic Resource Use*, edited by C. J. Ellis and J. C. Lothrop, pp. 1–9. Westview Press, Boulder, Colorado.

González Echegaray, J., and I. Barandiarán

1981 *El Paleolítico Superior de la cueva del Rascaño*. Monografías 3. Centro de Investigación y Museo de Altamira, Santander, Spain.

González Morales, Manuel R.

1990 La prehistoria de las marismas: Excavaciones en el abrigo de la Peña del Perro (Santoña, Cantabria). *Cuadernos de Trasmiera* 2:13–28.

1999 Varilla decorada del Magdaleniense de la cueva de la Fragua. *Sautuola* 6:177–183.

González Morales, Manuel R., and Yolanda Diez Casado

1992 Excavaciones en los abrigos de la Peña del Perro (Santoña, Cantabria): Estratigrafía, cronología y comentario preliminar de sus industrias. *Veleia* 8–9:43–64.

González Morales, Manuel R., and Lawrence Straus

2000 La cueva del Mirón (Ramales de la Victoria, Cantabria): Excavaciones, 1996–1999. *Trabajos de Prehistoria* 57:121–133.

González Sainz, César

1991 Algunas reflexiones sobre las materias primas líticas y la variabilidad técnica y tipológica, a termino de Paleolítico Superior de la región Cantábrica. In T*ecnología y cadenas operativas líticas*, edited by R. Mora, X. Terradas, A. Parpal, and C. Plana, pp. 173–191. Ballaterra, Spain.

Gould, Richard

1978 The Anthropology of Human Residues. *American Anthropologist* 80:815–835.

Gould, Richard, and Sherry Saggers

1985 Lithic Procurement in Central Australia: A Closer Look at Binford's Idea of Embeddedness in Archaeology. *American Antiquity* 50: 117–136.

Ibáñez Estévez, Juan José, and Jesús Emilio González Urquijo

1998 The Production and Use of Lithic Tools at the End of the Upper Paleolithic in the Basque

Country. In *The Organization of Lithic Technology in Late Glacial and Early Postglacial Europe*, edited by Sarah Milliken, pp. 17–38. BAR International Series 700. British Archaeological Reports, Oxford.

Kantner, John

2004 Geographical Approaches for Reconstructing Past Human Behavior from Prehistoric Roadways. In *Spatially Integrated Social Science*, edited by Michael Goodchild and Donald Janelle, pp. 323–348. Oxford University Press, Oxford.

Kelly, Robert L.

1983 Hunter-Gatherer Mobility Strategies. *Journal of Anthropological Research* 39:277–306.

1988 The Three Sides of a Biface. *American Antiquity* 53:717–734.

Longley, Paul A., Michael F. Goodchild, David J. Maguire, and David W. Rhind

2005 *Geographical Information Systems and Science*. 2nd ed. Wiley and Sons, West Sussex, United Kingdom.

Marín Arroyo, Ana Belén

2004 Análisis arqueozoológico, tafonómico y de distribución espacial de la cueva de La Fragua (Santoña-Cantabria). *Munibe (Antropología-Arkeología)* 56:19–44.

2008 El yacimiento paleolítico de la cueva del Mirón: Resultados de la aplicación de nuevas metodologías arqueozoológicas. In *Zooarqueología hoy*, edited by C. Díez, pp. 69–87. Encuentros Hispano-Argentinos, Servicio de Publicaciones de la Universidad de Burgos, Burgos, Spain.

McBryde, Isabel

1997 "The landscape is a series of stories": Grindstones, Quarries and Exchange in Aboriginal Australia; A Lake Eyre Case Study. In *Siliceous Rocks and Cultures*, edited by A. Ramos-Millan and Ma. Bustillo, pp. 587–607. Universidad de Granada, Granada, Spain.

Rissetto, John D.

2005 Using Lithic Procurement Strategies to Define Magdalenian Hunter-Gatherer Mobility Patterns in the Asón Valley of Eastern Cantabria, Spain. In *Proceedings of the 4th Archaeological Congress of the Iberian Peninsula*, pp. 481–492. Faro, Portugal.

2009 Late Pleistocene Hunter-Gatherer Lithic Exploitation and Mobility Patterns in Eastern Cantabria, Spain. Unpublished Ph.D. dissertation, University of New Mexico, Albuquerque.

Sarabia Rogina, Pedro M.

1999 Aprovechamiento y utilización de materias primas líticas en los tecnocomplejos del Paleolítico en Cantabria. Ph.D. dissertation, Servicio de Publicaciones, Universidad de Cantabria, Santander, Spain.

Schwendler, Rebecca

2004 Hunter-Gatherer Social Interaction in Magdalenian Western Europe. Unpublished Ph.D. dissertation, University of New Mexico, Albuquerque.

Sonneville-Bordes, D. de, and J. Perrot

1954 Lexique typologique du Paléolithique Supérieur. *Bulletin de la Société Préhistorique Française* 327–333.

Straus, Lawrence

1980 The Role of Raw Materials in Lithic Assemblage Variability. *Lithic Technology* 9:68–72.

1986 Late Würm Adaptive Systems in Cantabrian Spain: The Case of Eastern Asturias. *Journal of Anthropological Archaeology* 5:330–368.

1991 The Role of Raw Materials in Upper Paleolithic and Mesolithic Stone Artifact Assemblage Variability in Southwest Europe. In *Raw Material Economies among Prehistoric Hunter-Gatherers*, edited by A. Montet-White and S. Holen, pp. 169–185. Publications in Anthropology 19. University of Kansas, Lawrence.

1992 *Iberia before the Iberians*. University of New Mexico Press, Albuquerque.

Straus, Lawrence, and Geoffrey Clark

1986 *La Riera Cave: Stone Age Hunter-Gatherer Adaptations in Northern Spain*. Anthropological Research Papers 36. Arizona State University, Tempe.

Straus, Lawrence, and Manuel González Morales

1998 Report on the Initial Excavations in El Mirón Cave (Ramales de la Victoria, Cantabria, Spain) with Emphasis on the Magdalenian Occupations. *Journal of Iberian Archaeology* 0:173–188.

2007 Early Tardiglacial Human Uses of El Mirón Cave (Cantabria, Spain). In *On Shelter's Ledge: Histories, Theories and Methods of Rockshelter Research (Proceedings of the XVI Congress of the UISPP, Lisbon)*, edited by M. Kornfeld, S. Vasil'ev, and L. Miotti, pp. 83–93. BAR International Series S1655. British Archaeological Reports, Oxford.

Straus, Lawrence, Manuel González Morales, Miguel Fano Martínez, and Mari Paz García Gelabert

2002 Last Glacial Human Settlement in Eastern Cantabria (Northern Spain). *Journal of Archaeological Science* 29:1403–1414.

Straus, Lawrence, Manuel González Morales, and Elizabeth Stewart

2008 Early Magdalenian Variability: New Evidence

from El Mirón Cave, Cantabria, Spain. *Journal of Field Archaeology* 33:197–216.

Straus, Lawrence, Jesús Ortea, Luis Suárez del Río, and Rosa Esbert Alemany

1986 Patterns of Lithic Raw Material Variation at La Riera. In *La Riera Cave: Stone Age Hunter-Gatherer Adaptations in Northern Spain*, edited by L. Straus and G. Clark, pp. 189–208. Anthropological Research Papers 36. Arizona State University, Tempe.

Taçon, Paul

1991 The Power of Stone: Symbolic Aspects of Stone Use and Tool Development in Western Arnhem Land, Australia. *Antiquity* 65:192–207.

Tarriño Vinagre, Andoni

2006 *El sílex en la Cuenca Vasco Cantábrica y Pirineo Navarro: Caracterización y su aprovechamiento en la prehistoria*. Monografías 21. Museo Nacional y Centro de Investigación de Altamira, Santander, Spain.

Utrilla Miranda, Pilar

1981 *El Magdaleniense Inferior y Médio en la costa Cantábrica*. Centro de Investigación y Museo de Altamira, Santander, Spain.

Wheatley, David, and Mark Gillings

2002 *Spatial Technology and Archaeology: The Archaeological Applications of GIS*. Taylor and Francis, London.

Zipf, G. K.

1949 *Human Behavior and the Principle of Least Effort: An Introduction to Human Ecology*. Addison-Wesley, Cambridge, Massachusetts.

CHAPTER 3

Connecting the Dots

Least Cost Analysis, Paleogeography, and the Search for Paleoindian Sites in Southern Highland Peru

Kurt Rademaker, David A. Reid, and Gordon R. M. Bromley

3.1. Introduction

This chapter provides an example of how predictive GIS-based least cost analysis (LCA) can be situated effectively within a larger interdisciplinary research program. Our project examines links between Late Glacial (LG) period environmental change and human settlement of the high Andes. Paleoindians entered South America during the LG period, a time of major climatic, glacial, and ecologic shifts. How did early settlers successfully adapt to such changes and rapidly colonize a continent of diverse landscapes — the hyperarid coasts, the rugged, high-altitude Andes Mountains, and the vast Amazon Basin? While these environments are all extreme in some way, they also have abundant resources, making them favorable places for people with the correct adaptations.

Paleoindian sites such as Quebrada Tacahuay (Keefer et al. 1998; deFrance et al. 2001) and Quebrada Jaguay (QJ-280) (Sandweiss et al. 1998) indicate that some of South America's earliest people were settling the Pacific coast very early, taking advantage of rich fisheries and littoral zones (Plate 1). Quebrada Jaguay's well-dated Paleoindian component also contains material evidence of a connection with interior zones. This evidence includes artifacts made of petrified wood, which crops out ~20 km to the north of the site (Tanner 2001), and seeds from *Opuntia* cf. *ficus-indica* (prickly pear cactus) (Sandweiss 2003), found today at elevations between 2400 and 3600 m. The most dramatic material evidence of contact with interior zones is provided by a small bifacial tool fragment and associated debitage made of obsidian (Tanner 2001), found in directly dated LG and Early Holocene (EH) contexts at Quebrada Jaguay. Radiocarbon dates from these deposits span 9270±75 to 11,088±120 ^{14}C BP (~10,600–13,000 cal BP), indicating multiple procurement events.

Using instrumental neutron activation analysis (INAA) on 26 pieces of debitage, Glascock initially sourced Quebrada Jaguay's obsidian to a small geologic outcrop Burger and others (1998) had identified and characterized previously near the town of Alca in the Cotahuasi Canyon, some 150 km distant at 2850 m (Sandweiss et al. 1998). Jennings and Glascock (2002) later reported additional small, geochemically similar Alca obsidian deposits within the Cotahuasi Canyon, also near the town of Alca. Rademaker (2006) suggests that the Quebrada Jaguay obsidian more likely was obtained at the conspicuous ~412-ha. obsidian deposit at Cerro Condorsayana, located on the east rim of the Cotahuasi Canyon between 4000 and 4820 m, or from a small deposit located on the west side of the Pucuncho Basin at 4365 m.

Since no known geologic mechanism can transport Alca obsidian to the coast, either the same group of Paleoindian foragers was moving between coastal and highland ecozones or separate coastal and highland Paleoindian groups were trading. Under either scenario, highland sites contemporary with Quebrada Jaguay should occur near Alca source deposits. This connection between Quebrada Jaguay and Alca obsidian deposits is the only positively demonstrated material link between specific coastal and highland locales in all of South America during the LG period, and it provides an opportunity for us to find and investigate a series of linked Paleoindian sites situated in multiple environmental zones.

The first step in doing so involves locating one or more additional Paleoindian sites in the intervening ~150 km between the two known locales (Plate 1), particularly near the Alca volcanic field, an area characterized by high site visibility and good preservation potential. A comparative study of related Paleoindian sites within this corridor may help us determine the nature of the early coast-highland relationship and ultimately shed light on initial colonization routes within South America.

We used LCA, along with mapping and geochemical characterization of obsidian deposits and local paleoenvironmental research, to identify and survey a predictive least cost corridor linking Quebrada Jaguay with the Alca obsidian source. Along predictive least cost paths (LCPs), we discovered a number of highland open-air and rockshelter sites containing coastal lithic materials and diagnostic projectile points dating from the LG and EH. Some of these sites may contain substantial and directly datable Paleoindian evidence, and we anticipate further archaeological investigations.

3.2. Identifying the Destination(s): Provenance of Alca-1 Obsidian

LCA is particularly applicable to archaeological sites containing exotic botanical, faunal, and mineral resources, especially if the specific provenance of exotic materials can be determined (see discussion by Anderson this volume). As lithic raw materials generally are not transported great distances from parent outcrops via geologic mechanisms in the project area, human dispersal is the most likely process. Therefore, the geographic distribution of artifacts of a given raw material can provide a measure of past peoples' ranges, mobility, or social networks. Since geologic deposits can be considered stationary resources over human time scales, it is a straightforward process to calculate LCPs connecting lithic raw material sources with the workshop or habitation sites where lithic materials were used and discarded (Risetto this volume; Taliaferro et al. 2010). However, a given raw material may crop out over large geographic areas or its overall distribution may be unknown, limiting the applicability of LCA.

Before modeling routes between Point A (the coastal site Quebrada Jaguay) and Point B (the source deposit of Quebrada Jaguay's obsidian), we first had to determine the specific location of Point B. Our first objective was to map and geochemically characterize Alca obsidian deposits to identify all potential geologic sources for the obsidian artifacts at Quebrada Jaguay.

Until recently, insufficient data hindered our understanding of the spatial extent and geochemical variability of geologic deposits of Alca obsidian and the prehistoric exploitation of these deposits. This problem stemmed from a lack of extensive and systematic geologic mapping, sampling, and geochemical characterization, advocated by Glascock et al. (1998) for successful study of obsidian sources. Previous datasets (Burger and Asaro 1993; Burger et al. 1998; Jennings and Glascock 2002) included samples collected from geologic outcrops near the bottom of the Cotahuasi Canyon. Only small deposits of Alca obsidian had been located, and no prehistoric extraction or processing sites had been found, despite substantial archaeological evidence for the prehistoric use of Alca obsidian at sites throughout Peru (Burger et al. 2000).

In 2004 we began systematic geoarchaeological study of Alca obsidian deposits. Six spatially and geochemically distinct Alca obsidian deposits are exposed between 2850 and 4820 m over ~1200 km^2 of the highlands east of the Cotahuasi River. The largest and most conspicuous is a ~412-ha. obsidian deposit at Cerro Condorsayana, located on the east rim of the canyon (Rademaker 2006). We compared geochemical analyses of Alca obsidian source deposits with same-technique analyses of a sample of obsidian debitage from the Quebrada Jaguay lithic assemblage. The coastal artifacts originated from a single geochemical subgroup, designated Alca-1. Matching Alca-1 deposits occur in three locations: near the town of Alca (2850 m), Cerro Condorsayana (4000–4820 m), and on the western margin of the Pucuncho Basin (4365 m). All three of these areas are considered possible sources for the Quebrada Jaguay Paleoindian artifacts and are incorporated in archaeologically testable LCPs.

3.3. Optimal Routes and Central Places

Once we had determined specific Alca-1 destinations for Quebrada Jaguay, we could build a predictive model to identify likely locations for contemporary early sites in the intervening area. Our predictive model is based on two primary concepts of optimal foraging theory (Winterhalder and Smith 1981; Kelly 1983; Kelly and Todd 1988), discussed below.

In a high-relief region such as the Quebrada Jaguay–Alca-1 corridor, forager routes should follow paths of least topographic resistance (Anderson and Gillam 2000; Kelly 2003). Areas of gentle topography obviously facilitate movement more readily than steeper, rugged terrain, but low-slope areas offer additional benefits. For a colonizing Paleoindian group entering a new or unfamiliar landscape for the first time, large-scale movements should occur along easily traceable geographic features, such as river valleys or canyons, linear mountain chains, or ecologic corridors such as coasts, a pattern consistent with landscape learning principles (Kelly 2003). Aldenderfer (1999, 2006) has suggested that LG high-altitude environments presented extreme physiological, behavioral, and demographic challenges for forager groups attempting to colonize these zones from lower-elevation staging areas. If this hypothesis is correct, no highland forager group could have been resident near the Alca-1 obsidian source ~11,000–13,000 cal BP, and in general, early forays into the Andes would have followed corridor drainages leading from the Pacific coast into the highland interior. These drainages would have offered dependable fresh water and attractive biotic resources relative to the hyperarid intervalley coast. *Opuntia* cf. *ficus-indica* (prickly pear cactus) seeds in LG deposits at Quebrada Jaguay (Sandweiss 2003) provide evidence of contact with perennially watered valley habitats between 2400 and 3600 m.

In high-altitude, arid mountain environments such as the Andean plateau, where water and faunal resources are clustered in patches (Winterhalder and Thomas 1978), foragers should situate their residential sites in optimal central places that afford easy access to multiple high-quality resources (Kaplan and Hill 1992; Morgan 2008; Marín Arroyo 2009). Such a strategy would minimize the high energetic costs of movement and subsistence in high-altitude mountain terrain (Aldenderfer 1998). This central-place foraging strategy would mitigate the risk of failure in an extreme, heterogeneous, or perhaps new and unfamiliar environment and would be expected of a colonizing or transient explorer population (Beaton 1991; Meltzer 2002; Kelly 2003). While the behavior of a group exploring new and unfamiliar territory might not be perfectly optimal, it would be conservative and risk-averse (Kelly 2003; Steele and Rockman 2003). In an extreme environment such as the LG high-altitude Andes, there would have been little room for error, and rapid landscape learning would have been necessary for successful colonization.

3.4. Coast-Highland Route Explorations

Assuming that interior Paleoindian sites related to Quebrada Jaguay should be located along optimal coast-highland routes, we defined three possible corridors linking Quebrada Jaguay and Alca-1 obsidian deposits (Plate 1): travel north to the highlands via the Quebrada Jaguay–Manga intermittent drainage, travel west along the coast and north to the highlands via the perennial Ocoña-Cotahuasi River, and travel east along the coast and north to the highlands via the perennial Majes River.

In 2007 we personally explored the Quebrada Jaguay–Manga canyon from its mouth at 40 m to ~4000 m, suggested by Sandweiss et al. (1998) as the most likely route connecting Quebrada Jaguay with the Alca source. However, while the Quebrada Jaguay–Manga canyon provides the shortest route to the Andean plateau, it probably was not used by people. About 15 km north of Quebrada Jaguay, the canyon becomes an impassable bedrock slot; the only way to reach the rim is to make a circuitous and difficult climb in an area of absolute desert. A lack of archaeological sites on the rim above the coastal and *lomas* ecozones (or >1200 m elevation) provides further support that prehistoric people made only limited interzonal movements along this route, although a definitive answer must await a more complete survey.

The second possible corridor, the Ocoña-Cotahuasi River, also is an unlikely route to Alca-1 obsidian deposits, judging by our exploration of a portion of this route. Travel north through the Ocoña-Cotahuasi canyon is possible to an area north of Nevado Solimana (6095 m), about 30 km southwest of the small Cotahuasi Canyon Alca-1 deposits. In this section of the steep-walled canyon, the Cotahuasi River passes through a ~10-m-wide slot and plunges 150 m over the Sipia Cataract. Due to highly seasonal stream flux and frequent tectonic activity and associated landslides, no transportation routes have been constructed in this extremely rugged section of the Ocoña-Cotahuasi drainage, either historically or in modern times. A modified version of this route could head east up the Arma River canyon to pass south of Nevado Solimana and access the high-elevation Alca-1 deposits. We explored this route in 2008, but the canyon forms a precipitous bedrock slot several kilometers upstream of the Ocoña confluence and is unlikely to have been used prehistorically.

Our explorations, though subjective, suggest that the third option, the Majes River, was a potential route to the highlands. The Majes is a large, resource-rich perennial corridor drainage and primary travel route linking southern Peruvian coastal communities to the highlands today. The only modern highway accessing the Cotahuasi Canyon from the coast follows the Majes River north-northwest to the town of Chuquibamba, ascends to the plateau rim via a series of switchbacks, and then crosses the plateau west of Nevado Coropuna (6425 m) before descending to the Cotahuasi Canyon. Many segments of Inca Road have been found along this same route (B. Owen, personal communication 2005; A. Cardonas, personal communication 2008).

3.5. Least Cost Analysis and Colonizing Foragers

To evaluate these coast-highland corridors quantitatively, we conducted LCA between Quebrada Jaguay and Alca-1 obsidian deposits. Before presenting the methods and results of the LCA, we first must address the theoretical applicability of using this technique for archaeological prediction of Paleoindian site locations, since we assume these foragers were a colonizing population. A common objection to using LCA in the study of colonization behavior is that people first entering a new region would have had no idea what lay before them. LCA, which works from defined sources and destinations, assumes an overall objective, and thus directionality, to movement. Yet it is probable that colonizing groups would have had no such ultimate destination or objective, at least on the first movement(s) into a given area. We might expect that as a result of landscape learning, foragers would rapidly configure optimal routes across a landscape.

Admittedly, LCA for prediction of forager routes is limited by many unknowns. Archaeologists are unlikely to know completely the extent of forager group territories (especially for colonizers of empty landscapes), the specific reasons for discrete movements, and how site locations articulated within larger settlement systems. We cannot assume that we know the true starting and ending points for forager routes in the past, particularly for colonizing groups that were entering new and unfamiliar territory and whose small-scale movements may have been somewhat stochastic (Field and Mirazon Lahr 2005).

As Anderson and Gillam (2000) point out, colonizers cannot be expected to have followed cumulative optimal LCPs since they did not know where they were going. Instead, foragers probably followed sequentially optimal LCP segments, exploring new areas along routes characterized by relatively gentle topography and valuable resources. Whether a single forager group was moving between Quebrada Jaguay and Alca-1 deposits, or two separate groups were meeting at some intermediate location, the ~150-km distance between Alca-1 deposits and the Quebrada Jaguay site probably would have been traversed in stages, not in one continuous trip. The net rate and direction of the cumulative coast-highland journey would have been dependent on the goals of shorter-distance movements and the nature of the origin, objective destination, and intervening areas. Movement rates probably were highly variable, depending on whether a group was in the middle of a rich resource patch it was in no hurry to leave, or whether there was some need to get through a resource-poor zone more quickly, or other reasons (see in-depth discussion by Anderson this volume).

Since no early sites have been identified previously in the Quebrada Jaguay–Alca-1 corridor, we must begin with the site data we have, and we must use any predictive technique we can, however imperfectly suited to the task. We also must acknowledge that our predictive LCA will provide, at best, a rough estimate of the location of prehistoric routes within a portion of a potentially much larger settlement system (Risetto this volume; Surface-Evans this volume). We do not assume that our LCPs will represent entire forager path systems any more than we assume that the *archaeological* settlement patterns we detect through regional surveys will represent entire forager settlement systems (Moseley 1983). LCA is an appropriate technique for prediction of forager and colonizer routes because by employing LCA within the archaeological survey strategy, we improve our chances of intersecting segments of actual forager routes and sites if those routes and sites even roughly approximated an optimal configuration within the landscape.

What is the best way to measure the cost of movement for a colonizing Paleoindian group? Rugged terrain is a major challenge to overland travel, so slope can be an appropriate currency, provided it is scaled correctly (but see discussion by Kantner this volume). The simplest push-button algorithms scale cost values according to a scale of slope values calculated from the elevation dataset. Time is unlikely to have been a major issue for early foragers, as there is no archaeological evidence suggesting that multiple competing populations needed to race for the same resources. We generally assume low population densities for LG foragers, based on global anthropological (Kelly 1995) and Central Andean archaeological (Rick 1988) research. Energy probably was the most significant limiting resource for most hunter-gatherer populations (Jenike 2001). Energy expended in walking over varied terrain has been parameterized via human physiological experiments (McDonald 1961; Soule and Goldman 1972; Pandolf et al. 1977; Santee et al. 2001), and White (2007 and this volume) has developed a method for integrating these parameters in LCA.

3.6. Methods

We calculated LCPs using ArcGIS 9.1 software. For all models, we specified the Quebrada Jaguay (QJ-280) site as the source and defined three alternate Alca-1 destinations, using GPS-obtained coordinates for Alca-1 obsidian deposits at Cerro Condorsayana (4820 m), the Pucuncho Basin (4365 m), and the town of Alca in the Cotahuasi Canyon (2850 m).

The elevation dataset was a single 90-m cell resolution Shuttle Radar Topography Mission (SRTM) v.4 DEM tile (22_16). While the horizontal cell resolution of the 90-m SRTM DEM is poorer than available 30-m Advanced Spaceborne Thermal Emission and Reflection Radiometer (ASTER) DEMs, high-resolution GPS accuracy assessment of both DEM sets at Nevado Coropuna indicates that the SRTM is the more reliable elevation dataset and does not contain the significant vertical bias of the ASTER DEM (Racoviteanu et al. 2007). The specified SRTM accuracy standard is ±16 m for global coverage (Rabus et al. 2003), and Racoviteanu et al. (2007) determined an absolute vertical accuracy of 23.4 ± 10 m for the SRTM 22_16 tile.

We used the ArcGIS Spatial Analyst Slope function to construct a raster based on percent slope and the ArcGIS Path Distance function to produce a cumulative cost raster. How well are slope-derived cost values scaled with respect to the true energetic cost to humans moving across a topographic landscape? To address this issue, we calculated a series of LCPs, allowing comparison of results derived from a push-button, slope-based cost raster with a set of energy expenditure–based cost rasters.

Following methodology suggested by White (personal communication 2009 and this volume), we used the Pandolf et al. (1977) formula for constructing energy-based cost surfaces. We chose this approach for three reasons: (1) difficulty in applying two formulas to separate portions of a single raster, (2) the fact that travel from Quebrada Jaguay to any of the Alca-1 obsidian deposits would require substantially greater level and uphill movement relative to downhill movement, and (3) Santee et al.'s (2001:565) evaluation that "for input into a simulation model, it is far more important to accurately estimate uphill and level load-carriage costs than downhill costs."

TABLE 3.1. Parameter Combinations Used in Energy-Based Cost Rasters and Model Results.

Model #	Model Description	Weight (W) in kg	Walking speed (V) in m/s	Burden weight (L) in kg	Model Result
1	Lightweight forager, average speed, zero burden	50	1.2	0	1
2	Heavyweight forager, average speed, zero burden	70	1.2	0	1
3	Lightweight forager, fast speed, zero burden	50	2.4	0	2
4	Heavyweight forager, fast speed, zero burden	70	2.4	0	2
5	Lightweight forager, slow speed, zero burden	50	0.4	0	3
6	Heavyweight forager, slow speed, zero burden	70	0.4	0	3
7	Lightweight forager, average speed, heavy burden	50	1.2	25	1
8	Heavyweight forager, average speed, heavy burden	70	1.2	25	1
9	Lightweight forager, fast speed, heavy burden	50	2.4	25	2
10	Heavyweight forager, fast speed, heavy burden	70	2.4	25	2
11	Lightweight forager, slow speed, heavy burden	50	0.4	25	3
12	Heavyweight forager, slow speed, heavy burden	70	0.4	25	3

We chose a single terrain coefficient based on categories reported in Soule and Goldman (1972). Sediments vary widely throughout the study area and include sandy slopes near the Pacific coast, alluvium within river valleys, and desert pavements, compact volcanic sediments, and bedrock at higher elevations. Most of the arid western Andean slope and plateau are unvegetated or very sparsely vegetated, corresponding to Soule and Goldman's dirt road (η = 1.0) and light brush (η = 1.2) categories. We used a mean terrain coefficient (η) of 1.1.

Anthropometric studies of male and female hunter-gatherers throughout the world provided a range of values for forager weight (W). Kelly (1995:102, Tables 3–6) reports 43–66 kg and 38–56 kg as ranges of mean weights for male and female foragers, respectively. Jenike (2001:223–224, Table 5) lists 42.9–59.9 kg as the range for males and 37.8–51.8 kg as the range for females. We used two weight values, 50 kg and 70 kg, for our models. We were unable to find any anthropological data on the weight of burdens carried by hunter-gatherers, so we propose a 0–25-kg range of burden values as realistic for Paleoindian foragers. We calculated LCPs using minimum and maximum burden weight values, 0 kg and 25 kg, respectively.

Physiological walking experiments (McDonald 1961; Soule and Goldman 1972; Pandolf et al. 1977; Santee et al. 2001) and observational studies of modern city pedestrians throughout the world (Knoblauch et al. 1996; http://www.richardwiseman.com/quirkology/pace_home.htm) provide a range of values for average human walking speeds. Experiments normally tested walking speeds between 0.69 m/s and 2.4 m/s. This 0.69 m/s lower limit of walking speed "is seldom reached while an individual is walking under normal conditions," while at walking speeds greater than 2.4 m/s "the efficiency of walking becomes lower than running" (Pandolf et al. 1977). McDonald (1961) determined that the least cost walking speed was 1.3 m/s.

We used three speed values—2.4 m/s, the upper speed limit for walking; a medium-range/optimal value of 1.2 m/s; and an extreme low-speed value of 0.4 m/s—in our LCA. To construct the energy-based cost rasters, we incorporated grade (G) values from the ArcGIS-produced percent slope raster and values for the other parameters using the ArcGIS Raster Math functions. We constructed a series of cost surfaces using various possible combinations of forager weights, burden weights, and mean walking speeds, shown in Table 3.1.

3.7 Results

Push-button LCPs based on slope suggest that the Majes River plateau is the optimal route to *both* the lower-elevation and high-elevation Alca-1 deposits. The slope-based Path Distance

LCPs for the Alca-1 destinations are shown on the hillshaded SRTM DEM (Plate 2 A). Regardless of whether the Alca-1 destination is at the bottom of the Cotahuasi Canyon, at the Pucuncho Basin, or at the top of Cerro Condorsayana, the optimal route follows the Majes River Valley to the plateau west of Nevado Coropuna. The paths diverge just southwest of the Pucuncho Basin.

Energy-based LCPs were quite similar to all three Alca-1 destinations, despite varying input parameters in the twelve models. All Path Distance LCPs follow the east or west rim of the Quebrada Manga to ascend the high-altitude plateau (Plate 2 B, C, D). No energy-based LCPs follow the Ocoña-Cotahuasi drainage, even to access the lower-elevation Alca-1 obsidian deposit at the bottom of the Cotahuasi Canyon.

Both energy-based LCPs and slope-based LCPs connect Alca-1 deposits to the edge of the plateau west of the town of Chuquibamba. However, the energy-based LCPs are different from the slope-based LCPs for the southern portion of the project area, which encompasses the ~4000 m of relief from the edge of the plateau south to the Pacific coast. This difference indicates that the ArcGIS Spatial Analyst's automated scaling of cost values from slope values does not well approximate the actual energetic cost to humans moving over rugged terrain. We suggest that the cost of navigating steeper terrain is exaggerated when the ArcGIS Spatial Analyst push-button feature is used to produce a cost surface from slope values alone, and we caution against reliance on this user-friendly, but perhaps oversimplified, function. We believe researchers may obtain more reliable results by experimenting with the scaling of the slope cost values and comparing the resulting solutions or by comparing a slope-based LCP with LCPs based on energy expenditure, time, or some other appropriate and accurately quantified measure of cost.

The twelve models result in three path variations to each of the Alca-1 destinations, and comparison of these results allows the identification of active parameter(s) affecting path variance. Result 1 (Plate 2 B) and Result 2 (Plate 2 C) are only slightly different, whether the input values for W (forager weight, in kg) are at the lower or upper ends of the ethnographic ranges for modern hunter-gatherers, whether L (forager burden, in kg) is 0 or 25 kg, or whether V (walking speed) is close to the observed mean value of 1.3 m/s (McDonald 1961) or at the upper limit of human walking speed, 2.4 m/s (Pandolf et al. 1977). Forager walking speed is the active variable determining the model result, not forager weight or burden weight.

LCPs using a V of 0.4 m/s, a value even lower than the lowest recorded human walking speed of 0.58 m/s (http://www.richardwiseman.com/quirkology/pace_home.htm), are quite different from the other solutions, regardless of changes in the other input parameters. This Result 3 (Plate 2 D) approximates a straight line between source and destinations irrespective of topography, indicating that there is some velocity threshold below which sloped terrain fails to influence the LCA algorithm. For this reason, Result 3 is unlikely to represent a realistic optimal path. Rerunning the analyses with the $V = 0.58$ m/s value results in a LCP nearly identical with Results 1 and 2.

The high degree of similarity among energy-based LCPs indicates that variations in forager weight, burden weight, and to some extent walking speed do not greatly alter model results, supporting the reliability of the method. On the other hand, the LCPs differ slightly, suggesting that an archaeological survey strategy should encompass the variation among LCPs by targeting an area that encloses all the modeled pathways. We can compare LCPs to the three Alca-1 deposits in terms of overall distance (Table 3.2). Energy expenditure–based LCPs to all three Alca-1 sources are much shorter in overall distance than their slope-based counterparts (see Kantner this volume). Comparison of all LCPs shows that the Pucuncho Basin is the closest of the three Alca-1 deposits to Quebrada Jaguay, at a distance of ~225 km or ~145 km, according to respective slope and realistic (Results 1 and 2) energy-based LCPs. The Cotahuasi Canyon and Cerro Condorsayana Alca-1 deposits are about equidistant from Quebrada Jaguay, at ~30–40 km additional distance.

Table 3.3 compares the metabolic rates (measures of the total energetic expenditures) of the LCPs to the three Alca-1 deposits. LCPs from Quebrada Jaguay to the Pucuncho and Cerro Condorsayana Alca-1 obsidian deposits involve only uphill or level movement, allowing accurate

TABLE 3.2. Distances (in km) of LCPs from Quebrada Jaguay to Alca-1 Deposits.

Model Result	Model Description	QJ-280 to Cotahuasi Alca-1	QJ-280 to Condorsayana Alca-1	QJ-280 to Pucuncho Alca-1
Slope	Push-button path distance	266.3	265.5	223.7
Energy Result #1	Average speed	177.8	181.9	147.1
Energy Result #2	Fast speed	174.7	180.1	145.7
Energy Result #3	Slow speed	163.1	165.8	137.3

TABLE 3.3. Metabolic Rates (in megawatts) of LCPs from Quebrada Jaguay to Alca-1 Deposits.

Model Result	Model Description	QJ-280 to Cotahuasi Alca-1	QJ-280 to Condorsayana Alca-1	QJ-280 to Pucuncho Alca-1
Energy Result # 1	Average speed	>63.328	58.881	48.725
Energy Result #2	Fast speed	>155.241	147.132	121.212
Energy Result #3	Slow speed	>10.961	11.009	9.057

modeling of energetic expenditure to those destinations (but see next paragraph). Because the Cotahuasi Canyon Alca-1 LCP involves ~2500 m of elevation loss from the plateau to the base of the canyon, our approach cannot accurately calculate the cost of travel on the steep, downhill segment of that route since it does not take into account the energetic cost of braking. Without a consideration of this braking action, the reported value for the LCP to the Cotahuasi Alca-1 deposit is an *underestimate* of the total cost of reaching that deposit. LCPs based on realistic walking speeds (Results 1 and 2) consistently indicate that the Pucuncho Alca-1 deposit is the least costly of the three obsidian deposits to access and that the Cotahuasi Alca-1 deposit is the most costly. Not surprisingly, LCP results based on slower walking speeds require less overall energy expenditure than those based on faster walking speeds.

Pandolf et al. (1977) derived their equation from experiments at low elevation, and it therefore does not take into account the additional energy required for work at altitudes above 2500 m (Aldenderfer 1998). Our calculations of total energy expenditures to the high-elevation Alca-1 deposits (Table 3.3) are probably *underestimates.* However, the LCP reported value for the Cotahuasi Canyon Alca-1 deposit is also an underestimate, and the Cotahuasi Canyon Alca-1 deposit remains the farthest of the three from Quebrada Jaguay in terms of total distance (Table 3.2). Moreover, the energy expenditure method we use produces results that compare favorably with those produced by the Tobler hiking function method (see Kantner this volume), suggesting that were we to use time as a cost currency, the resulting LCP to the Cotahuasi Canyon Alca-1 deposit would be more costly than LCPs to the high-altitude Alca-1 deposits. To calculate the energetic cost of navigating high-altitude terrain more accurately, one could gather experimental data quantifying the additional energy expenditure needed to perform the same work at altitude relative to sea level, add an altitude modifier to the Pandolf et al. (1977) equation, and rerun the analysis using a new energy-based cost raster.

3.8. Testing the Models: Archaeological Investigations and Paleogeography

LCA has served as a useful predictive tool to guide archaeological survey of the coast-highland corridor. We consider areas where LCPs intersect potentially attractive forager habitats (e.g., perennial streams, wetlands) to have a high potential for containing early archaeological sites, as these locales satisfy optimal foraging theory predictions for both optimal routes and central-place site locations. Since we are investigating prehistoric use of LG landscapes, it has been important to determine the paleogeography of the corridor at 10,600–13,000 cal BP, as well as any post-LG landscape changes that may have affected preservation of Paleoindian archaeological evidence.

Although the energy-based LCPs are optimal with respect to topography and are theoretically more robust than the slope-based LCPs, the energy-based LCPs are non-optimal in light of our current understanding of local hydrology. The energy-based LCPs follow the east rim of the intermittent Quebrada Manga drainage to the plateau west of the town of Chuquibamba. We explored much of this route in 2006, and although we found Inca Road segments that followed the east rim of the canyon toward the highlands, we failed to find potentially early sites. Even if this route is used during the rainy season, as we did in our exploration, the east rim of the Manga Canyon currently contains no springs and is ~600 m above the river bed. Travel between the rim and the base of the canyon is impossible or difficult along much of the sheer-walled route, and continuous coast-highland travel within the canyon is not possible due to steep bedrock ledges and slots. Above ~1200 m elevation along the canyon rim is a zone of absolute desert extending for ~50 km northward. Above 4000 m elevation there are perennial stream flows, plant and animal resources, and potentially early archaeological sites.

In contrast, the slope-based LCP follows the Majes River, a large, resource-rich perennial corridor drainage. Paleoindian foragers probably would have discovered and inhabited the Majes drainage, especially since the intermittent availability of fresh water at the Quebrada Jaguay locale would have precluded year-round occupation there (Sandweiss et al. 1998). Sites could have been located on Pleistocene terraces along the Majes Valley from the coast at modern-day Camaná north to Chuquibamba.

Given the dramatic ecologic differences between the slope-based and energy-based LCPs, the study area might seem ideal for testing whether topographic vs. resource optimality influenced the location of early forager coast-highland routes in southern Peru. Both coast-highland corridors would have to be surveyed systematically if potential Paleoindian-age sites were to be identified. However, due to a number of geomorphic processes and subsequent prehistoric and modern occupation of some areas, segments of the Majes Valley LCP are unlikely to contain intact Paleoindian evidence.

Profound landscape alteration within the Majes Valley probably has obscured or destroyed any potential LG-age archaeological sites in open-air settings. Fed by an extensive 17,400-km^2 highland catchment, the Majes River is a wide, braided stream with a high discharge and no intact terraces below the Capiza River confluence at ~950 m elevation (Steffen et al. 2009). Geomorphic study and radiocarbon dating of fluvial landforms of the Moquegua River, ~200 km southeast of the Majes River, revealed that 80 percent of the Moquegua River floodplain is younger than 550 ^{14}C BP, a result of periodic El Niño–Southern Oscillation (ENSO) floods (Manners et al. 2007). The same situation may apply to the Majes River floodplain. Torrential ENSO rainfall, episodic in this region, would likely erode unvegetated canyon slopes, causing colluvial sediment deposition and burial of the proximal surfaces of fans (Keefer and Moseley 2004). If any Pleistocene landforms with early archaeological sites survived these geomorphic changes, at least 4000 years of irrigation agriculture have modified them. Any arable and habitable land surfaces have been plowed, canalized, terraced, or occupied by more recent settlements and roads, probably destroying or obscuring any early archaeological sites that might exist, at least in the heavily populated portions of the Majes Valley.

Consequently, we have focused our survey efforts on the northern portion of the coast-highland corridor, corresponding to the plateau between Chuquibamba and the three Alca-1 obsidian deposits (Plate 3). All LCPs pass through this area, and the relative geomorphic stability makes it the most likely area for extant Paleoindian sites. Relative to corridor drainages such as the Majes River, the plateau is conducive to archaeological investigations, for several reasons. First, low sedimentation rates and limited fluvial erosion result in good landform preservation and site visibility. Second, limited human disturbance of archaeological sites has occurred, since there are no major population centers, agriculture, or looting. Third, rockshelters are abundant here, and rockshelters were commonly used as campsites by early foragers elsewhere in the Central Andes (Rick 1980; Santoro and Núñez 1987; Lavallee et al. 1995; Núñez et al. 2002) and exhibit great potential for preservation of datable organics (Goldberg and Macphail 2006).

We restricted our survey coverage to an area that encompassed one or two valleys to either side of the LCPs between Chuquibamba and Alca-1 obsidian deposits. We especially focused survey efforts on intersections of LCPs with perennial streams and wetlands, as previous work has shown these habitats to be foci for early Central Andean foragers (Aldenderfer 1998; Grosjean et al. 2005; Quade et al. 2008).

From our surveys, we have identified rockshelters in andesite exposures and open-air archaeological sites on alluvial fans surrounding the Pucuncho Basin (Plate 3). Additional shelters occur along the Río Blanco and other drainages southwest of Nevado Coropuna. These open-air sites and rockshelters likely preserve datable evidence of early human occupations. Several of these sites contain artifacts made of coastal materials, substantiating Pacific coast–Pucuncho connections. In the Río Blanco drainage we discovered a Type 4A projectile point (~9400–11,000 cal BP) (Klink and Aldenderfer 2005) made of pink chalcedony, a material that crops out on the intervalley coast. At Pampa Colorada, just west of Quebrada Jaguay, McInnis (2006) identified six Type 4A specimens, one made of petrified wood and five of obsidian. The obsidian specimens have not yet been sourced geochemically, but we speculate that they are made of Alca obsidian. On the western margin of the Pucuncho Basin where LCPs intersect the Pucuncho Alca-1 deposit, we identified an open-air site containing a weathered fine-grained andesite Fishtail projectile point (Plate 3). This point is the first of its kind to be found in the high Andes (above 4000 m elevation) and indicates that people were at Pucuncho between ~11,700 and 12,800 years ago, the well-established age range for Fishtail points (Jackson 2006).

The Pucuncho Basin contains one of the largest and most productive wetlands in the western Peruvian Andes (Peru Ministry of Agriculture, personal communication 2008), supporting ~300 permanent pastoralists and ~10,000 domesticated camelids, in addition to wild vicuña. Vegetation is dominated by *Distichia muscoides* mats. These *bofedal* communities are the preferred forage for camelids (Koford 1957; Webster 1973; Franklin 1981), which, in turn, have been the most important animal resource for prehistoric Andean human populations (Wheeler et al. 1976; Flores-Ochoa 1979; Santoro and Núñez 1987; Rick 1988; Lavallee et al. 1995; Núñez et al. 2002; Aldenderfer 2008). In summary, the Pucuncho Basin meets all the criteria for an optimal central place for Andean foragers, as defined by Rick (1980) and Aldenderfer (1998) for the Central Andean highlands.

Preliminary glacial geologic research indicates that glaciers on Nevados Coropuna, Solimana, and Firura advanced during the LG (Bromley et al. 2009; Bromley, Hall, Rademaker et al. 2011; Bromley, Hall, Schaefer et al. 2011) in response to an atmospheric cooling of as much as 3°C. Although the extent of this advance was small relative to maximum ice-age conditions (~20,000–25,000 years ago; Bromley et al. 2009) and did not present a physical barrier to foragers (Plate 3), it is likely that the regional 0°C isotherm was depressed from its modern elevation of ~4900 m (Dornbusch 1998) to as low as ~4400 m. Such a cooling would place large portions of the study area in the zone of discontinuous (seasonal) permafrost (0 to -2°C) and much of the land above ~4800 m in continuous permafrost. Since the Pucuncho Basin lies at an altitude of ~4400 m or less, surrounding rockshelter and open-air sites likely were just below the permafrost zone. Early foragers may have been exploring this environment as local wetlands were thawing and becoming productive.

With the exception of extensive lithic workshop activity at the Cerro Condorsayana Alca-1 obsidian deposit, our systematic survey efforts failed to find any potentially early archaeological sites along the LCPs to the Cerro Condorsayana and Cotahuasi Canyon Alca-1 deposits, even in a zone of many lakes surrounding Nevado Firura (Plate 3). This absence of early evidence makes sense in light of LG paleoenvironmental data, as the plateau north of the Pucuncho Basin likely was permafrozen and not biologically productive when people were first exploring this area of the Andes.

3.9. Concluding Thoughts

Our discovery of highland sites containing coastal lithic materials and diagnostic projectile points dating from the LG and EH attests to the potential of an interdisciplinary research design combining optimal foraging theory, predictive LCA, paleoenvironmental study, and systematic archaeological survey to find early forager archaeological sites. The explicit assumption of forager optimality has guided our investigations, and the resulting discovery of a Paleoindian projectile point and other potentially early evidence at the Pucuncho Basin suggests that this assumption is correct. However, LCP generation and site discovery can tell us nothing about the nature of the relationships among the sites—only archaeological excavations and comparative material analyses can do that. We plan further archaeological and paleoecologic investigations of the Pucuncho area to determine more precisely the timing of initial settlement of this area and the environmental conditions under which settlement occurred. Because the methodology we have described is particularly suited for site assemblages containing exotic materials, a feature of many Paleoindian sites, it will be applicable elsewhere in the Americas, especially in arid zones with good site preservation and visibility. Further use and refinement of LCA in survey efforts should help reveal many new Paleoindian sites.

Certain features of our study area or the period we are investigating may contribute to the success of LCA. LCA may be most useful when one is dealing with high-relief regions characterized by extremely rugged topography, whereas areas with many possible similar-cost solutions may be more difficult for the technique to resolve (Kantner this volume). Forager mobility may be amenable to optimal path modeling, particularly for groups colonizing empty landscapes, since there are simply fewer variables potentially affecting route choice that archaeologists must consider. On the other hand, successful investigation of forager land use behavior must take into account paleogeographic changes and the ephemeral nature of many early forager sites, as well as the effects of long-term geomorphic processes and land use and modification by subsequent human populations.

When LCA is used for archaeological prediction, it is important to consider other factors beyond terrain difficulty, such as the availability of fresh water and biotic resources, particularly in arid environments, where resources are unequally distributed. For most projects, time and economic constraints preclude systematic survey of the entire extent of LCPs, which may cover large geographic areas. One way to focus site discovery efforts is to target areas where LCPs intersect one another and intersect potentially attractive forager habitats, such as perennial streams, wetlands, or other ecologically productive zones. Such intersections satisfy optimal foraging theory predictions for both optimal routes and central-place site locations. It is critical to remember that the geographic distribution of ecologically optimal zones cannot be assumed to have been the same in the distant past as today. When properly constrained by local paleogeographic data and tested through systematic survey efforts, LCP modeling offers the archaeologist a powerful predictive tool for connecting the dots.

Acknowledgments

We thank Sarah Surface-Evans and Devin White for organizing a thought-provoking Society for American Archaeology session on LCA and for inviting our contribution to this book. Session discussants John Kantner and David Anderson provided helpful reviews and advice. Devin White's expertise in LCA was instrumental to the success of the modeling work reported here. Fieldwork was made possible through generous support from the Dan and Betty Churchill Exploration Fund. We especially thank all the fine people who helped us during our coast-highland explorations: Saúl Cerón of Inca Tours Peru, Arequipa; the residents of the Chagenoc oasis, Pachana, and Chanchallay; and the Zuñiga and Huisacaina families of the Pucuncho Basin.

References

Aldenderfer, M. S.

1998 *Montane Foragers: Asana and the South-Central Andean Archaic.* University of Iowa Press, Iowa City.

1999 The Pleistocene/Holocene Transition in Peru and Its Effects upon Human Use of the Landscape. *Quaternary International* 53–54: 11–19.

2006 Modeling Plateaux Peoples: The Early Human Use of the World's High Plateaux. *World Archaeology* 38(3):357–370.

2008 High Elevation Foraging Societies. In *Handbook of South American Archaeology*, edited by H. Silverman and W. H. Isbell, pp. 131–144. Springer, New York.

Anderson, D. G., and J. C. Gillam
2000 Colonization of the Americas: Implications from an Examination of Physiography, Demography, and Artifact Distribution. *American Antiquity* 65:43–66.
Beaton, J. M.
1991 Colonizing Continents: Some Problems from Australia and the Americas. In *The First Americans: Search and Research*, edited by T. D. Dillehay and D. J. Meltzer, pp. 209–230. CRC Press, Boca Raton, Florida.
Bromley, G. R. M., B. L. Hall, K. M. Rademaker, C. E. Todd, and A. E. Racoviteanu
2011 Late Pleistocene Snowline Fluctuations at Nevado Coropuna (15°S), Southern Peruvian Andes. *Journal of Quaternary Science* 26(3): 305–317.
Bromley, G. R. M., B. L. Hall, J. M. Schaefer, G. Winckler, C. E. Todd, and K. M. Rademaker
2011 Glacier Fluctuations in the Southern Peruvian Andes during the Late-Glacial Period, Constrained with Cosmogenic ^{3}He. *Journal of Quaternary Science* 26(1):37–43.
Bromley, G. R. M., J. M. Schaefer, G. Winckler, B. L. Hall, C. E. Todd, and K. Rademaker
2009 Relative Timing of Last Glacial Maximum and Late Glacial Events in the Tropical Andes. *Quaternary Science Reviews* 28(23–24):2514–2526.
Burger, R. L., and F. Asaro
1993 La distribución y procedencia de artefactos de obsidiana durante el período inicial y horizonte temprano. In *Emergencia de la civilización en los Andes: Ensayos de interpretación*, edited by R. L. Burger, pp. 180–231. Universidad Nacional Mayor de San Marcos, Lima, Peru.
Burger, R. L., F. Asaro, P. Trawick, and F. Stross
1998 The Alca Obsidian Source: The Origin of Raw Material for Cuzco Type Artifacts. *Andean Past* 5:185–202.
Burger, R. L., K. L. Mohr Chávez, and S. J. Chávez
2000 Through the Glass Darkly: Prehispanic Obsidian Procurement and Exchange in Southern Peru and Northern Bolivia. *Journal of World Prehistory* 3:267–362.
Collins, M. B.
1991 Rock Shelters and the Early Archaeological Record in the Americas. In *The First Americans: Search and Research*, edited by T. D. Dillehay and D. J. Meltzer, pp. 157–175. CRC Press, Boca Raton, Florida.
deFrance, S. D., D. K. Keefer, J. B. Richardson III, and A. Umire Alvarez
2001 Late Paleo-Indian Coastal Foragers: Specialized Extractive Behavior at Quebrada Tacahuay, Peru. *Latin American Antiquity* 12:413–426.
Dornbusch, U.
1998 Current Large-Scale Climatic Conditions in Southern Peru and Their Influence on Snowline Altitudes. *Erdkunde* 52:41–54.
Field, J. S., and M. Mirazon Lahr
2005 Assessment of the Southern Dispersal: GIS-Based Analyses of Potential Routes at Oxygen Isotopic Stage 4. *Journal of World Prehistory* 19(1):1–45.
Flores Ochoa, J. A.
1979 *Pastoralists of the Andes: The Alpaca Herders of Paratia*. Institute for the Study of Human Issues, Philadelphia.
Franklin, W. L.
1981 Biology, Ecology, and Relationship to Man of the South American Camelids. In *Mammalian Biology in South America*, edited by M. A. Mares and H. H. Genoways, pp. 457–489. Special Publication Series, Vol. 6. Pymatuning Laboratory of Ecology, University of Pittsburgh.
Glascock, M. D., G. E. Braswell, and R. H. Cobean
1998 A Systematic Approach to Obsidian Source Characterization. In *Archaeological Obsidian Studies*, edited by M. S. Shackley, pp. 15–65. Plenum Press, New York.
Goldberg, P., and R. I. Macphail
2006 *Practical and Theoretical Geoarchaeology*. Blackwell, Malden, Massachusetts.
Grosjean, M., L. Núñez, and I. Cartajena
2005 Palaeoindian Occupation of the Atacama Desert, Northern Chile. *Journal of Quaternary Science* 20:643–653.
Jackson, L. J.
2006 Fluted and Fishtail Points from Southern Coastal Chile: New Evidence Suggesting Clovis- and Folsom-Related Occupations in Southernmost South America. In *Paleoindian Archaeology: A Hemispheric Perspective*, edited by J. E. Morrow and C. Gnecco, pp. 105–122. University of Florida Press, Gainesville.
Jenike, M. R.
2001 Nutritional Ecology: Diet, Physical Activity and Body Size. In *Hunter-Gatherers: An Interdisciplinary Perspective*, edited by C. Panter-Brick, R. H. Layton, and P. Rowley-Conwy, pp. 205–238. Cambridge University Press, New York.
Jennings, J., and M. D. Glascock
2002 Description and Method of Exploitation of the Alca Obsidian Source, Peru. *Latin American Antiquity* 13(1):107–118.

Kaplan, H., and K. Hill
1992 The Evolutionary Ecology of Food Acquisition. In *Evolutionary Ecology and Human Behavior*, edited by E. A. Smith and B. Winterhalder, pp. 167–201. Aldine de Gruyter, New York.

Keefer, D. K., S. D. deFrance, M. E. Moseley, J. B. Richardson III, D. R. Satterlee, and A. Day-Lewis
1998 Early Maritime Economy and El Niño Events at Quebrada Tacahuay, Peru. *Science* 281:1833–1835.

Keefer, D. K., and M. E. Moseley
2004 Southern Peru Desert Shattered by the Great 2001 Earthquake: Implications for Paleoseismic and Paleo–El Niño–Southern Oscillation Records. *Proceedings of the National Academy of Sciences* 101:10878–10883.

Kelly, R. L.
1983 Hunter-Gatherer Mobility Strategies. *Journal of Anthropological Research* 39:277–306.
1995 *The Foraging Spectrum*. Smithsonian Institution Press, Washington, D.C.
2003 Colonization of New Land by Hunter-Gatherers: Expectations and Implications Based on Ethnographic Data. In *Colonization of Unfamiliar Landscapes: The Archaeology of Adaptation*, edited by M. Rockman and J. Steele, pp. 44–58. Routledge, New York.

Kelly, R. L., and L. C. Todd
1988 Coming into the Country: Early Paleoindian Hunting and Mobility. *American Antiquity* 53:231–244.

Klink, C. J., and M. S. Aldenderfer
2005 A Projectile Point Chronology for the South-Central Andean Highlands. In *Advances in Titicaca Basin Archaeology, Vol. 1*, edited by C. Stanish, A. B. Cohen, and M. S. Aldenderfer, pp. 25–54. Cotsen Institute of Archaeology at UCLA, Los Angeles.

Knoblauch, R. L., M. T. Pietrucha, and M. Nitzburg
1996 Field Studies of Pedestrian Walking Speed and Start-Up Time. *Transportation Research Record: Journal of the Transportation Research Board* 1538:27–38.

Koford, C. B.
1957 The Vicuña and the Puna. *Ecological Monographs* 27(2):153–219.

Lavallee, D., M. Julien, J. Wheeler, and C. Karlin
1995 *Telarmachay: Cazadores y pastores prehistóricos de los Andes*. Instituto Frances de Estudios Andinos, Lima, Peru.

Manners, R. B., F. J. Magilligan, and P. S. Goldstein
2007 Floodplain Development, El Niño, and Cultural Consequences in a Hyperarid Andean Environment. *Annals of the Association of American Geographers* 97(2):229–249.

Marín Arroyo, A. B.
2009 The Use of Optimal Foraging Theory to Estimate Late Glacial Site Catchment Areas from a Central Place: The Case of Eastern Cantabria, Spain. *Journal of Anthropological Archaeology* 28(1):27–36.

McDonald, I.
1961 Statistical Studies of Recorded Energy Expenditure of Man. *Nutrition Abstracts and Reviews* 31(3):739–762.

McInnis, H.
2006 Middle Holocene Culture and Climate on the South Coast of Peru: Archaeological Investigation of the Pampa Colorada. Unpublished Ph.D. dissertation, University of Oregon, Eugene.

Meltzer, D. J.
2002 What Do You Do When No One's Been There Before? Thoughts on the Exploration and Colonization of New Lands. In *The First Americans*, edited by N. G. Jablonksi, pp. 27–58. Memoirs of the California Academy of Sciences 27. San Francisco.

Morgan, C.
2008 Reconstructing Prehistoric Hunter-Gatherer Foraging Radii: A Case Study from California's Southern Sierra Nevada. *Journal of Archaeological Science* 35:247–258.

Moseley, M.
1983 White Holes and Warps. In *Prehistoric Settlement Patterns: Essays in Honor of Gordon R. Willey*, edited by Evon Z. Vogt and Richard M. Leventhal. University of New Mexico Press, Albuquerque.

Núñez, L., M. Grosjean, and I. Cartajena
2002 Human Occupations and Climate Change in the Puna de Atacama, Chile. *Science* 298:821–824.

Pandolf, K. B., B. Givoni, and R. F. Goldman
1977 Predicting Energy Expenditure with Loads while Standing or Walking Very Slowly. *Journal of Applied Physiology* 43(4):577–581.

Quade, J., J. A. Rech, J. L. Betancourt, C. LaTorre, B. Quade, K. A. Rylander, and T. Fisher
2008 Paleowetlands and Regional Climate Change in the Central Atacama Desert, Northern Chile. *Quaternary Research* 69:343–360.

Rabus, B., M. Eineder, A. Roth, and R. Bamler
2003 The Shuttle Radar Topography Mission — A New Class of Digital Elevation Models Acquired by Spaceborne Radar. *ISPRS Journal of Photogrammetry and Remote Sensing* 57(4):241–262.

Racoviteanu, A. E., W. F. Manley, Y. Arnaud, and M. W. Williams
2007 Evaluating Digital Elevation Models for Glaci-

ologic Applications: An Example from Nevado Coropuna, Peruvian Andes. *Global and Planetary Change* 59:110–125.
Rademaker, K.
2006 Geoarchaeological Investigations of the Wayñuna Site and the Alca Obsidian Source, Peru. Unpublished Master's thesis, University of Maine, Orono.
Rick, J. W.
1980 *Prehistoric Hunters of the High Andes*. Academic Press, New York.
1988 The Character and Context of Highland Preceramic Society. In *Peruvian Prehistory*, edited by R. W. Keatinge, pp. 3–40. Cambridge University Press, New York.
Sandweiss, D. H.
2003 Terminal Pleistocene through Mid-Holocene Archaeological Sites as Paleoclimatic Archives for the Peruvian Coast. *Palaeogeography, Palaeoclimatology, Palaeoecology* 194:23–40.
Sandweiss, D. H., H. McInnis, R. L. Burger, A. Cano, B. Ojeda, R. Paredes, M. Sandweiss, and M. Glascock
1998 Quebrada Jaguay: Early Maritime Adaptations in South America. *Science* 281:1830–1832.
Santee, W. R., W. F. Allison, L. A. Blanchard, and M. G. Small.
2001 A Proposed Model for Load Carriage on Sloped Terrain. *Aviation, Space, and Environmental Medicine* 72(6):562–566.
Santoro, C. M., and L. Núñez
1987 Hunters of the Dry Puna and the Salt Puna in Northern Chile. *Andean Past* 1:57–109.
Soule, R. G., and Goldman, R. F.
1972 Terrain Coefficients for Energy Cost Prediction. *Journal of Applied Physiology* 32(5): 706–708.
Steele, J., and M. Rockman
2003 "Where Do We Go from Here?" Modeling the Decision-Making Process during Exploratory Dispersal. In *Colonization of Unfamiliar Landscapes: The Archaeology of Adaptation*, edited by M. Rockman and J. Steele, pp. 130–143. Routledge, New York.
Steffen, D., F. Schlunegger, and F. Preusser
2009 Late Pleistocene Fans and Terraces in the Majes Valley, Southern Peru, and Their Relation to Climatic Variations. *International Journal of Earth Sciences* 99(8):1975–1989.
Taliaferro, M. S., B. A. Schriever, and M. S. Shackley
2010 Obsidian Procurement, Least-Cost Path Analysis, and Social Interaction in the Mimbres Area of Southwestern New Mexico. *Journal of Archaeological Science* 37:536–548.
Tanner, B. R.
2001 Lithic Analysis of Chipped Stone Artifacts Recovered from Quebrada Jaguay, Peru. Unpublished Master's thesis, University of Maine, Orono.
Webster, S.
1973 Native Pastoralism in the South Andes. *Ethnology* 12(2):115–133.
Wheeler Pires-Ferreira, J. C., E. Pires-Ferreira, and P. Kaulicke
1976 Preceramic Animal Utilization in the Central Peruvian Andes. *Science* 194(4264):483–490.
White, D.
2007 Transportation, Integration, Facilitation: Prehistoric Trails Network of the Western Papaguería. Unpublished Ph.D. dissertation, University of Colorado, Boulder.
Winterhalder, B. P., and E. A. Smith
1981 *Hunter-Gatherer Foraging Strategies: Ethnographic and Archaeological Analyses*. University of Chicago Press, Chicago.
Winterhalder, B. P., and R. B. Thomas
1978 *Geoecology of Southern Highland Peru: A Human Adaptation Perspective*. Occasional Paper 27. Institute of Arctic and Alpine Research, University of Colorado, Boulder.

CHAPTER 4

Wandering the Desert

Least Cost Path Modeling for Water Transport Trails in the Jornada Mogollon Region, Fort Bliss, South-Central New Mexico

Shaun M. Phillips and Phillip O. Leckman

4.1. Introduction

Roads, trails, and footpaths present both challenges and opportunities for archaeologists. As a record of a patterned "landscape of movement" (Snead 2008:113), trail systems offer direct evidence for the network of interconnections linking sites, activity areas, and other archaeological phenomena in ways not immediately suggested by the distributions of these more circumscribed phenomena alone. The routes taken by archaeological trail systems across the landscape may reveal patterns of habitual movement or linked action (Tilley 1994:28–31), hint at the meaning invested in particular landscape features (Becker and Altschul 2008:436–438; Darling and Eiselt 2003; Darling 2006, 2009; Kantner 1997; Snead et al. 2009), or provide clues to political or social boundaries (Snead 2002, 2008, 2009). Recent research in the American Southwest (Altschul et al. 2005; Becker and Altschul 2008; Kantner 1997; Kludt 2007; Snead 2002, 2008, 2009; Snead et al. 2009) and elsewhere has demonstrated that the identification and delineation of prehistoric trail systems can offer productive insights into past movement, interconnection, and social organization in ways not easily afforded by more circumscribed archaeological phenomena.

While the significance of trail systems as a record of interconnection and meaning above the level of individual sites is thus an increasingly important and frequently discussed topic in contemporary landscape archaeology (e.g., Barrett 1994; Tilley 1994; Snead 2002, 2008; Snead et al. 2009), the identification and documentation of archaeological trails and linear features continue to pose significant challenges to archaeological inquiry. In most cases, the identification of prehistoric trail networks is dependent on the firsthand physical identification of sections of the linear networks themselves, although remote sensing and aerial imagery have been used effectively for the preliminary identification and delineation of segments of large, intensively constructed trail networks like the Chaco roads (e.g., Lyons and Hitchcock 1977; Ware and Gumerman 1977). However, the physical remnants of trails are often extremely difficult to detect because the continued visibility of individual trail segments is highly contingent on local geology and geomorphology. Often, physical evidence is absent entirely, leaving the routes taken by trail systems linking archaeological manifestations to be inferred or hypothesized based on "topography and . . . the presence of associated monuments" (Snead 2002:756). Even when physical evidence is present, however, it may be difficult or impossible to distinguish trails made by prehistoric humans from those created by livestock or other animals (Becker and Altschul 2008: 429–436).

One potential area for improvement in the identification of archaeological paths and trails may be an increased focus on the artifacts and other material culture associated with the use of trail systems, rather than the physical traces of trails themselves. The association of prehistoric material culture, particularly such sites or features as rock art or trailside cairns, with visible trail segments is often used to establish a particular trail's prehistoric origin or to provide general date ranges for trail use (e.g., Becker and Altschul 2008:422–426; Davidson 2009:27–40; Dore and McElroy 2006:1; Johnson and Johnson 1957:23–25; Snead 2002:761; Weaver 1967), but trails are rarely identified on this basis alone. In part, this may reflect the difficulties posed by traditional "site-based" pedestrian survey, in which artifacts associated with a site are intensively recorded at the site level while low-density scatters of artifacts are often disregarded as isolated occurrences and reduced in consideration to a general location and brief description in a report appendix. This immediate segregation into site and nonsite contexts and the difference in recording depth and precision it typically entails make a holistic examination of artifact distributions within a project area much more difficult, reducing the likelihood that linear patterning present within a dataset will be identified as the possible traces of a road or trail. The difficulty of identifying linear features is magnified in areas where alluvial and colluvial forces are active or where the feature was not formalized by construction, maintenance, or trail-side activity sites. Even where physical traces of potential prehistoric trails remain, site-based archaeology may also bias the identification of associated cultural materials, as archaeologists following a trail across a landscape may record archaeological deposits that intersect the trail while leaving other areas unsurveyed (Becker and Altschul 2008:430–431).

In contrast, if all identified archaeological manifestations are mapped and recorded at a comparable level of detail without an immediate determination of whether a site is present, the low archaeological visibility of linear features can be greatly enhanced, permitting the identification of potential trails even where their artifactual correlates are relatively low-density. Methods of survey that utilize this approach are also less likely to fall victim to "false positives" created by survey bias focused on the physical traces of trails, since it is the linear distribution of artifacts across a landscape that signals a potential trail, rather than any physical remnant of the trail itself. As this chapter demonstrates, potential linear patterns identified during intensive siteless survey can then be explored and evaluated via least cost path analysis, giving us a more nuanced understanding of the interconnections linking sites and other cultural manifestations in a particular local context.

In the remainder of this chapter, we explore a series of potential trails identified by recent archaeological survey and testing work at Fort Bliss, a large U.S. Army installation located in the Tularosa Basin/Hueco Bolson region of southern New Mexico and western Texas (Figure 4.1). First identified through the observation of linear distributions of ceramic artifacts within datasets derived from large-scale siteless survey, these trails appear to link areas of intensive prehistoric habitation with potential water sources, a key concern in this arid environment. By generating least cost paths modeled to explore a variety of potential trail end points and evaluate these paths in terms of the artifactual evidence, we may be able to more fully contextualize the potential trails suggested by artifact data in terms of the local cultural and physical landscape. The correlation or lack thereof between modeled and artifactual pathways may also serve to indicate the situations in which the cost of travel across a landscape was a central concern in path making and where other factors, such as lines of sight or the proximity of nearby areas of habitation or habitual activity, may have been of more importance. While these additional factors cannot be explored exhaustively here, we employ viewshed analysis as a way to begin to address the role played by local landscape features as potential landmarks influencing trail formation. Especially in open, level settings where relatively little separates "optimal" paths from less efficient routes across the landscape, our research suggests that continued line of sight with a potential destination or other physical landmarks may have significantly influenced where trails develop.

Before entering into this discussion, however, we briefly summarize the prehistory of the Tularosa Basin region, emphasizing changes in patterns of mobility and resource utilization over time, and provide context regarding the history of the siteless survey method used to identify the potential trail networks.

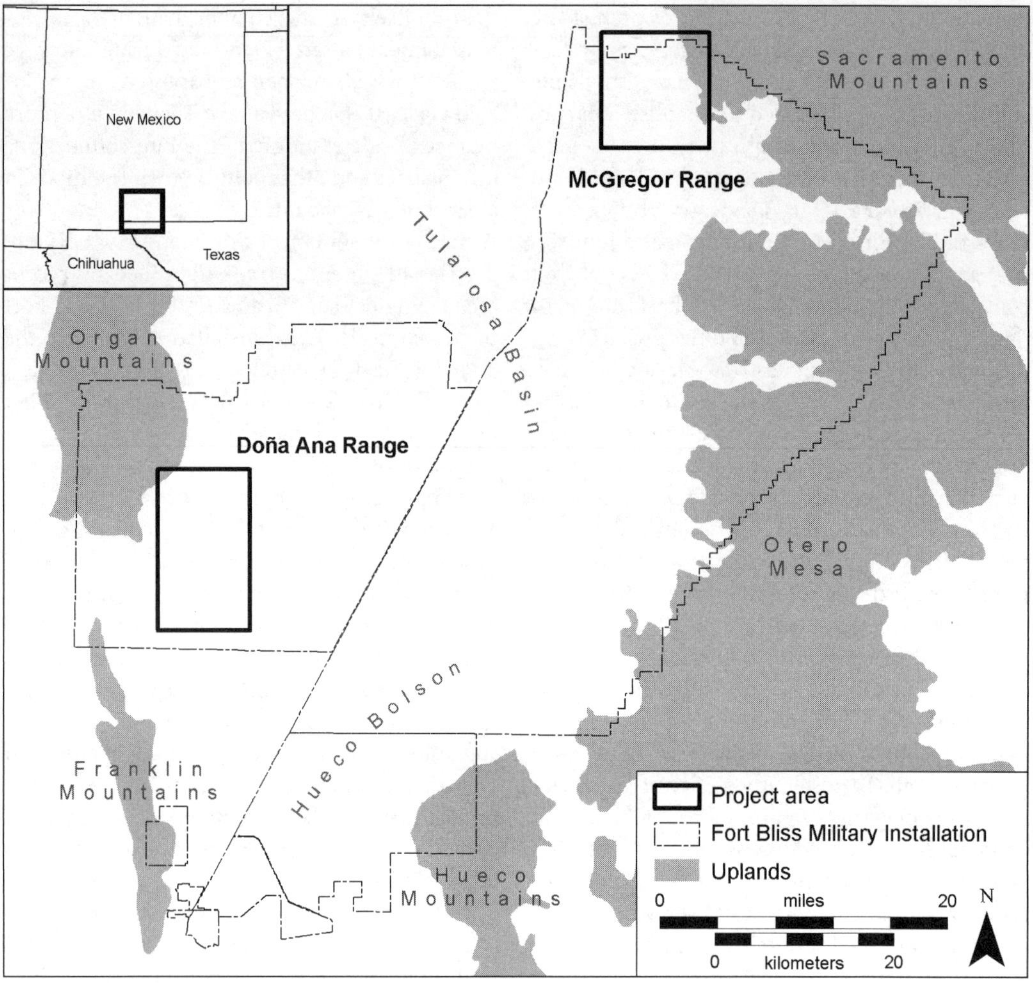

FIGURE 4.1. Overview of Tularosa Basin and location of project areas.

4.2. The Jornada Mogollon Sequence and the Prehistory of the Tularosa Basin

Located near the international boundary between the United States (Texas and New Mexico) and Mexico (Chihuahua), southern New Mexico's Tularosa Basin is among the largest of the broad, interiorly drained, north-south-trending basins that characterize western North America's Basin and Range geological province (Hawley and Kottlowski 1969; Kottlowski 1958; Schmidt et al. 2009). The basin lies within the arid Chihuahuan Desert and receives an average of only 8 to 10 inches (195–254 mm) of precipitation annually (Brown 1994; Schmidt et al. 2009). Although large alluvial fans fed by intermittent drainages originating in the Sacramentos, the Organs, and other surrounding mountain ranges ring the periphery of the basin, it contains no major permanent streams or bodies of water beyond a few saline lakes and ephemeral playas. Due to extremely deep sediments and minimal historic rainfall, the basin has a very deep, limited aquifer (Healy et al. 1978; Mattick 1967; Strain 1966). Accordingly, the scarcity of water has always structured and constrained human activity within the basin.

Good evidence for human occupation in the vicinity of the Tularosa Basin dates to the early Paleoindian period (Beckett 1983; Krone 1976; Vierra et al. 2009; Weber and Agogino 1968). By the late Paleoindian period, artifact distributions suggest an increased focus on playas, ephemeral drainages, and other water sources as a drier precipitation regime broadly similar to the current climate came into effect. The succeeding Early Archaic period (6000–4000 BC) remains poorly documented in the basin, suggesting the region may have been sparsely populated at this time (Vierra et al. 2009). Use of the area intensified during the Middle Archaic (4000–1200 BC) (Carpenter et al. 2005; Miller and Shackley 1998; Vierra et al. 2009) and continued to increase during the Late Archaic (1200 BC–AD 250), perhaps spurred to some degree by the introduction of maize into the region at the beginning of the period (Hard and Roney 2005; Vierra et al. 2009; Tagg 1996; Upham et al. 1987). Residential mobility is hypothesized to have decreased across the Late Archaic, however, coupled with the increasing use of maize agriculture focused on alluvial fan zones at the basin edge (Anderson 1993).

The Formative period in the Tularosa Basin begins with the onset of the Mesilla phase (AD 250–1000), marked by the initial appearance of ceramics, somewhat larger and more elaborate domestic structures, and, perhaps, an increasing reliance on maize agriculture (Vierra et al. 2009). However, lifeways generally resembled those of the preceding Late Archaic in many respects, and wild plant foods such as mesquite and cactus continued to dominate the diet. A somewhat different pattern of seasonal mobility has been proposed for this phase (Hard 1986; Whalen 1994) in which small-scale basin sites were used during summer and fall and larger sites in riverine settings or on the alluvial fans along the basin edge during winter and spring. The ubiquity of maize continued to increase during the Middle Formative Doña Ana phase (AD 1000–1275/1300), accompanied by growing residential stability focused on settings with high agriculture potential such as the distal ends of large alluvial fans, as well as increased architectural formalization (Vierra 2009; Vierra et al. 2009). By the Late Formative El Paso phase (AD 1275/1300–1450), larger, more aggregated populations were concentrated around areas with sufficient water resources to support the relatively intensive agriculture practiced during the period, particularly on the margins of large playas and at the distal ends of alluvial fans in locations with access to nearby fan surfaces for runoff farming (Leckman et al. 2009; Vierra 2009).

4.3. Transect Recording Unit Survey

Since the late 1980s, all archaeological contractors working on Fort Bliss have been required to use the Transect Recording Unit (TRU) methodology when conducting pedestrian survey (Kludt et al. 2007). This method is a nonsite survey strategy in which attribute and positional information are recorded at comparable levels of effort for all identified cultural materials (Miller et al. 2009:1.12). During fieldwork, all data are digitally recorded at the level of the 15-m-by-15-m cell (Vierra et al. 2008). Sites are not defined during initial survey but are aggregated from the resulting TRU dataset based on defined site criteria (Miller et al. 2009:1.12). Collected under identical procedures, both site and nonsite TRU data are viewed as part of a single, continuous whole: the dataset remains intact for the entire survey, and nonsite TRUs are not discarded. The ability and flexibility of TRU datasets for identifying nontraditional cultural resources result from this intrinsic siteless nature; during the initial survey, all cells are recorded and managed identically. TRU survey also overcomes many potential biases imposed by the traditional survey process (Kludt et al. 2007), in which sites are differentiated from isolates and site boundaries are defined in the field based on a supervisor's subjective interpretation of site criteria; in contrast, TRU-based sites are defined and expanded empirically on the basis of adjacent positive cells. While individual contractor's recording systems vary, the overall methodology remains the same and produces a result that can be compared and combined across separate surveys conducted by separate companies.

4.4. TRUs and Trails

While the TRU methodology is useful for recording nonsite artifacts and features, this utility is limited for the purposes of identifying and delineating linear trail networks without a proper understanding of how a linear feature might be manifested archaeologically in the Tularosa Basin environment. Most of the direct evidence for prehistoric trails and footpaths, such as the depressions, notches, or vegetation patterns that mark prehistoric Hopi trails (Ferguson et al. 2009:30–31), would likely have disappeared quickly after abandonment, due to its ephemeral character: nothing remotely approaching the formal constructed pathways seen in northern New Mexico (e.g., Ware and Gumerman 1977; Snead 2002, 2009) has been documented in the Tularosa Basin region. Even the more ephemeral linear swales or cleared alignments observed on stable desert settings elsewhere in the American Southwest (e.g., Darling 2006, 2009; Ferguson et al. 2009; White this volume) have to date not been documented at Fort Bliss. The dominant local geomorphic regime, with its shifting, fairly shallow sand deposits and dunal areas overlying caliche hardpan without intervening desert pavement, limits the potential for preserving remnants of actual trail segments themselves. Many regions in the basin exhibit relatively stable geomorphic contexts, however, suggesting that indirect evidence, such as trail- or path-side artifact scatters or features, may provide clear clues by which prehistoric linear features can be identified, especially when archaeological data are collected at a minimum spatial resolution of 15 m by 15 m.

Because trails and footpaths are most often associated with local logistic activities or daily use, one should expect to see evidence generated by these activities along the trail (Kludt et al. 2007). Activities such as collecting water, gathering plant resources, and tending agricultural fields typically occur on a daily basis, and the primary artifact class shared across these activities is ceramics. Thus, the artifact class most likely associated with daily activities and, therefore, trails or footpaths is ceramics. Indeed, archaeological and ethnographic studies confirm the association between these activities and ceramic breaks. In northern New Mexico, for instance, archaeological work along a Chacoan road segment in the vicinity of Lobo Mesa repeatedly documented the presence of "numerous pot fragments along the road," suggesting to some researchers that the roadways were used to transport water (Kantner 1997). Similarly, numerous isolated ceramics and "occasional large spreads of sherds" were frequently noted along an extensive prehistoric trail system documented in the Mohave Desert during the 1950s (Johnson and Johnson 1957:29–30), as well as other Mohave trails (Davidson 2009; Rogers 1966; Von Werlhof 1988). Artifact scatters located "on or immediately adjacent" to trail traces are also noted as corroborating evidence for the identification of prehistoric and historic Hopi trails (Ferguson et al. 2009:31).

Ethnographic and ethnoarchaeological sources provide additional support for this linkage between water transportation and vessel breakage. For example, an ethnohistoric account from the Mohave region singles out jars for carrying water as necessary accoutrements for foot trail travel (Fowler 2009:90), and ethnoarchaeological research conducted by Beck (2006) among traditional communities in Kalinga Province in the Philippines documents a strong association between water gathering and ceramic breakage: "because transport to water sources is heavily associated with vessel breakage, vessels tend to accumulate near water sources or along pathways to water sources" (Beck 2006:46). Within the Tularosa Basin, recent mitigation work at LA 30116, a large site in the vicinity of the current Doña Ana project area, encountered a concentration of ceramics during surface investigations (MacWilliams 2009). On further investigation, the small topographic rise where dense accumulations of sherds were located was identified as an earthen berm surrounding an El Paso phase reservoir associated with several nearby residential structures. A similar relationship was found at Hot Well Pueblo, where ceramics were concentrated around a reservoir associated with an El Paso phase residential site (Scarborough 1988). By extension, other activities involving the use of ceramic containers, such as gathering wild plant resources or pot-watering agricultural fields, might also be expected to correlate with accidental breakage during transit to and from activity sites.

Besides artifacts, several types of features are often associated with trails or paths. As mentioned, cairns, trailside camps or shelters, and rock art have been noted in direct association with trails and paths in several places, including the Pajarito Plateau in north-central New Mexico (Snead 2002). Furthermore, both cairns and rock art are noted in association with prehistoric trail

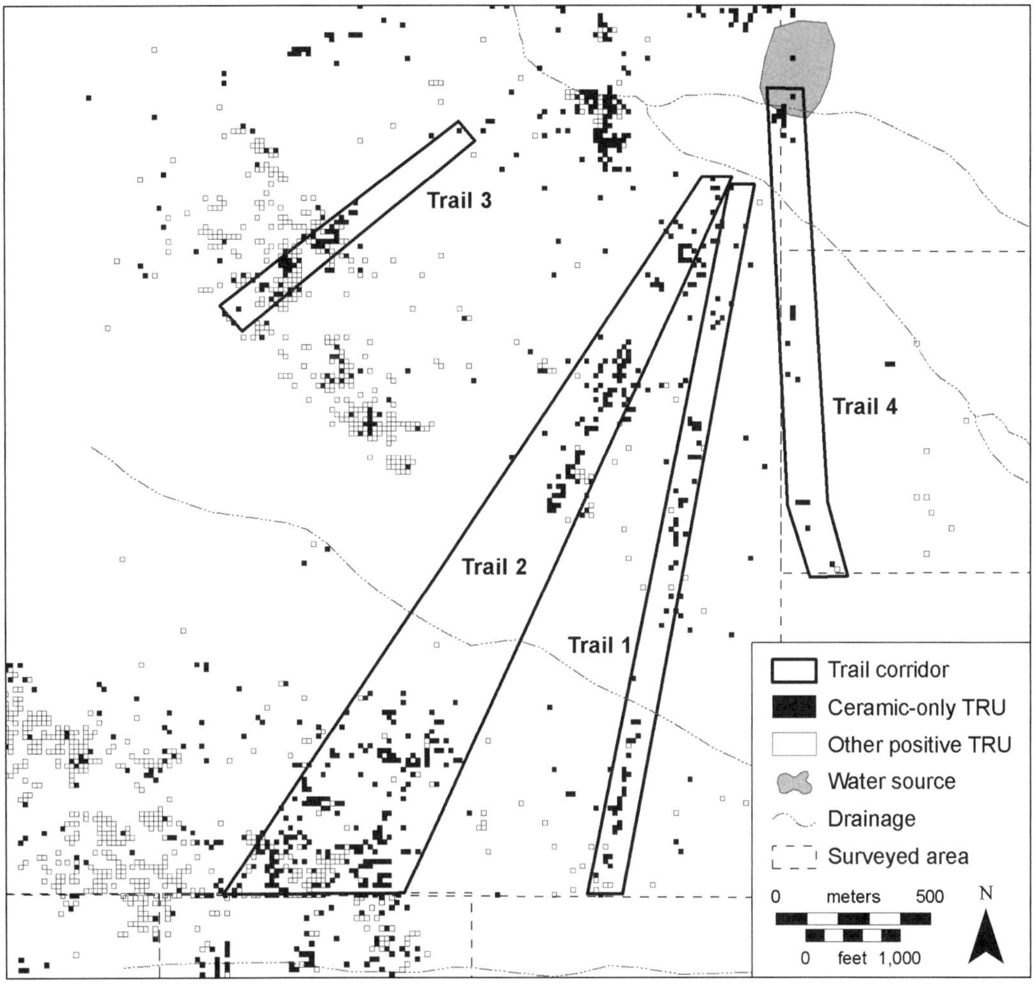

FIGURE 4.2. North McGregor trail system.

systems identified in southeastern California's Mohave Desert (Becker and Altschul 2008:422–426; Davidson 2009; Fowler 2009; Johnson and Johnson 1957; Rogers 1966; Von Werlhof 1988; Weaver 1967). To date, few unambiguously prehistoric examples of such potential trail markers have been noted on the basin floor or alluvial fans of the Tularosa Basin, although cairns and markers associated with the region's later history of extensive ranching and military activity are frequently documented on survey. Given this later record, prehistoric artifacts are likely a better proxy for prehistoric travel corridors across the basin. This is especially true when artifacts associated with travel corridors can potentially leave a distinctive signature that differentiates them from other accumulations of material culture.

4.5. Archaeological Trails within the Tularosa Basin

4.5.1. North McGregor Range System

Between 2004 and 2006, archaeologists conducting an 8,000-acre survey in the Tularosa Basin on Fort Bliss's northern McGregor Range (Kludt et al. 2007) identified three linear arrangements of TRUs that consisted almost entirely of ceramic artifacts, crossed relict alluvial deposits, and originated at the end of a large drainage descending from the Sacramento Mountains (Figure 4.2). Based on the artifact assemblage, the stable geomorphic setting, and their association with a potential water source, these alignments were interpreted as water transport trails connecting a potential water source and several large Formative period residential sites to the south. The methods employed to identify these features closely match ethnographic and other archaeological examples and clearly demonstrate the facility of a nonsite-based survey system such as the TRU method for identifying and tracing prehistoric linear features on the basis of linear distributions of artifacts.

As noted previously, one of the key advantages of the TRU method and the standardization it imposes on data collection is that datasets from differing projects and contractors are consistent and can be combined into a more robust dataset. A basic inspection of the area surrounding the potential water source was undertaken with an expanded TRU dataset and with the same criteria used to identify the original three trails: a linear artifact assemblage dominated by ceramics, a stable geomorphic context, and an approximate alignment toward the potential water source. One additional potential trail, trail 4, was identified to the east of the current three on this basis (Figure 4.2). While the linear ceramic assemblage in this area is not as dense as that associated with the other three trails, this fourth potential trail possesses all the same characteristics. Of the 24 TRUs potentially composing this trail, all contain at least one ceramic sherd and only three have lithic materials. The large gaps in this potential trail correspond to several drainages that have created similar gaps in the three original trails. The end closest to the water source contains several TRUs tightly clustered with high ceramic counts, including a ceramic concentration. While reports and survey data from the vicinity of the potential fourth trail do not reference anything that suggests a prehistoric reservoir, it is interesting to note that the large concentration of ceramics near a suspected water source closely matches findings at other prehistoric reservoir sites such as LA 30116 and Hot Well Pueblo.

4.5.2. Doña Ana Trail

In 2009, 25 previously recorded sites were evaluated for National Register of Historic Places (NRHP) eligibility near the base of the Organ Mountains (Figure 4.3). During the initial TRU recording on one of these sites (LA 91357), positive cells containing large quantities of ceramics with relatively few other artifact classes present were noted to be aligned roughly north-south. Although this was not a survey, we investigated sufficiently wide buffers to the east and west to be confident that this linear pattern of positive cells was not a product of imbalanced coverage. The north edge of this investigation was arbitrarily cut off, as it was not within the project's guidelines to resurvey the entire area. However, an informal reconnaissance north of the recorded portion showed that linear distributions of ceramics continued at least a hundred meters farther. A modern military road and associated activities heavily impact the alignment immediately to the south of this site.

On analyzing other sites that had already been evaluated, we found two additional sites with similar characteristics: LA 91070 and LA 91162. Collectively, the TRUs composing these three sites are aligned north-south and stretch over 2.5 km, cutting diagonally across an alluvial fan. Unfortunately, the geomorphic dataset used to determine landform stability on the McGregor Range does not extend to this part of the military installation. Areas near several of the evaluation sites contained evidence of sheet erosion; however, the artifacts were generally found above the sheet-washed areas, on stable dune and alluvial deposits. Also, the degree of alluvial erosion present suggested it was primarily caused by slow-moving water and thus would not have affected the location of artifacts severely (David Kuehn, personal communication 2009).

Continuing the north-south line formed by these ceramic-dominated sites in both directions leads to two possible end points: a large cluster of residential sites approximately 4.5 km to the north of the northernmost recorded segment of the possible trail and a large fault-trough depression 1 km to the south.

Recent excavations at several of the large sites in the residential cluster demonstrate that their primary period of occupation was between AD 500 and 1000, during the Mesilla phase (Ward et al. 2009; Condon et al. 2010). Significantly, ceramic vessel forms recovered from one of the largest residential sites in the cluster are dominated

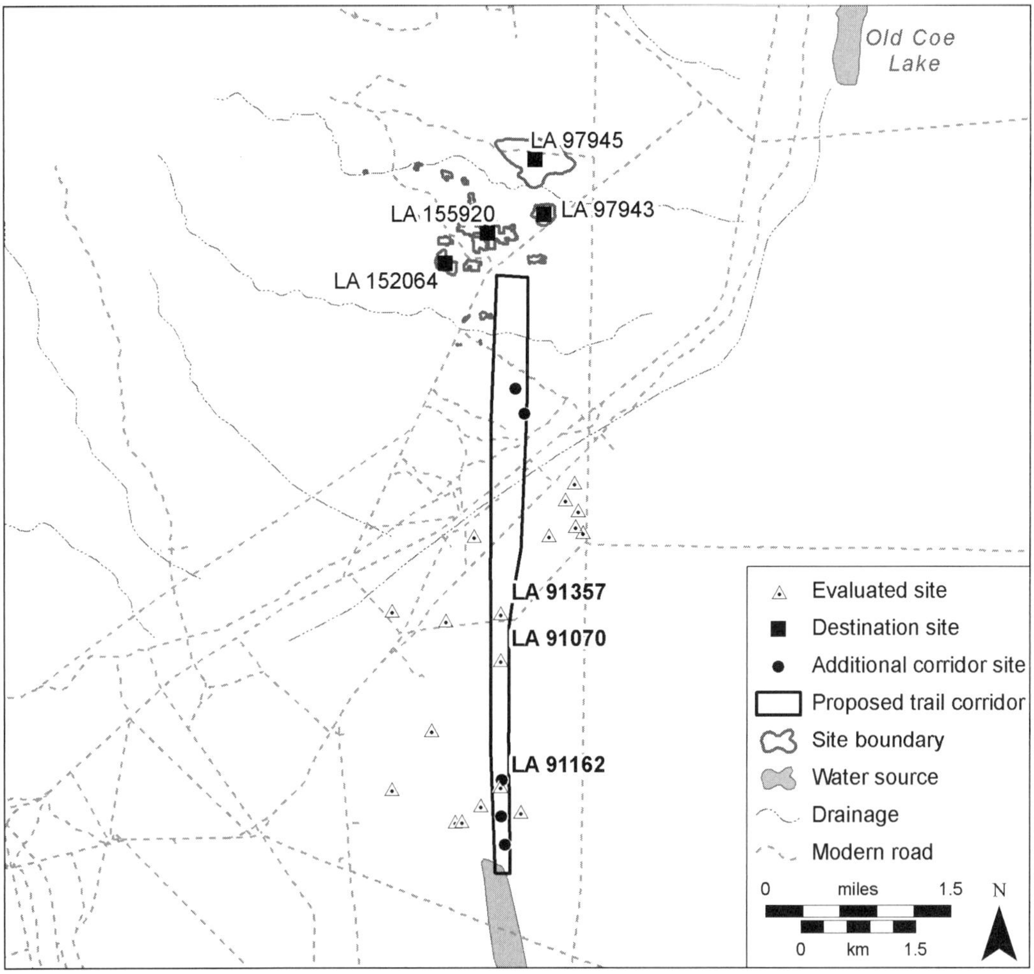

FIGURE 4.3. Doña Ana trail and associated sites.

by small necked or neckless jars suitable for water transport or storage (Unruh et al. 2009:218; Myles Miller, personal communication 2009). The analysis of pollen and macrobotanical remains from excavations at four of the sites in the residential cluster revealed several species that either require or thrive in a mesic environment, including desert willow and cotton (Adams 2009). This analysis indicates that the residents at these sites were collecting and utilizing plants from wetter areas beyond their immediate surroundings. The repeated use and length of occupations at these sites and the nature of Mesilla subsistence practices provide the necessary "stability of destination and departure points" for a trail to form (Kludt et al. 2007). The majority of the diagnostic ceramics found along the trail also date from the Mesilla phase, indicating the possibility that the residents of those villages used this trail for water collection or plant resource gathering.

Auger testing conducted in the fault trough during the evaluation project suggests that the sediments present would have retained at least subsurface water, indicating that it might have supported a rich variety of plant resources. Previously recorded archaeological data that were not part of the project, but that fell within a 250–500-m-wide corridor centered on the potential trail, were incorporated into this study in an attempt to fill in gaps between the evaluated sites and the possible end points. Several sites containing relatively high densities of ceramics similar to the evaluated sites were noted in the corridor, as well as several isolated ceramic artifacts. These instances tentatively fill in the gaps and provide linkages to the possible end points. Other gaps are explained by modern military activities associated with road construction and use and historic ranching activities along a drainage leading to Old Coe Lake. Finally, the majority of the potential trail runs perpendicular to the slope of the fan, deviating only near the residential sites. Together these arguments of artifact types and densities, length and continuity of the alignment, and geomorphic integrity satisfy the criteria used to identify the original three McGregor trails and make a strong case that this alignment represents a trail as well.

The initial observation and recognition of a trail is an important first step in the investigation of the people who used it. But what do these trails reveal about the decisions that led to the creation and use of the trails? As we noted in our introduction, prehistoric trail systems are often most useful for the information they reveal about the interconnection between sites, resource areas, and other loci of past human activity and the patterned human activity that occurred between these loci. In the remainder of this chapter, we examine the trail systems via a variety of GIS-based analytical methods in an effort to unravel the processes behind their formation. Do the trail networks follow predictable corridors across the landscape suggested by local terrain? Or are other social or economic factors at work? Are additional unrecognized transportation routes within the various datasets revealed when GIS-based methods are combined with the methodology and landscape characteristics represented by the existing trails? Finally, what other sites and activity areas, if any, are connected by the trail systems and what do these connections suggest about the context of trail use?

4.6. Least Cost Path Analysis

After completing the visual inspection of the data, we created a least cost path model for each study area using Tobler's hiking function as the basis for the cost surface (Tobler 1993). This function incorporates slope into the calculations of how fast a person can travel over the terrain but does not account for other obstacles such as vegetation and vertical cuts in the landscape. The hiking function, although not overly complex, has been shown to be accurate in archaeological and ethnographic settings and therefore is an appropriate model on which to base the cost surface (Kantner 1997).

Cost surfaces generated using Tobler's hiking function are anisotropic, meaning, as articulated elsewhere in this book, that the cost associated with traversing a particular slope varies, depending on the direction of travel across it. For both areas in the study, the USGS National Elevation Dataset's 10-m-resolution digital elevation model (DEM) was used to derive the slope values. Besides the cost surface, least cost paths require a set of sources and destinations. While the Doña Ana trail has a set of presumed end points, the north McGregor trail system has only a known source (the water source); the destinations for these trails are not known.

In order to choose appropriate destinations, we investigated an expanded dataset of 14,000 TRUs, encompassing approximately 55 km^2 (22 mi^2) with the water source near the northeast corner of the area (Figure 4.4). This region of Fort Bliss is characterized by numerous large archaeological palimpsests, each consisting of thousands of positive TRUs and representing repeated occupations over much of prehistory. Concentrations of TRUs containing very high densities of ceramics, ground stone, and features were seen as an appropriate proxy measure for residential sites. Eleven of these high-concentration areas were chosen as "destination sites," including two that are located at the end of trails 2 and 3. Only two destinations were chosen outside these guidelines: a small concentration of TRUs representing a primarily preceramic site (LA 163580; Murrell et al. 2009) and an isolated point located in the middle of an unsurveyed area, representing a possible unsurveyed residential location.

This water source may have been used during the Paleoindian and Archaic periods as well, but prior use of the trail system is unknown since the trail locations are clearly signaled only by ceramic artifacts. Water containers during previous periods would likely have been perishable, and the broad, low-density scatter of lithic artifacts that spans the project area makes the identification of aceramic pathways on an artifactual basis extremely difficult. However, if the modeled paths to known ceramic sites are shown to be reflective

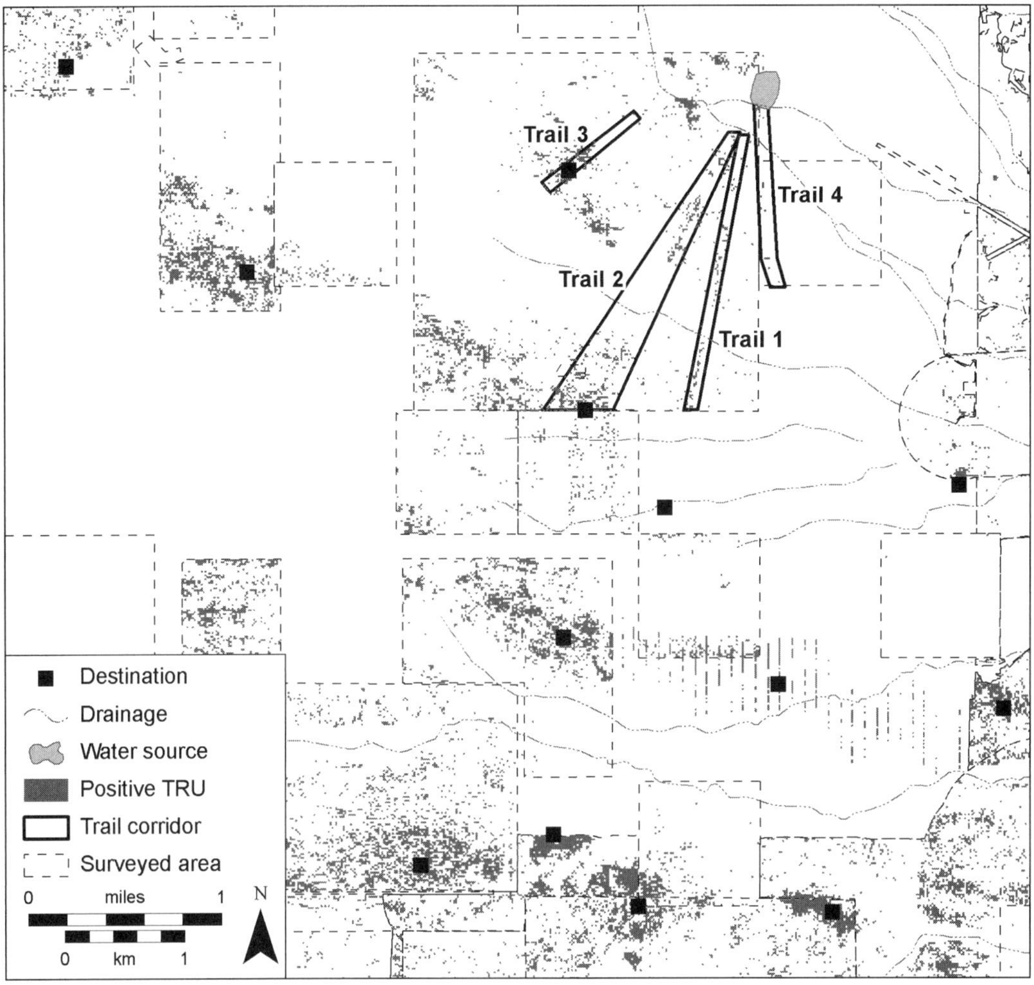

FIGURE 4.4. North McGregor trail system and selected destinations.

of the archaeological data, a modeled path to a preceramic site, such as LA 163580, could provide insights into earlier land use and suggest temporal assignments for other nonceramic sites along it (Rademaker et al. this volume).

The second location was chosen because it is approximately 1 km south of and in alignment with trail 1. Also, there is evidence in adjacent surveyed areas of a large concentration of TRUs centered in the unsurveyed region. The possible water source was used as the source for all paths for the McGregor trails.

For the Doña Ana trail, in addition to the fault-trough depression discussed above, Old Coe Lake was used as a possible comparison source (Figure 4.3). Old Coe Lake, a large playa fed by summer monsoons, was an important focus of later El Paso phase settlements and agricultural activities (Miller and Graves 2009). As Old Coe Lake is a closer potential water source for the residential sites than the fault trough at the terminus of the trail, it was hypothesized that the route to the depression might be less costly in terms of time and energy expended and thus might have been favored over Old Coe Lake. Four of the largest sites in the residential complex, representing a variety of locations within the cluster, were chosen as the other end point of the Doña Ana trail.

Owing to the anisotropic nature of the hiking function, we generated cost surfaces both toward and away from the water source for the McGregor trails and in both directions for both sources and a central destination for the Doña Ana trail, using the Path Distance tool in ArcGIS

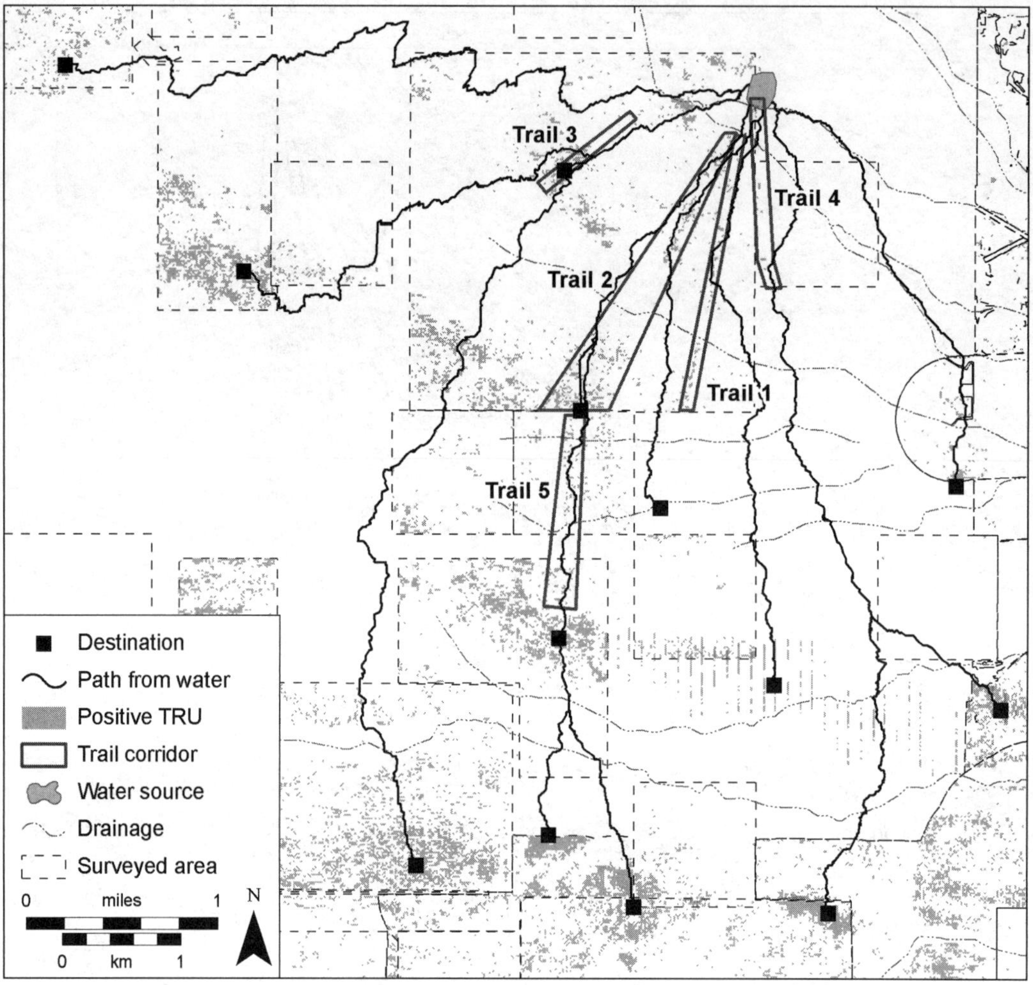

FIGURE 4.5. Least cost paths generated for North McGregor trails.

9.3. For the McGregor system, the two surfaces resulted in similar least cost paths, so for simplicity, only the "away from" model is discussed. For the Doña Ana trail, although there were some differences between the "toward" and "away from" models, they had relatively little impact on the overall model, and thus only the "away from" trails are discussed for this area as well. In the following sections, which describe the results of the analysis, we use the term "path" to refer to the results of the least cost path analysis and "trail" to refer to archaeologically attested travel routes.

4.6.1. McGregor Range Trail Network Least Cost Paths

The paths generated for the McGregor trail system radiate out in several directions from the source (Figure 4.5). Each originally identified trail (1–3) corresponds with modeled paths in at least some areas. The corridor of trail 2 closely matches the model, as two least cost paths follow this corridor for approximately 1 km before diverging and exiting the corridor. The portion of trail 2 corresponding to both modeled paths has an especially dense concentration of positive cells with ceramics. This may indicate that the model's suggestion of a heavily used corridor is accurate. The north half of trail 1 also matches well with a modeled trail. It is interesting to note that at the point where each of the three modeled paths diverges from trails 1 and 2, the evidence for the trails also stops. In the case of one of these paths, it later reenters the corridor for trail 2 approximately 2.5 km from the water source. These simultaneous departures occur at a large drainage that has washed away any archaeological evidence for trails. The least cost paths suggest that in this case the south half of trail 1 and portions of the north half of trail 2 may be segments of the same overall trail rather than portions of distinct separate trails. On the other hand, the downcutting and erosion caused by the drainage since the trails were used could have also affected the elevation model enough to cause the least cost path to diverge from where it would have gone originally.

Trail 3 was the shortest of the recognized trails and suffered the most from erosion at either end, but a least cost path roughly parallels it for its entire length. Interestingly, despite there being a possible destination at the end of this trail, two additional paths split at this point and continue to other destinations, indicating that this could have also been a high-traffic area, representing the trample zone of a destination and departure point and thus being responsible for the large accumulation of artifacts (Kludt et al. 2007). While trail 4 is the least dense of any of the trails, two separate paths run along it for short distances and two more originate at its north end. The large departure from the modeled central portion highlights a possible alternate route since it crosses two isolated TRUs, each containing a single sherd.

Our application of least cost paths also identified several possible previously unrecognized trails. Trail 5 is one example of a pathway that was highlighted by one of the least cost paths. It consists of approximately 100 TRUs in a roughly north-south linear alignment. All the cells contain at least one ceramic sherd, with several potential pot drops, and only eight contain other artifact classes. The only gap in the archaeological data for this trail is also a gap in survey coverage. Like the previously identified trails, this one runs parallel to the topographic contours. However, unlike the others, this potential trail is located in a geomorphically active area. Several medium to large arroyos cut across this possible trail, along with several smaller washes. Also, a relatively dense scatter of lithic artifacts is found in two dune areas upstream from this possible trail; these ceramics could simply be washing downstream from a more stable context. Conversely, the upstream lithic scatters also contain ceramics, possibly indicating that few artifacts have shifted because of erosion. Further, if the artifacts have shifted, the uniformity of the displacement would be remarkable, with different erosional channels flowing at different strengths and moving artifacts roughly the same distance. Despite these factors, this segment could be a continuation of the originally identified trail 2, making this the only archaeologically identified trail that connects two probable residential complexes separated by approximately 2 km.

Because these modeled paths match up fairly well with the known archaeological trails, it is reasonable to assume that the destination choices are broadly appropriate and accurate. Therefore, we can begin to make statements relating the trails to individual destinations and purposes. With the near alignment between modeled paths and archaeological trails, the destinations of trail 2/5 seem to be well founded. As noted by previous researchers, the dense, wide scatter of artifacts at the southern end of trail 2 could represent the trample zone created as individuals disperse on reaching a destination (Kludt et al. 2007). As with trail 3, the southern end of trail 5 could represent the trample zone at both a destination and a departure point, splitting into two paths and heading farther south. The lack of artifacts south of the destination of trail 5 is due to an active alluvial zone and lack of survey coverage, and thus interpretations are limited beyond this point.

The path leading from trail 4's southern terminus is interesting, despite the lack of survey coverage to confirm its route. The destination for its eastern fork is a very large, very dense cluster of TRUs containing primarily lithics and features. This was interpreted to be an area representing an Archaic logistic locale for plant processing, due to its high density of flaked and ground stone tools and burned rock features and its very low quantity of ceramics (Stowe 2008; Murrell et al. 2009). If this destination is the remains of an Archaic locale, it would make sense that the trail leading from it to the water source would have the least ceramics of any of the trails, as the primary use life of the trail would have been before the introduction of ceramic technology in the area. The few ceramics located within the trail corridor may be from a later, less intensive use of the same path. Also, the direction of this trail suggests a different purpose than that of the other trails. While the other three trails head from the water source toward dense residential areas within the basin, this trail points toward the mountains and fans, where more logistically based activities occurred. This orientation, combined with less intensive use during the Formative period, could indicate that this is a trail connecting logistic sites to each other instead of to a residential camp. Due to the infrequent nature of this type of movement (logistic site to logistic site), the trail may not become as well defined as one connecting higher-traffic areas (Kludt et al. 2007). Because ceramic types are often used to date the use life of a trail, and various terrains were favored during different time periods in the Tularosa Basin, the trail's turn toward the upper fans, coupled with the lack of late Formative ceramics, could also indicate a Mesilla phase use and larger occupation nearby (Harner 1957; Johnson and Johnson 1957; Vierra 2009).

Finally, the modeled path leading to the destination chosen within the small preceramic lithic scatter in the northeast provides an interesting result. While much of the path is unsurveyed, the portion near the destination aligns with several smaller clusters of primarily lithic TRUs, indicating that this path was a possible travel corridor for lithic procurement or for other logistic activities not involving ceramics. There is also the possibility that these sites represent preceramic activities in the foothills of the Sacramento Mountains. The evidence from LA 163580 suggests both possibilities: the majority of the artifacts found on the site are related to lithic tool production, and possible Paleoindian and Late Archaic components were recorded at the site.

4.6.2. Doña Ana Trail Least Cost Path Analysis

One benefit of using the hiking function as the basis for the cost surface is that the value at any single place in the resulting surface is the total time it would take to reach that point from the source. Thus it is a simple matter to calculate the total time of a trip and the overall average time. Within the Doña Ana trail area, if one looks only at the travel time away from the site (toward the water), Old Coe Lake is a significantly shorter route, requiring approximately 17 minutes of travel versus 40 minutes to the fault trough. On the return trip, however, the route from the depression would be much shorter and easier, at only 27 minutes compared with 42 minutes to return from Old Coe Lake. These differences occur because Old Coe Lake is located downslope of the residential sites, while the depression is roughly equivalent in elevation to the sites on the alluvial fan. While the model accounts for slope and elevation, it does not account for the increase in weight from carrying water, so although both return trips would likely increase in time, we could expect the trip from Old Coe Lake to increase more due to a larger change in elevation and greater slope. Unless the increase was dramatically more than expected, the route to Old Coe Lake would constitute an easier path in terms of total distance and time. However, because a return journey with fully laden water containers or collected food resources would presumably be more tiring than traveling to the water source with empty vessels, a shorter, less steep trip back might have been viewed as advantageous and chosen more often, leading to the formation of the archaeologically attested trail from the sites to the depression. Until the area between the residential cluster and Old Coe Lake is surveyed at the same resolution as adjacent areas, however, the existence of a trail from the residential sites to Old Coe Lake is purely conjectural, precluding any additional analysis.

Least cost paths modeled from the Doña Ana trail's destination and source points provide several interesting insights. The paths created by the least cost model vary greatly depending on whether the residential complex or the water sources are used as the source for the cost surface (Figure 4.6). This could indicate two different routes: one heading away from the villages and a second, different route leading away from the water source. The archaeologically known trail most closely matches this second route, leading away from the depression and toward the residential sites. This could indicate that the trail represents the return trips, when the ceramic vessels would be full of water. It is not implausible to think that the combination of a full load of water and a 15–17-km round trip would be tiring and contribute to accidental breakage of vessels. While this certainly does not preclude the presence of dense ceramics along the path away from

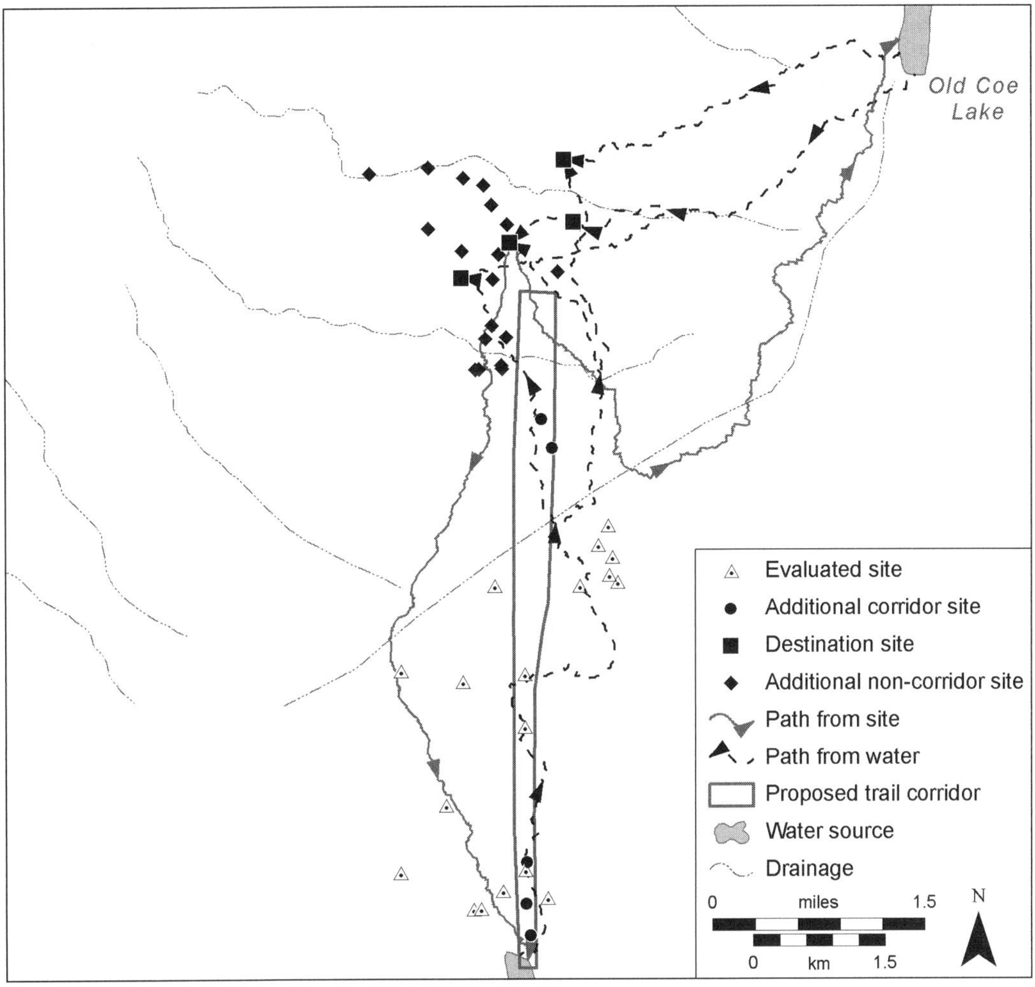

FIGURE 4.6. Least cost paths generated for the Doña Ana trail.

the village, it might be expected that a higher number of vessels would be dropped on the return trip, since a full vessel was presumably more difficult to carry and maneuver than an empty one.

Although the major deviations of the least cost paths from archaeologically documented trails were initially somewhat confounding, further review suggests that these deviations may offer insights into the locations of other nearby sites falling outside the proposed trail corridor but near the modeled paths. Only two of the seventeen sites evaluated during the project are more than 300 m away from either the modeled paths or the known trail corridor. While none of these sites was densely occupied, they might imply the potential use of trails in the area for logistic activities not associated with water collection, such as plant gathering or hunting. If the modeled paths are compared with an expanded site dataset, several other sites, including smaller residential sites outside the primary residential cluster, also fall along possible routes. The strong correlation between sites and modeled paths in this area suggests that habitation and other prehistoric activities in the trail vicinity may relate to traffic along the suggested prehistoric travel corridor, much as modern cities are spatially and economically tied to major highways. This connection between travel corridors and logistic sites has been noted ethnographically as well (Albert and Le Tourneau 2007:584).

In the north McGregor trail system, the directness of the trails to the water sources also indicates a potential openness in the landscape. If the physical landscape did not shape the paths, social landscapes may have. As seen in other areas of the southwestern United States, social relations and time depth in use can alter trails beyond what would be optimal (Snead 2002, 2009). The straightness of these trails is a possible indication that the landscape was socially as well as physically open. There was no need to avoid your neighbors; in fact, having your neighbors come along on the trip could be beneficial in terms of both physical safety and social relations (Kludt et al. 2007). This is potentially what is occurring with trail 2/5 in the north McGregor system, which links two residential sites before continuing to the water source. Similarly, the frequent occurrence of forking or widening as paths approach destination locations might represent individuals separating to their various residences. A similar explanation may hold for the relationship between least cost paths and site locations in the Doña Ana example. People were not only taking the easiest route to the water; they were visiting neighbors and strengthening social ties along the way.

4.7. Viewshed Analysis

The various physical locations and geomorphic settings of the trails present different challenges for safe navigation to and from the water source. Within a basin, beyond the alluvial fans, a person's immediate field of view is often blocked by dune ridges or other topographic features. This can be disorienting, as one dune often looks like the next. Without a focus point on the horizon, perception research suggests, a disoriented person will often walk in circles, increasing the confusion. When given a fixed point at a distance, the same person can successfully navigate a straight line over long distances (Souman et al. 2008). Visually restricted settings are common across the Tularosa Basin and present a challenge even for current-day navigation. Whereas dune ridges can block the line of sight in a basin setting, an alluvial fan setting generally provides a broader view, encompassing a wider field and greater distance of vision. Thus a direct line of sight from source to destination is more likely in an alluvial fan setting than in the dunal environments of the basin. While the two landscapes have different challenges for local navigation, they share the same overall characteristics: relatively level, open terrain with mountains in the distance to the east and west. Despite their differing settings, the straightness of both sets of trails discussed here suggests that the people using them had firm points of navigation to direct them to their destinations. In order to identify and quantify the correlation between the routes of the modeled paths and archaeologically attested trails and the field of view of the individuals who would have traversed them, we conducted a viewshed analysis from the destinations and sources of each area.

A viewshed analysis of the destinations in the McGregor trail network reveals that the water source cannot be seen from any of the southern destination points (Plate 4). Only the three destinations due east of the source are intervisible with it at this point; people departing from all the southern destinations would be walking toward the water source without visible local clues to its location. The absence of line of sight between source and destination suggests that people may have maintained their orientation by sighting on peaks, canyons, or other prominent landmarks in the Sacramento Mountains or adjacent mountain ranges. These landmarks may even have included major rockshelters along the Sacramento escarpment, as some of these are prominently visible over considerable distances.

The Doña Ana trail is in a very different setting from the McGregor trails, being situated on the distal edge of an alluvial fan rather than on the basin floor. A viewshed analysis from the source and destination points of the Doña Ana trail reveals the alluvial fan setting to be much more open (Plate 4). Within 100 m of the fault trough, the cluster of villages, which sits on a prominent ridge, would be clearly visible to the north. Additional visual cues, such as smoke from fires, would have enhanced the visibility of the villages. The reverse is also true; as one looks out from the villages, the edge of the depression is visible in the distance. The change in vegetation, from the creosote and mesquite of the well-drained alluvial fans to a mesic environment with desert willow and other water-dependent plants, would be easily recognizable. While additional landmarks may have been used for small distances, the nearly constant line of sight between the two end points would have provided the majority of the necessary navigational signals. Given the potential costs of getting lost, the ease of navigation afforded by this near-constant line of sight could have overridden the benefits gained from taking the least costly path and may help explain the deviations from the modeled paths. In fact, combining the least cost path away from the depression with the results of the viewshed reveals a major disadvantage of the least cost route: anyone traveling that route would lose sight of the villages for approximately 1 km. The discontinuity between the modeled "optimal" path and archaeologically attested routes in this area is thus a clear suggestion that intervisibility with a destination was more important than optimal travel time in this case of prehistoric path making.

4.8. Conclusion

This chapter asks and answers several questions about two possible trail systems identified at Fort Bliss: where they were located, what were they used for, and why they followed the routes they did. In addition to three previously identified trail segments in the Tularosa Basin, two trails were identified that are either a continuation of a previously identified trail or in close proximity, and probably related to, an integrated trail network. We also identified a trail near the base of the Organ Mountains which serves a similar purpose but exhibits somewhat different characteristics from the basin trails. In both cases, the comparison of archaeological evidence for trails against least cost paths provides insights into the purpose and placement of the trails, as well as suggesting possible destinations for each trail system.

By combining least cost paths with viewshed analyses from the destinations and sources of the trails, we also investigated the potential role of line of sight in path making. Several previous cost-path studies have noted that least cost routes across flat, open landscapes frequently deviate dramatically from archaeologically attested prehistoric trails (e.g., Becker and Altschul 2008:443; Kantner 1997). Explanations for these deviations have typically suggested that the difference in efficiency between cost paths and less optimal routes in such settings is sufficiently minimal that no single route is significantly better (Becker and Altschul 2008:443). Our results support that contention but also suggest that within such a landscape, intervisibility between departure and destination locations plays an important role in path making. In the absence of dramatic topographic shifts or other major physical constraints on movement, the potential for continued line of sight may outweigh any minimal benefit gained by following an "optimal" route in terms of energy expenditure. In relatively flat settings, at least, visibility analysis thus provides a powerful tool for use alongside least cost modeling for the interpretation of the routes followed by archaeologically attested prehistoric trails. A preference for continued line of sight provides a compelling explanation for several deviations from the least cost paths modeled for the Doña Ana project setting.

Conversely, within a locally restricted-view environment, the viewshed analyses present future opportunities for investigating prehistoric navigation questions. The analysis of the McGregor project area suggests that peaks or other distant landscape features might have played an important role in navigation and that they could have been imbued with special significance. Finally, we propose that the placement of the trails relative to adjacent residential and logistic sites provides insights into the interconnectedness and openness of the basin in both a physical and social sense.

Acknowledgments

Special thanks to Fort Bliss DPW-E staff, especially Belinda Mollard, for processing our multitude of data requests to build our expanded datasets. Thank you to several of our colleagues at SRI—Christine Ward, Arthur MacWilliams, William Hayden, and Nicholas Reseburg—for valuable insight into the archaeology of the area, ideas on cost modeling, and reviews of early drafts. Finally, thank you to our editors, whose idea to organize a least cost paths session at the 2009 Society for American Archaeology meeting and to use that session as the basis for this book gave us the opportunity to expand on these ideas.

References

Adams, Karen R.

2009 Flotation Analysis. In *Archaeological Investigations at Five Prehistoric Sites on the Eastern Flank of the Organ Mountains, Fort Bliss, Doña Ana County, New Mexico*, edited by C. G. Ward, B. J. Vierra, and K. M. Schmidt, pp. 283–304. Historic and Natural Resources Report No. 07-53. SRI Technical Report No. 09-43. Submitted to the United States Army Fort Bliss Directorate of Public Works, Environmental Division, Conservation Branch, Fort Bliss, Texas. Prepared by Statistical Research, Inc., El Paso, Texas.

Albert, Bruce, and Francois-Michel Le Tourneau

2007 Ethnogeography and Resource Use among the Yanomami: Toward a Model of "Reticular Space." *Current Anthropology* 48:584–592.

Altschul, Jeffrey H., Terry H. Klein, and Lynne Sebastian

2005 *A Workshop on Predictive Modeling and Cultural Resource Management on Military Installations, Santa Fe, New Mexico, November 15–18, 2004: Legacy Resource Management Program, Project #03-167*. Foundation Preservation Research Series 4. SRI Foundation, Rio Rancho, New Mexico.

Anderson, Sally

1993 Archaic Period Land Use in the Southern Tularosa Basin, New Mexico. In *Preliminary Investigations of the Archaic in the Region of Las Cruces, New Mexico*, edited by R. MacNeish, pp. 48–67. Historic and Natural Resources Report No. 9. Environmental Management Office, Directorate of Engineering and Housing, U.S. Army Air Defense Artillery Center, Fort Bliss, Texas.

Barrett, J. C.

1994 *Fragments from Antiquity: An Archaeology of Social Life in Britain, 2900–1200 BC*. Blackwell, Oxford.

Beck, Margaret E.

2006 Midden Ceramic Assemblage Formation. *American Antiquity* 71:27–51.

Becker, Kenneth H., and Jeffrey H. Altschul

2008 Path Finding: The Archaeology of Trails and Trail Systems. In *Fragile Patterns: The Archaeology of the Western Papaguería*, edited by Jeffrey H. Althschul and Adrianne G. Rankin, pp. 419–446. SRI Press, Tucson, Arizona.

Beckett, Patrick H.

1983 The Paleo-Indian Prehistory of the Tularosa Basin. In *The Prehistory of Rhodes Canyon, New Mexico*, edited by P. L. Eidenbach, pp. 95–103. Human Systems Research, Tularosa, New Mexico.

Brown, David E.

1994 Chihuahuan Desertscrub. In *Biotic Communities: Southwestern United States and Northwestern Mexico*, edited by D. E. Brown. University of Utah Press, Salt Lake City.

Carpenter, John P., Guadalupe Sánchez, and María Elisa Villalpando C.

2005 The Late Archaic/Early Agricultural Period in Sonora, Mexico. In *The Late Archaic across the Borderlands: From Foraging to Farming*, edited by B. J. Vierra, pp. 13–40. Texas Archaeology and Ethnohistory Series, Thomas R. Hester, general editor. University of Texas Press, Austin.

Condon, Peter C., Willi Hermann, Lillian M. Ponce, Javier Vasquez, Seth Sampson, Grant D. Smith, Sarah N. Cervera, and Luis Sierra

2008 *Assessing Organizational Strategies during the Late Mesilla Phase (AD 600 to 1100): A Data Recovery of Four Prehistoric Sites along the Organ Mountain Alluvial Fans, Doña Ana Firing Complex, Range 48, Fort Bliss Military Installation, Doña Ana County, New Mexico.* Historic and Natural Resources Report No. 03-04. TRC Report No. 103874.0100. Submitted to the United States Army Fort Bliss Directorate of Public Works, Environmental Division, Conservation Branch, Fort Bliss, Texas. Prepared by TRC Environmental, Inc., El Paso, Texas.

Condon, P. C., M. E. Hroncich, B. G. Bury, W. Hermann, D. D. Kuehn, J. A. Jacobson, L. S. Cummings, E. Hickey, K. Puseman, and A. Romero

2010 *Cultural Coalescence and Economic Diversity in the Formative Period: A Data Recovery of 14 Prehistoric Sites Adjacent to U.S. Highway Loop 375, Maneuver Area 1B, Fort Bliss Military Installation, El Paso County, Texas.* Report No. 153885.0001. TRC Environmental, Inc., El Paso, Texas. Fort Bliss Cultural Resources Report No. 07-45. Prepared for the Directorate of Public Works, Environmental Division, Conservation Branch, Fort Bliss, Texas.

Darling, J. Andrew

2006 Pima Song and the Archaeology of Space. Draft. Manuscript on file, Cultural Resource Management Program, Gila River Indian Community, Sacaton, Arizona.

2009 O'odham Song and the Archaeology of Space. In *Landscapes of Movement: Trails, Paths, and Roads in Anthropological Perspective*, edited by James E. Snead, Clark L. Erickson, and J. Andrew Darling, pp. 61–83. University of Pennsylvania Museum of Archaeology and Anthropology, Philadelphia.

Darling, J. Andrew, and B. Sunday Eiselt
2003 Trails Research in the Gila Bend Area. In *Trails, Rock Features, and Homesteading in the Gila Bend Area: A Report on the State Route 85, Gila Bend to Buckeye Archaeological Project*, edited by John Czarzasty, Kathleen Peterson, and Glen Rice, pp. 209–242. Draft. Anthropological Field Studies No. 43, Vol. 2. Cultural Resource Management Program, Arizona State University, Tempe.

Davidson, Nancy J.
2009 Making the Connections: An Archaeological Survey of Prehistoric Trails and Trail Markers along the Lower Colorado River. Unpublished Master's thesis, California State University, Fullerton.

Dore, Christopher D., and Stephen A. McElroy
2006 *Automated Trail Identification and Mapping: An Experiment in Archaeological Spectral-Image Analysis Using Commercial High-Resolution Satellite and Remote-Sensing Data*. Technical Report No. 06-02. Statistical Research, Inc., Tucson, Arizona. Prepared for U.S. Army Proving Ground, Yuma, Arizona.

Ferguson, T. J., G. Lennis Berlin, and Leigh J. Kuwanwisiwma
2009 Kukhepya: Searching for Hopi Trails. In *Landscapes of Movement: Trails, Paths, and Roads in Anthropological Perspective*, edited by James E. Snead, Clark L. Erickson, and J. Andrew Darling, pp. 20–41. University of Pennsylvania Museum of Archaeology and Anthropology, Philadelphia.

Fowler, Catherine S.
2009 Reconstructing Southern Paiute–Chemehuevi Trails in the Mojave Desert of Southern Nevada and California. In *Landscapes of Movement: Trails, Paths, and Roads in Anthropological Perspective*, edited by James E. Snead, Clark L. Erickson, and J. Andrew Darling, pp. 84–105. University of Pennsylvania Museum of Archaeology and Anthropology, Philadelphia.

Hard, Robert J.
1986 Ecological Relationships Affecting the Rise of Farming Economies: A Test from the American Southwest. Unpublished Ph.D. dissertation, University of New Mexico, Albuquerque.

Hard, Robert J., and John R. Roney
2005 The Transition to Farming on the Rio Casas Grandes and in the Southern Jornada Mogollon Region in the North American Southwest. In *The Late Archaic across the Borderlands: From Foraging to Farming*, edited by B. J. Vierra, pp. 141–186. University of Texas Press, Austin.

Harner, Michael J.
1957 Appendix: Potsherds and the Tentative Dating of the San Gorgonio–Big Maria Trail. *Reports of the University of California Archaeological Survey* 37:35–37.

Hawley, John W., and Frank E. Kottlowski
1969 Quaternary Geology of the South-Central New Mexico Border Region. In *Border Stratigraphy Symposium*, edited by F. E. Kottlowski and D. V. Lemone, pp. 89–115. Circular 104. New Mexico Bureau of Mines and Mineral Resources, Socorro.

Healy, D. L., Ronald R. Wahl, and F. E. Currey
1978 *Gravity Survey of the Tularosa Valley and Adjacent Areas, New Mexico*. Open File Report No. 78-309. U.S. Geological Survey, Reston, Virginia.

Johnson, Francis J., and Patricia H. Johnson
1957 An Indian Trail Complex of the Central Colorado Desert: A Preliminary Survey. *Reports of the University of California Archaeological Survey* 37:22–34.

Kantner, John
1997 Ancient Roads, Modern Mapping: Evaluating Prehistoric Chaco Anasazi Roadways Using GIS Technology. *Expedition* 39(3):49–62.

Kludt, Trevor, Mike Stowe, Tim Church, and Scott Walley
2007 *Pathfinding on McGregor Range: Archaeological Survey of Approximately 8,000 Acres on Fort Bliss, New Mexico*. Historic and Natural Report No. 04-12. Lone Mountain Report No. 560-008. Submitted to the United States Army Fort Bliss Directorate of Public Works, Environmental Division, Conservation Branch, Fort Bliss, Texas. Prepared by Lone Mountain Archaeological Services, Inc., El Paso, Texas.

Kottlowski, Frank E.
1958 Geologic History of the Rio Grande Near El Paso. In *Franklin and Hueco Mountains, Texas*, edited by A. Young, H. L. Williams, J. P. Salisbury, H. N. Freznel, P. D. Hull Jr., and G. L. Evans, pp. 46–54. Guidebook 1958 Field Trip. West Texas Geological Society, Midland.

Krone, Milton F.
1976 A Clovis Point from the El Paso Area. *Artifact* 14(2):45–48.

Leckman, Phillip O., William H. Doleman, Bradley J. Vierra, Monica L. Murrell, and Shaun M. Phillips
2009 Spatial and Aspatial Patterning within Identified Archaeological Assemblages. In *Results*

of a 10,000-Acre Cultural Resource Survey on Northern McGregor Range, Fort Bliss Military Reservation, Otero County, New Mexico, edited by Christine G. Ward, Bradley J. Vierra, and Kari M. Schmidt, pp. 241–299. Submitted to the Directorate of Public Works, Environmental Division, Conservation Branch, Fort Bliss, Texas, Fort Bliss Report No. 08-09. Technical Report 09-30. Prepared by Statistical Research, Inc., El Paso, Texas.

Lyons, Thomas R., and Robert K. Hitchcock

1977 Remote Sensing Interpretation of an Anasazi Land Route System. In *Aerial Remote Sensing Techniques in Archeology*, edited by T. R. Lyons and Robert K. Hitchcock, pp. 111–134. Reports of the Chaco Center Vol. 2. Chaco Center, Albuquerque, New Mexico.

MacWilliams, A. C., Brad Vierra, and Kari Schmidt

2009 *Archaeological Mitigation at FB 17 (LA 91017) and FB 9122 (LA 30116) on the Doña Ana Range, Fort Bliss, Doña Ana County, New Mexico.* Historic and Natural Resources Report No. 07-49. SRI Technical Report No. 09-13. Submitted to the United States Army Fort Bliss Directorate of Public Works, Environmental Division, Conservation Branch, Fort Bliss, Texas. Prepared by Statistical Research, Inc., El Paso, Texas.

Mattick, R. E.

1967 A Seismic and Gravity Profile across the Hueco Bolson, Texas. In *Geological Survey Research 1967*, pp. 85–91. Professional Paper 575-D. U.S. Geological Survey, Government Printing Office, Washington, D.C.

Miller, Myles R., and Tim B. Graves

2009 *Madera Quemada Pueblo: Archaeological Investigations of a 14th Century Jornada Mogollon Pueblo.* Historic and Natural Resources Report No. 03-12. Geo-Marine Report No. 679EP. Submitted to the United States Army Fort Bliss Directorate of Public Works, Environmental Division, Conservation Branch, Fort Bliss, Texas. Prepared by Geo-Marine, Inc., El Paso, Texas.

Miller, Myles R., N. A. Kenmotsu, and M. R. Landreth (editors)

2009 *Significance and Research Standards for Prehistoric Archaeological Sites at Fort Bliss (Revised 2008): A Design for the Evaluation, Management, and Treatment of Cultural Resources.* Report submitted to the Environmental Division, Garrison Command, Fort Bliss, Texas. Submitted by Geo-Marine, Inc., El Paso, Texas.

Miller, Myles R., and M. S. Shackley

1998 Obsidian Procurement and Distribution in the Jornada Mogollon Region of West Texas, Southern New Mexico, and Northern Chihuahua. Paper presented at the meetings of the Texas Archaeological Society, Waco.

Murrell, Monica L., Bradley J. Vierra, and Kari M. Schmidt

2009 Results of a 9,314-Acre Cultural Resource Survey on Northern McGregor Range, Fort Bliss Military Reservation, Otero County, New Mexico. Historic and Natural Resources Report No. 08-28. SRI Technical Report No. 09-25. Submitted to the United States Army Fort Bliss Directorate of Public Works, Environmental Division, Conservation Branch, Fort Bliss, Texas. Prepared by Statistical Research, Inc., El Paso, Texas.

Rogers, Malcolm J.

1966 *Ancient Hunters of the Far West.* Union-Tribune Publishing, San Diego, California.

Scarborough, Vernon L.

1988 A Water Storage Adaptation in the American Southwest. *Journal of Anthropological Research* 44(1):21–40.

Schmidt, Kari M., Monica L. Murrell, and Tim Mills

2009 Environmental Context. In *Results of a 10,000-Acre Cultural Resource Survey on Northern McGregor Range, Fort Bliss, Otero County, New Mexico*, edited by C. G. Ward, B. J. Vierra, and K. M. Schmidt, pp. 2.1–2.22. Historic and Natural Resource Project No. 08-09. Prepared for the United States Army Fort Bliss, Directorate of Public Works, Environmental Division, Conservation Branch, Fort Bliss, Texas. Technical Report No. 09-43. Submitted by Statistical Research, Inc., El Paso, Texas.

Snead, James E.

2002 Ancestral Pueblo Trails and the Cultural Landscape of the Pajarito Plateau. *Antiquity* 76(293):756–765.

2008 *Ancestral Landscapes of the Pueblo World.* University of Arizona Press, Tucson.

2009 Trails of Tradition: Movement, Meaning, and Place. In *Landscapes of Movement: Trails, Paths, and Roads in Anthropological Perspective,* edited by James E. Snead, Clark L. Erickson, and J. Andrew Darling, pp. 42–60. University of Pennsylvania Museum of Archaeology and Anthropology, Philadelphia.

Snead, James E., Clark L. Erickson, and J. Andrew Darling (editors)

2009 *Landscapes of Movement: Trails, Paths, and Roads in Anthropological Perspective.* University of Pennsylvania Museum of Archaeology and Anthropology, Philadelphia.

Souman, J. L., I. Frissen, M. N. Sreenivasa, and M. O. Ernst
2008 Walking in Circles: The Role of Visual Information in Navigation. Paper presented at the 31st European Conference on Visual Perception, Utrecht, Netherlands.

Stowe, Michael
2008 *In the Zone: Evaluation of Three Sites for Off Limits Area (Red Zone) Nomination in Maneuver Areas 10 and 12, Fort Bliss Military Reservation, Otero County, New Mexico.* Historic and Natural Resources Report No. 07-07. Geo-Marine Report No. 746EP. Submitted to the United States Army Fort Bliss Directorate of Public Works, Environmental Division, Conservation Branch, Fort Bliss, Texas. Prepared by Geo-Marine, Inc., El Paso, Texas.

Strain, William S.
1966 *Blancan Mammalian Fauna and Pleistocene Formations, Hudspeth County, Texas.* Bulletin 10. Texas Memorial Museum, University of Texas, Austin.

Tagg, M. D.
1996 Early Cultigens from Fresnal Shelter, Southeastern New Mexico. *American Antiquity* 61:311–324.

Tilley, Christopher
1994 *A Phenomenology of Landscape: Places, Paths and Monuments.* Berg, Oxford.

Tobler, Waldo
1993 *Three Presentations on Geographical Analysis and Modeling: Non-isotropic Modeling.* Technical Report 93-1. National Center for Geographic Information and Analysis, Santa Barbara, California.

Unruh, David T., Robert A. Heckman, and Lisa C. Atkinson
2009 Ceramic Analysis. In *Archaeological Investigations at Five Prehistoric Sites on the Eastern Flanks of the Organ Mountains, Fort Bliss, Dona Ana County, New Mexico*, edited by Christine G. Ward, Bradley J. Vierra, and Kari M. Schmidt, pp. 201–224. Historic and Natural Resource Report No. 07-53. Prepared for the United States Army Fort Bliss, Directorate of Public Works, Environmental Division, Conservation Branch, Fort Bliss, Texas. Technical Report No. 09-43. Submitted by Statistical Research, Inc., El Paso, Texas.

Upham, S., R. S. MacNeish, W. C. Galinat, and C. M. Stevenson
1987 Evidence Concerning the Origin of Maize de Ocho. *American Anthropologist* 89:410–419.

Vierra, Bradley J.
2009 Research Design. In *Results of a 10,000-Acre Cultural Resource Survey on Northern McGregor Range, Fort Bliss, Otero County, New Mexico*, edited by C. G. Ward, B. J. Vierra, and K. M. Schmidt, pp. 51–56. Submitted to the Directorate of Public Works, Environmental Division, Conservation Branch, Fort Bliss, Texas. Contract No. W911SG-07-D-0012. Statistical Research, Albuquerque, New Mexico.

Vierra, Bradley J., and Richard I. Ford
2007 Foragers and Farmers in the Northern Rio Grande Valley, New Mexico. *Kiva* 73:117–130.

Vierra, Bradley J., P. O. Leckman, and R. A. Heckman
2008 *Plan of Work for National Register of Historic Places Evaluations of Sites and Inventory of 5,000 Acres on the McGregor Range, Fort Bliss, Otero County, New Mexico.* Technical Report 08-58. Statistical Research, Albuquerque, New Mexico.

Vierra, Bradley J., Kari M. Schmidt, and Carrie J. Gregory
2009 Archaeological Background for Fort Bliss. In *Results of a 10,000-Acre Cultural Resource Survey on Northern McGregor Range, Fort Bliss, Otero County, New Mexico*, edited by C. G. Ward, B. J. Vierra, and K. M. Schmidt, pp. 39–50. Submitted to the Directorate of Public Works, Environmental Division, Conservation Branch, Fort Bliss, Texas. Contract No. W911SG-07-D-0012. Statistical Research, Albuquerque, New Mexico.

Von Werlhof, Jay
1988 Trails in Eastern San Diego County and Imperial County: An Interim Report. *Pacific Coast Archaeological Society Quarterly* 24(1):51–75.

Ward, Christine G., Bradley J. Vierra, and Kari M. Schmidt
2009 *Archaeological Investigations at Five Prehistoric Sites on the Eastern Flank of the Organ Mountains, Fort Bliss, Doña Ana County, New Mexico.* Historic and Natural Resources Report No. 07-53. SRI Technical Report No. 09-43. Submitted to the United States Army Fort Bliss Directorate of Public Works, Environmental Division, Conservation Branch, Fort Bliss, Texas. Prepared by Statistical Research, Inc., El Paso, Texas.

Ware, J. A., and G. J. Gumerman
1977 Remote Sensing Methodology and the Chaco Canyon Prehistoric Road System. In *Aerial Remote Sensing Techniques in Archeology*, edited by T. R. Lyons and Robert K. Hitchcock, pp. 135–168. Reports of the Chaco Center Vol. 2. Chaco Center, Albuquerque, New Mexico.

Weaver, J. R.
1967 An Indian Trail near Needles, California. *Reports of the University of California Archaeological Survey* 70:151–157.

Weber, Robert H., and George A. Agogino
1968 Mockingbird Gap Paleo-Indian Site: Excavations in 1967. In *33rd Annual Meeting of the Society for American Archaeology*. Santa Fe, New Mexico.

Whalen, Michael E.
1994 *Turquoise Ridge and Late Prehistoric Residential Mobility in the Desert Mogollon Region*. Anthropological Papers No. 118. University of Utah Press, Salt Lake City.

CHAPTER 5

A Method for Multiple Cost-Surface Evaluation of a Model of Fort Ancient Interaction

Kevin C. Nolan and Robert A. Cook

5.1. Introduction

In this chapter we describe recent work on an evolutionary-ecological model of interaction and social change for the Middle Ohio River Valley during the Late Prehistoric period (AD 1000–1650). Our model is derived from expectations provided in the Winterhalder-Kelly (W-K) model for social interaction (Kelly 1995; Winterhalder 1986). The W-K model predicts changes in intercommunity relations given changes in resource abundance and variability. Briefly, the model predicts that communities will seek trading partners in regions that experience different patterns of resource abundance. Our original approach consisted of a time-series of modeled raster surfaces for temporal and spatial variation of summer moisture availability (the Palmer Drought Severity Index [PDSI] reconstructions of Cook et al. 2004) that predicted changes in social organization, settlement plan, and, most important for this chapter, patterns of exchange and intercommunity interaction. Fluctuations in moisture available for plants will produce fluctuations in harvests for agriculturalists. Therefore, patterns in moisture variability will allow us to apply the W-K model to predict the direction of trade that seeks to access partners with a different pattern of abundance or stress. Both this chapter and our original model provide new robust explanations for previous observations and make predictions to guide new problem-oriented research.

In addition to our contribution to the understanding Middle Ohio Valley Late Prehistory, our application in this chapter provides a unique application of the tools of least cost analysis (LCA) to solve anthropological problems. Three-dimensional space is not the only factor affecting interaction among groups of people, and our application just scrapes the surface of nonspace-derived factors conditioning movement.

Our original presentation (Nolan and Cook 2010a) was a broad-brush analysis conducted at the regional scale. From this high-level perspective, the expectations of the model largely duplicate the most widely accepted culture-historical narrative for Fort Ancient development (Pollack and Henderson 1992, 2000; see 5.1.1 below). The most important aspect of the original model is its ability to deductively generate predictions at multiple spatial scales and from multiple perspectives (regional, local, etc.) and in these ways to make predictions that may deviate from the accepted inductive, typological scenario. Problem-oriented research requires deductively generated predictions to seek to falsify explanations. Inductive accounts are not adequate to this task. The strength of our approach is in predicting cases that will draw a contrast against the accepted, empirically generalized, inductive narrative and in identifying how to test the prediction.

In any discipline, important patterns can be found at multiple scales. Explanations at each scale are important. Our initial efforts were at explaining large-scale, regional patterns in interaction. However, most archaeology, and for that matter human experience, is not at the regional level. Inverting the perspective of the original study allows for analysis of the factors conditioning decisions at the site level. While the same factors are evaluated in each case, not every location in the region will fit into the larger-scale patterning. Site-based analyses facilitate the exploration and explanation of patterns at the site and catchment level. Our primary goal in this chapter is to present a template for deriving site-based expectations from the original modeled surfaces for any point in the larger region.

Below we detail the procedures for developing a site-based model of interaction and discuss how this model provides robust explanations for previously observed similarities in Fort Ancient pottery between the Miami (southwestern Ohio) and Scioto (central Ohio) valleys during the Middle Late Prehistoric period (AD 1200–1400). Specifically, we describe how our model can be focused on local interactions for two sites: SunWatch (33MY57) and Reinhardt (33PI880). First, we derive a cost surface from a digital elevation model (DEM) for a large catchment around each focal site. The least cost paths (LCPs) derived from the topography serve as a first estimate of likely trading partners and trading routes and can be treated as a null hypothesis. Next we derive difference-of-PDSI (dPDSI) surfaces for each point of focal analysis for two of Nolan and Cook's (2010a) original 50-year periods that extant dates indicate have the highest probability of occupation for the focal sites. On the dPDSI raster surfaces, locations with moisture balance (measured by PDSI) similar to each site being analyzed will be on peaks and greater differences will be "downhill." Thus we create a faux topography with areas yielding similar patterns of resource abundance and stress located in the hilly uplands and areas experiencing different harvest patterns (therefore making ideal trading partners) located in the valleys. From this surface we derive cost surfaces and LCPs, allowing approximation of ecologically beneficial trade partners for each site. The direction of trade expected from each set of LCPs is compared with known site distributions and previously detected patterns of interaction. We find that, often, ecologically beneficial trade routes are not along the routes predicted by purely topographic LCA. Finally, we explore ways of combining the two predictions, as well as future directions and additional applications of this model.

5.1.1. The Late Prehistory of the Middle Ohio River Valley

The Fort Ancient "tradition" is located along the Ohio River in southern Ohio, northern Kentucky, southeastern Indiana, and western West Virginia (Figure 5.1). The "Fort Ancient" name for a prehistoric culture in the region was first used in the late nineteenth century, when it was applied to the remains of a "village-dwelling culture" that occupied southwestern Ohio (Moorehead 1892; Putnam 1886). The geographical extent of Fort Ancient was later broadened to include sites found throughout the remainder of the Middle Ohio River Valley, including the rest of southern Ohio (Griffin 1966; Mills 1904), northern Kentucky (H. Smith 1910), and neighboring parts of West Virginia and eastern Kentucky (Dunnell 1961; Mayer-Oakes 1955). The Fort Ancient peoples were initially thought to be builders of the eponymous hilltop enclosure (33WA2), though it became well known early on that the Hopewell (100 BC–AD 400) built that monument, and that Fort Ancient culture sites were confined to the Late Prehistoric period in the region. There are Fort Ancient villages both within the enclosure (South Fort) and on the floodplains below the large hilltop enclosure (Anderson, 33WA4), which caused the initial confusion.

In 1943, James Griffin (1966) wrote what is still the most comprehensive examination of Fort Ancient as a whole, albeit with far fewer sites than are now known. In *The Fort Ancient Aspect*, Griffin (1966) defined Fort Ancient as a cultural expression with a series of phases (composed of related sites) related to Mississippian cultures to the south and west (see Figure 5.1). While vast improvements have been made with respect to our understanding of chronology and additional phases, the whole system of classification according to the Midwestern Taxonomic System is very much alive, but it is not well suited for addressing

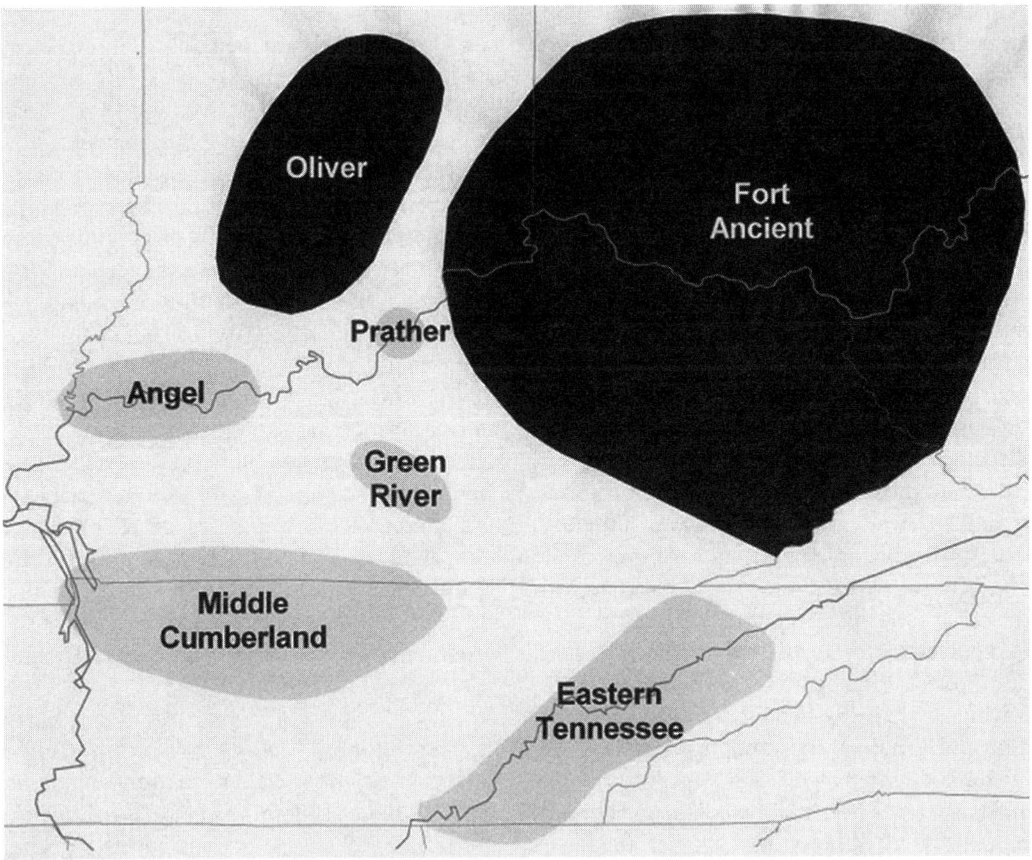

FIGURE 5.1. Location of culture-historic taxa in the Middle Ohio Valley and surrounding region.

dynamic questions now coming to the forefront of Fort Ancient research specifically and dynamic problems associated with human behavior more generally (for an overview pertaining to Fort Ancient, see Nolan and Cook 2010a). Relationships between Fort Ancient groups and Mississippians to the south and west have been debated on and off since Griffin's (1966) seminal work, toggling between local developments (Essenpreis 1978, 1982; Rafferty 1974) and migration scenarios (Carskadden and Morton 2000; Dunnell 1972; Prufer and Shane 1970), the most recent attempt being to bridge the gap in the so-called analogy-homology dilemma (Smith 1990). Recent examples of this bridge include works by both of us (e.g., Cook 2008; Nolan and Cook 2010a, b) and an ongoing research agenda.

The general lack of clear indicators of social hierarchy as expressed in elite mortuary treatments and monumental architecture led many researchers to the conclusion that Fort Ancient was "tribal" in terms of political composition (e.g., Henderson 1998; Redmond and McCullough 2000). While this conclusion still regarded as the general rule, it is becoming increasingly clear that human societies exhibit highly variable and shifting forms of organization not easily pigeonholed into neo-evolutionary types (Feinman and Neitzel 1984), and the Middle Ohio River Valley during the Late Prehistoric period is no exception.

Regional settlement patterning for Fort Ancient is relatively well known and clearly conditioned by subsistence concerns, dominated by a pronounced shift just before the development of Fort Ancient to maize horticulture (Greenlee 2002). This staple was supplemented with several other domesticates such as beans and squash as well as plants associated with the Eastern

Agricultural Complex (Martin 2009). Additionally, a large portion of the diet continued to come from a wide variety of wild plants and animals (for relevant reviews, see Henderson and Pollack 2001; Rossen 1992; Wagner 1987).

The larger settlements are most often located on floodplains or nearby terraces or bluffs for their proximity to fertile soils that underlay the focus on maize horticulture. These settlements are typified by circular villages, sometimes with one or more mounds. The presence of mounds and the formality of the village plan are more pronounced in the western Fort Ancient region, with the eastern cases seemingly less formally organized. Smaller sites used for a variety of purposes (seasonal gathering and hunting, farmhouses, burial) are clearly present in all parts of the Fort Ancient region, although they have only rarely been investigated and how they articulate with the larger settlements is poorly known.

Intrasite patterning is well known only for a few sites in the region due to limited excavations, leaving us with few general statements (see Nolan 2010 and 2011, and Cook and Burks 2011). These are limited to the following: villages are circular and have plazas of at least two sizes; one or more mounds or plaza poles or stockades are sometimes present; and site reuse is common either in the same locale or in very close proximity (e.g., Anderson, Madisonville, SunWatch) or in adjacent and sometimes slightly overlapping locales (e.g., Buffalo). At SunWatch, the one site where the issue has been examined in detail, households are clearly organized in localized corporate groups with village-level leadership positions emerging in relationship with growing population size and contacts with nonlocal Mississippians (Cook 2008).

During the 650-year Fort Ancient span, there is one clear change in settlement patterning: it occurred around AD 1400 and is perhaps related to the notable abandonment of the region bordering Fort Ancient to the west, the so-called Vacant Quarter, in the fifteenth century (Cobb and Butler 2002; Williams 1990). Fort Ancient groups became more geographically restricted in the extent of their settlement pattern, occupying a smaller region while exhibiting an increase in social complexity, most evident in the broadening of exchange networks and the development of incipient mortuary ranking (Cook 2008; Drooker 1997; Pollack and Henderson 1992, 2000). Fort Ancient site locations after circa AD 1400 were concentrated closer to the Ohio River and large floodplains (Kennedy 2000). This shift is at least partially relatable to the end of the Medieval Warm Period and the beginning of the Little Ice Age, a period of cool and dry conditions affecting many Northern Hemisphere populations. A related pattern is that sites are clearly most abundant between AD 1200 and 1400, which is also the time when neighboring related cultures form (e.g., Oliver). This "high-water mark" is followed by a notable attrition in site frequency and concentration on the main rivers after AD 1400. At the same time, there is a marked change in material culture, from a variety of local traditions (often valley specific) to a more singular one, typified by Madisonville Horizon pottery. Reasons for this shift are poorly known but include a variety of environmental and social variables (see Nolan and Cook 2010a).

5.1.2. Evolutionary Ecology and Fort Ancient Social Interactions

The W-K model (Figure 5.2) predicts that people and groups will engage in exchange with people and groups whose harvest returns are different from their own. For our application in this chapter, this is the single most important aspect of the model. We briefly discuss each of the four quadrants of the W-K model and its implications for group and individual behavior.

Warfare and other direct mechanisms of limiting outsider access to territories and resources are predicted when returns are highly variable (unpredictable from one harvest to the next; i.e., high standard deviation) for a group and highly correlated among neighbors (Case A in Figure 5.2). When returns among neighboring groups are uncorrelated (right side of Figure 5.2), exchange between groups becomes more likely. If the standard deviations for each group are high (Case B in Figure 5.2), then formal and intensive trade is predicted. This is likely to be represented by shared styles (sensu Dunnell 1978) among interacting groups. If standard deviations are lower (more consistent year to year; Case D in Figure 5.2), then exchange will be less formal and differentiated exchange will predominate; that is,

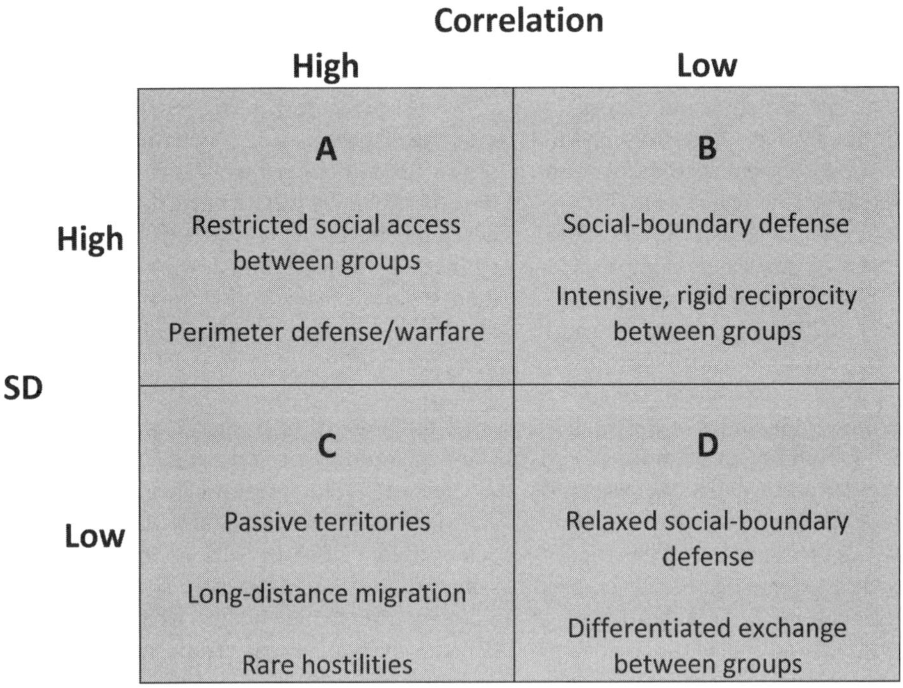

FIGURE 5.2. The Winterhalder-Kelly model. Figure modeled after Kelly (1995:Figure 5-6) and reproduced from Nolan and Cook (2010a:Figure 3).

what is locally abundant is traded with neighbors having different resources in abundance. If correlation among neighbors is high and standard deviations are low, then increased residential mobility is predicted with little to be gained by defense (Case C in Figure 5.2).

In this application, we are most concerned with the cases toward the upper right side of Figure 5.2 (high standard deviation and low intergroup correlation of harvests). Groups occupying this portion of the graph will be likely to exchange with each other and manipulate emblemic (sensu Weissner 1983) styles to identify eligible trading partners. We map out the trends and orientation of correlation in moisture balance during the summer months as a proxy for correlation in agricultural harvest returns. The trends in our dPDSI surfaces indicate the directions that yield the most ecologically beneficial trade for the occupants of the focal sites.

Our model uses the PDSI reconstructions of Cook et al. (2004) as the necessary background conditions for modeling hypothetical patterns of interaction among communities:

Cook et al.'s PDSI reconstructions are based on dendroclimatological analyses and provide data on moisture availability at an annual scale for all of North America. The PDSI is "a widely used measure of *relative* drought and wetness over the United States" (Cook et al. 2004:1016; emphasis added) and there is a growing archaeological literature that uses the index (e.g., Benson et al. 2007; Cooper 2008; Stahle et al. 2007). The PDSI scale ranges from -6 to +6 (arbitrary) using "[n]egative (positive) PDSI values [to] indicate dry (wet) periods, while those near zero presume a state near the average" (Mika et al. 2005:224). Cook et al. (1999:11477) found that tree ring data correlate best with summer PDSI values and therefore the reconstructed values are primarily estimations of summer precipitation for a given year. This is appropriate for our analysis in the temperate eastern United States as summer precipitation has the most direct impact on agricultural productivity. [Nolan and Cook 2010a:69]

When the W-K model was applied iteratively to the relative moisture availability data (PDSI)[1] for the Ohio River Valley, we were able to generate predictions for the direction and nature of interaction among Late Prehistoric period populations (e.g., Monongahela, Fort Ancient, Oliver, Prather, Angel; see Figure 5.1). Throughout most of the modeled period (AD 801–1400), PDSI values are least correlated along a generally east-west axis (i.e., more or less parallel to the Ohio River) with high correlation to the north and south (Nolan and Cook 2010a:Figures 5–10). This generally predicts an east-west orientation to exchange. The western portions of the modeled region (i.e., central Indiana through the American Bottom) consistently experience greater variability in returns (i.e., higher standard deviations of PDSI), applying simultaneous pressure for increased exchange and increased social complexity. Complex groups are better able to manage longer-distance trade and spread the risk of agricultural failure to minimize individual risk (Halstead and O'Shea 1989; Nolan and Cook 2010a, b).

During the period AD 1251–1300, the level of correlation increases drastically in the Lower Ohio Valley and decreases in southwestern to central Ohio (Nolan and Cook 2010a:Figure 9d). We predicted that the extant, politically complex Mississippian societies to the west would expand their trade networks toward the Middle Ohio Valley. Around AD 1350, the orientation of correlation begins to change. The high correlation with neighbors to the north and south disappears, therefore encouraging trade in a north-south direction. There is a continuation of low correlation along the east-west axis in most areas along the Middle and Lower (western) Ohio Valley, and the extant formal exchange relationships with neighbors to the east and west along the Ohio River are expected to have continued, but there is increasing benefit for those groups that establish north-south trade relationships (Nolan and Cook 2010a: 76). This is likely to be a major cause for the previously recognized Horizon (Madisonville Horizon) style, which occurs across the Middle Ohio Valley and has been recognized as starting there between AD 1400 and 1450 (e.g., Cowan 1986; Pollack and Henderson 1992).

As originally presented, the model's predictions are fairly coarse. However, there is enormous potential to zoom in to the modeled surfaces and apply the same framework. It is when viewed from a community perspective that our model is likely to make predictions that are at odds with and richer than the current culture-historical (typological) narrative for Fort Ancient development. In the remainder of this chapter, we discuss one way to drill down into the current model to examine interaction from a community-focused perspective. We examine two 50-year periods from the original model (AD 1251–1300 and AD 1301–1350) for the SunWatch (33MY57) and Reinhardt (33PI880) sites (for SunWatch, see Cook 2008 and Heilman et al. 1988; for Reinhardt, see Nolan 2009, 2010, 2011, and Nolan et al. 2008). The sites are located along the Great Miami and Scioto Rivers, respectively, at approximately the same latitude (Figure 5.3). Both sites have a solid cluster of dates in the Middle Fort Ancient period (Cook 2007; Feathers and Deppen 2009) with the highest mutual probability of occupation during the periods of analysis (AD 1251–1350).

5.1.3. Contribution to Least Cost Modeling

This chapter constitutes a unique application of LCA. While we make use of typical topographic cost paths to evaluate interaction over space, the uniqueness of our approach lies in our application of LCA to time-limited environmental data. This innovative use of the algorithms and technology of LCA highlights the great untapped potential of this type of analytical tool for addressing a variety of anthropologically significant problems. As our contribution highlights, these problems do not have to be limited to analyses in three-dimensional space. Our use of dPDSI surfaces is just one example of an application that considers only horizontal (two-dimensional) space, time, and a third variable (i.e., not vertical space). In this case, the third variable is correlation in moisture availability (as proxy for agricultural production); however, there are a host of behaviorally relevant attributes (correlated with two-dimensional space) that could substitute for the vertical dimension in traditional LCA. We are excited to consider the possibilities of these time-limited analyses, which we only begin to explore here.

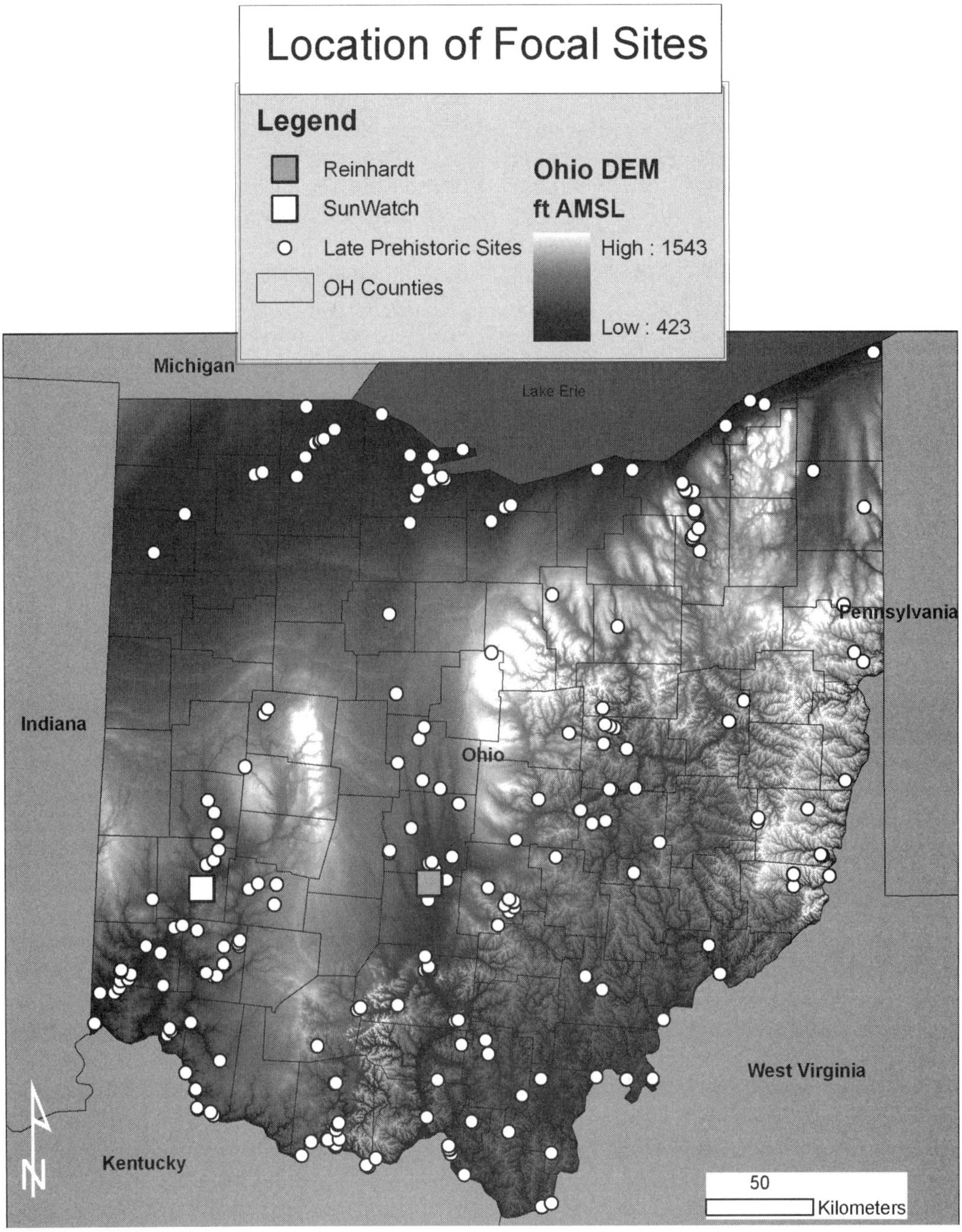

FIGURE 5.3. Locations of case study sites (SunWatch and Reinhardt) and other Late Prehistoric sites with fauna.

5.2. Methods

Our original model (Nolan and Cook 2010a) uses Cook et al.'s (2004) tree-ring-reconstructed Palmer Drought Severity Index (PDSI) as the environmental background. The PDSI reconstructions are available as a downloadable grid of 2.5° × 2.5° data points (NOAA Paleoclimatology's World Data Center for Paleoclimatology and the Applied Research Center for Paleoclimatology: http://www.ncdc.noaa.gov/paleo/pdsi.html) with yearly data for at least the last 1,200 years in the Middle Ohio Valley and as early as AD 342 for some grid points. We divided the annual data into a series of 50-year periods from AD 801 to 1400 (Nolan and Cook 2010a). Our somewhat arbitrary choice of 50-year periods was based on the need for groups of data points to be available to generate values for both axes of the Winterhalder-Kelly model (see Figure 5.2). There are many potential scales at which the data could be aggregated, but echoing the intent of Reid A. Bryson and colleagues (developers of the Macroclimatic Model [MCM], a.k.a. Archaeoclimatology; see Bryson and DeWall 2007), we are trying to get as close to the time scale of human experience as possible given the available data. Bryson's MCM details climate retrodictions for 100-year periods. Given the increasing precision offered by radiometric dating (and other dating methods), we selected 50 years to split the difference between the resolution of Cook et al.'s (2004) annual reconstructions and the MCM and also to approximate the 1σ values reported for most AMS dates (especially from BETA Analytic, Inc.). We are not alone in choosing this time scale for analysis of the potential effects of summer moisture availability on prehistoric populations (see Cooper 2008); however, this is not the only or even the best choice. There are a variety of ways that the PDSI data can be employed, and our periodization of the data is just one (see Benson et al. 2009 for an alternative approach). However, the results obtained in the original model and subsequent extensions (Nolan and Cook 2010b) demonstrate that there is heuristic value in this time scale for archaeological purposes.

For each of the 50-year time periods, we computed an average and standard deviation for the PDSI values for each grid point (Nolan and Cook 2010a). These statistics were then interpolated into raster surfaces for each period to allow analysis of the distribution of spatial and temporal variability in summer moisture balance. There is no guarantee that the interpolated surfaces are accurate for each individual spot on the maps; however, what is critical for our original purposes and our analysis in this chapter is the *orientation of trends* in the surface. This point is particularly important given the scale of the base dataset (2.5° × 2.5° grid). In this chapter we attempt to gain insights at a finer scale than the original model (i.e., the entire Middle Ohio Valley; Nolan and Cook 2010a). Even though our focus is at a finer scale, the major difference is the point of view taken in this analysis. We are still looking at environmental variability (and specifically moisture availability) at the original 2.5° × 2.5° scale. Given the continental climate of the region, the trends in the derived surfaces are very likely to be represented in the data compiled by Cook et al. (2004). The mean and standard deviation surfaces were used to derive the relative values for each of the axes in Winterhalder's model for social interaction (1986; modified by Kelly 1995:Chapter 5, Figure 5-6).

We need to point out that there is a significant difference in resolution between the digital elevation model (DEM) and the PDSI data. This disparity makes direct combination difficult if not impossible; however, the PDSI data are the best-resolution environmental data available. To avoid overinterpretation, we attempt to restrict ourselves to discussions of *trends* in the moisture data as compared with the orientation of topographically inexpensive avenues of travel. Quantitative, one-to-one comparisons are inappropriate given the current dataset. It is difficult to know the appropriate level at which ecological data and topography would be comparable. Our primary focus is on examining the effects of environmental variability over time on patterns of interaction. The addition of the DEM-derived cost paths offer another, atemporal influence on interaction and can serve as a source of alternative or null hypotheses in attempts to develop explanations.

We also develop a method of controlling for survey and recording bias in LCA. Many least cost analyses use known site locations as the targets of navigation. This can be problematic if survey and reporting has been uneven or unsystematic. Additionally, if time is not well controlled in the site records, then paths can be modeled that are temporally irrelevant to the question being addressed. To control for this bias, we created subcardinal catchment zones to serve as the targets of the LCA. The results yield atemporal least cost paths to regions of the catchment and not toward the potentially biased (spatially and temporally) site locations. The resultant atemporal LCPs can be used to address a number of questions and/or to guide problem-oriented fieldwork.

Below we detail how to zoom into the original modeled PDSI surfaces for each period to generate community-focused predictions. The specific steps described are for ArcMap 9.3, but they should be compatible with any 9.x version and easily translated into 8.x or 10.x versions. Our research integrates several types of data: topographic cost surfaces (derived from digital elevation models), archaeological site locations, and difference-of-PDSI (dPDSI) surfaces. We focus on the creation of the dPDSI surfaces because of the uniqueness of this application to archaeological research.

5.2.1. Topographic Cost Surfaces

The base map from which our topographic cost surfaces were derived is a mosaiced 10-m resolution DEM based on digitized versions of the USGS 7.5-minute quadrangles (Ohio EPA/USGS 2004; http://geodata.oit.ohio.gov/geodatadownload/data.aspx). For this stage of the analysis, we limited our geographic focus to the 40-km catchment (analysis mask) for each site. This distance was chosen to keep down computation time and file size but also to include the most proximate intensively investigated sites that will potentially have data available to help us evaluate the predictions based on the model.

5.2.2. Archaeological Sites Layer and Least Cost Path Analysis

Two different sets of paths were computed for each focal site. First, a path is derived to each site recorded in the Ohio Archaeological Inventory (OAI)[2] that has both diagnostics for the Late Prehistoric period (typically small triangular projectile points) and faunal remains. These sites are most likely to be habitation sites (camps or villages) and not special-purpose or procurement sites. There are 244 OAI records that meet these criteria (see Figure 5.3). A least cost path was computed for the cost surface of each 40-km catchment from the focal site to each site within that catchment.

Second, a type of least cost path was developed as a sort of "control" study to consider past human movement within the region. This portion of the analysis was directed at moving beyond the potential limitations of the archaeological site data, such as survey bias, site contemporaneity problems, or site destruction. We computed the control paths for sections of the 40-km catchment by dividing it into 5-km-wide rings and then cutting 45° wedges into the catchment, creating a wedge for each cardinal (north, south, east, and west) and subcardinal (northeast, southeast, southwest, northwest) direction. This resulted in 64 zones within the 40-km catchment (see Figures 5.4b and 5.5b). The 5-km directional zones were then used as the targets for the control least cost path analysis. Once the targets were established around a site, the same method (except target definition) described above was used to create least cost paths.

These control paths enable us to get a general picture of the avenues most likely to be traveled in any direction irrespective of the location of *known* sites. This prevents our analysis from being biased by extant distributional knowledge. Additionally, the directional catchment-zone paths are atemporal. That is, our analysis is not dependent on the assumption that the OAI sites are all contemporaneous. The cardinal catchment analysis will further allow us to assess the economy of placement (i.e., whether known sites occur along avenues of least cost) of known settlements. Assessment of contemporaneity is not currently possible.

In constructing our least cost surfaces for all three sets of least cost analyses, we used the default settings and only the required elements for the calculation found in ArcGIS 9.3. No optional limits or fields were used. To save on computation time, we set the cell size for the raster output to 20 m. Surface-Evans's LCP analyses of Archaic period sites in the Ohio valley showed little significant difference in path outcomes for 10-m- and 90-m-resolution DEMs (Surface-Evans 2007, 2009). Therefore, this compromise resolution is not expected to affect the resultant topographic-based cost paths significantly. We used the 10-m DEM or the dPDSI surface as the input raster for the calculation of the cost-distance raster. The focal site is identified in the "Distance to" field. A direction raster is also created in this step. The cost-distance raster records the "least-accumulative-cost" (ArcMap help files) for each cell from the source ("Distance to," or focal site). The cost-direction surface indicates the direction for each cell of the least-resistance travel given

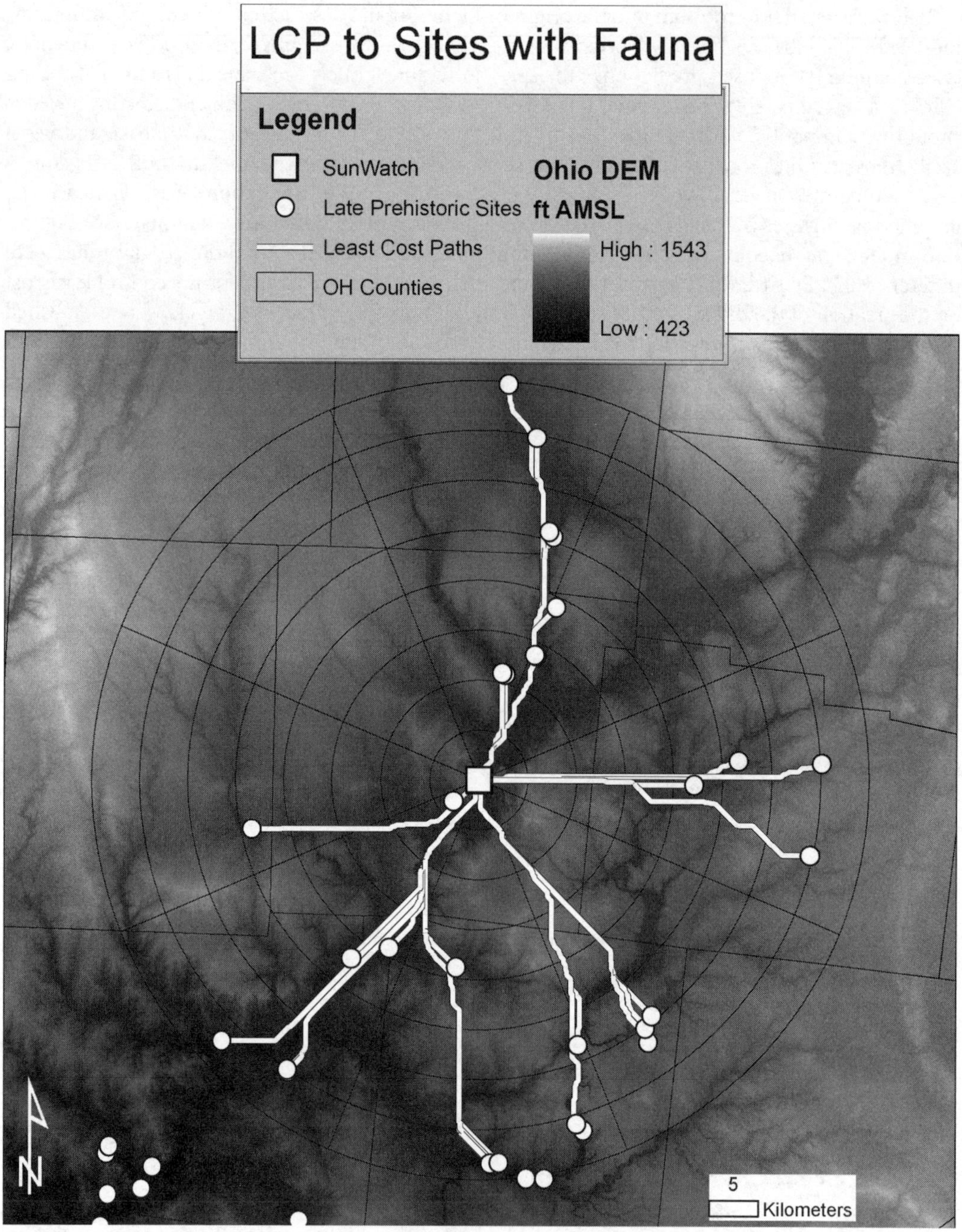

FIGURE 5.4A. Topographic least cost paths from SunWatch. Paths to sites with fauna within the 40-km catchment.

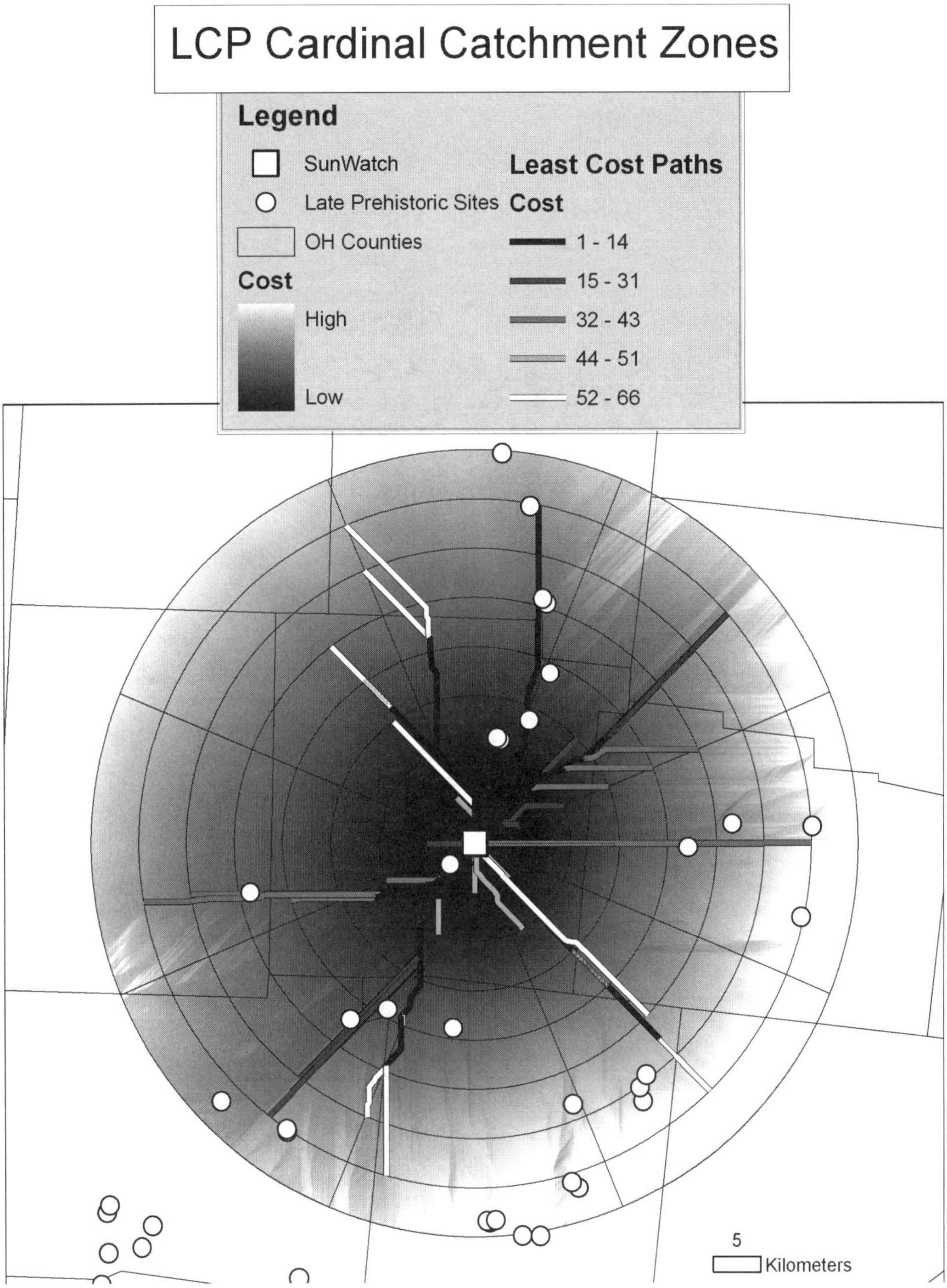

FIGURE 5.4B. Topographic least cost paths from SunWatch. Paths to each 5-km band for each direction within the 40-km catchment. Paths in B are ranked by cumulative cost (GRID_CODE).

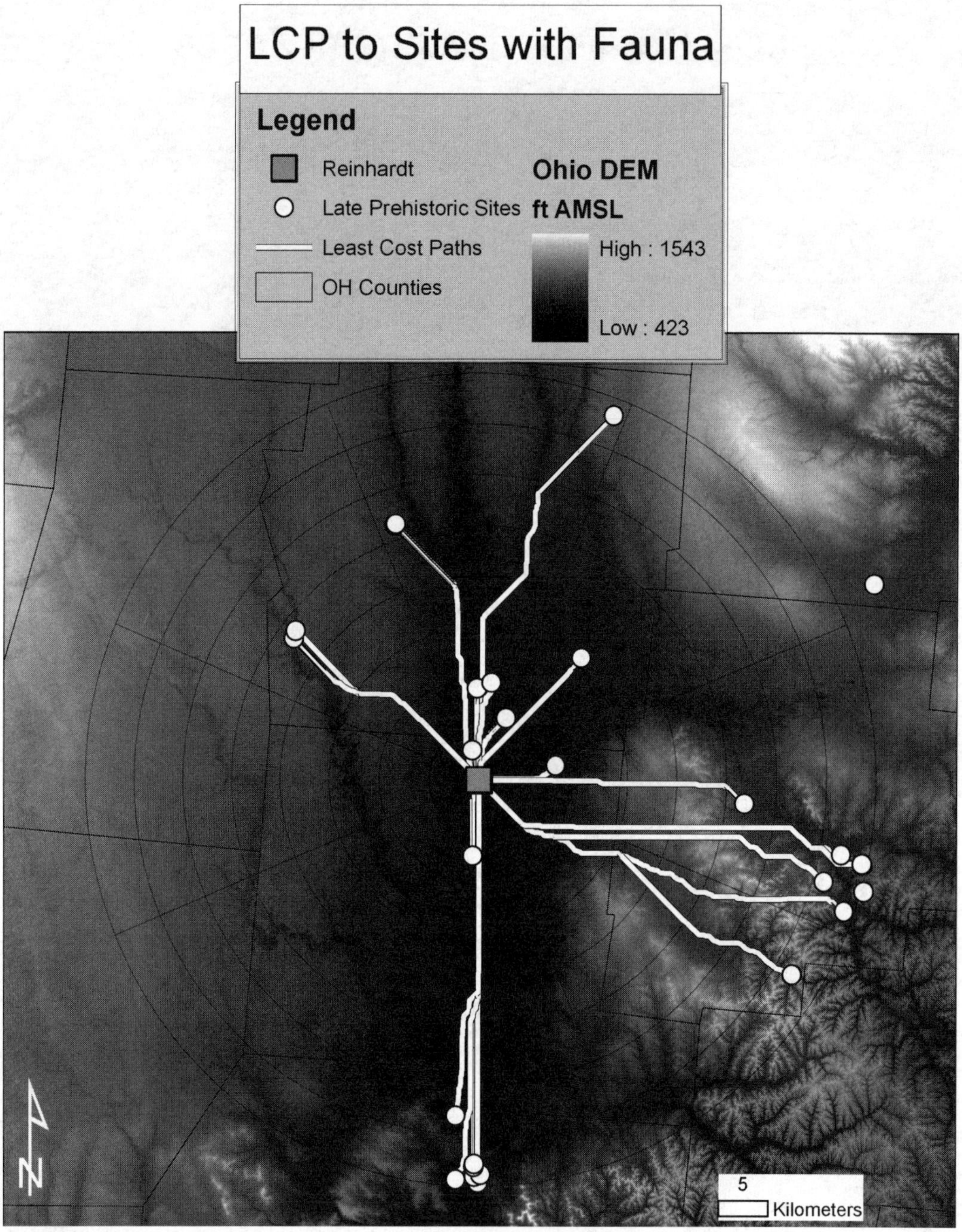

FIGURE 5.5A. Topographic least cost paths from Reinhardt. Paths to sites with fauna within the 40-km catchment.

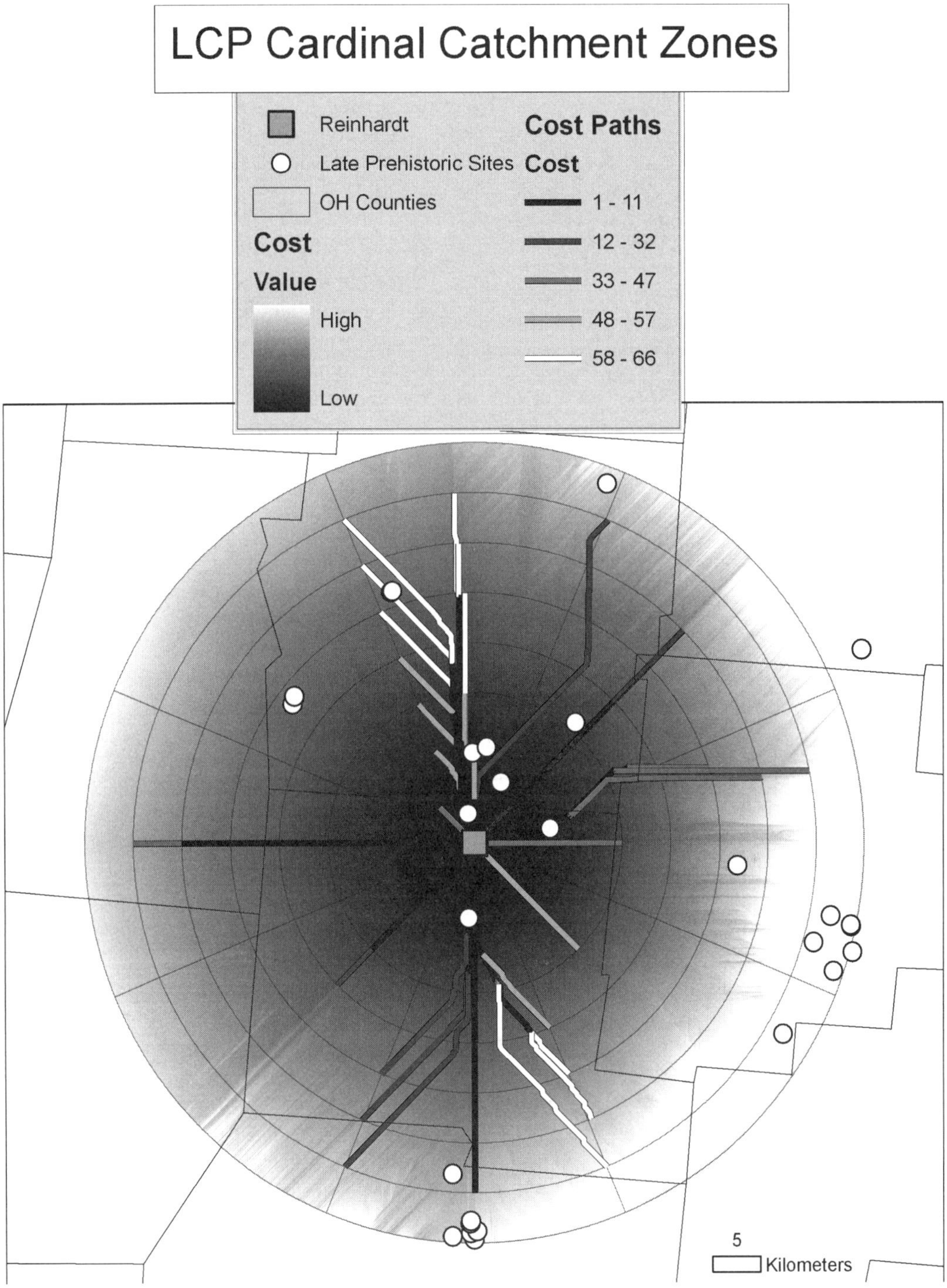

FIGURE 5.5B. Topographic least cost paths from Reinhardt. Paths to each 5-km band for each direction within the 40-km catchment. Paths in B are ranked by cumulative cost (GRID_CODE).

the input raster. Finally, one or more LCPs are calculated. From the Spatial Analyst menu, select Distance, then Shortest Path. The dialog box requests a "Path to," a cost-distance raster, and a cost-direction raster as input for the calculation. In the "Path to" field, we identified the point or polygon layer (OAI sites or subcardinal catchment zones, respectively) to which we wanted to navigate from the focal site for which the cost rasters were calculated.

5.2.3. Construction of Difference-of-PDSI Surfaces

The construction of the dPDSI surfaces for our analysis starts with the original mean PDSI surface for the period of interest (i.e., AD 1251–1300, AD 1301–1350). The first step is to assign a PDSI value for each point representing a Late Prehistoric (ca. AD 800–1650) site in the OAI database with faunal remains. The point shapefiles were originally projected in Universal Transverse Mercator (UTM) zones 16 (western Ohio; $N = 65$) and 17 (central and eastern Ohio; $N = 188$). To simplify the calculation of a single raster surface for the whole state, we projected the files for each zone into State Plane (Ohio, South, North American Datum 1927) coordinates and then combined them into a single coverage using the ArcToolbox Append function. This coverage was then clipped by a layer of the state of Ohio with county boundary polygons to eliminate all points in the OAI database that have either no coordinates or erroneous coordinates (i.e., placing the site outside the state). The clip operation left us with 244 OAI records out of the original 253.

Using the composite point coverage for all Ohio Late Prehistoric fauna sites, we extracted values from the respective mean PDSI surfaces for each period into a new column in the attribute table (PDSI1251, PDSI1301).[3] The next step was to calculate values for dPDSI from each of the focal sites for each period so that those sites with the greatest dPDSI from the focal site will have the lowest value; that is, the sites with least similar PDSI values (irrespective of direction of difference) will be downslope on the dPDSI surface. The formula used to calculate dPDSI (D) for each focal site (F_{xn}) for each period from each faunal site location (P_n) is:

$$P_n - F_{xn} = D,$$

where n can be either 1 or 2 for AD 1251–1300 or AD 1301 to 1350, respectively, and x can be either s or r for SunWatch or Reinhardt, respectively. For example, the calculation to populate each field in the dPDSI column for the AD 1251–1300 period for Reinhardt would have the notation:

$$P_1 - F_{r1} = D.$$[4]

The raw dPDSI results included negative numbers, but the Spatial Analyst cost calculation function cannot handle negative values. The raw differences were then subtracted from 1 to yield all positive numbers.[5] An additional hurdle is the maximum precision that the cost-surface calculator can handle. We eliminated all decimal points by multiplying the results by 1000 and storing the results as an integer in a new column. Each subsequent calculation was performed in a new column in the attribute table. This maintains the ability to use each subsequent stage in different forms of display and analysis if desired, as no calculation is overwritten by subsequent manipulation.

Now each point representing the Late Prehistoric fauna sites has a dPDSI value ranging from 0 to 1000. Sites with a value of 1000 have a raw PDSI value identical to the focal site for the period of analysis. The smaller the dPDSI values for the points, the greater the difference from the focal site's PDSI value for that period.

Next we interpolated the dPDSI surface using the newly populated dPDSI*1000 field for each period for each focal site as the Z value and the Late Prehistoric fauna sites as the source of the data. We chose to use a spline interpolation algorithm because the algorithm forces the surface through the known points and yields the smoothest surface possible. For these cost surfaces, we used the entire state of Ohio. We doubled the cell width to facilitate a quicker computation and smaller file sizes (i.e., 20 m).[6] From the dPDSI surface, we calculated a cost surface. This is exactly the same function used for regular topographic surfaces. Finally, we calculated the LCPs to each of the fauna sites. Unlike for the DEM LCPs, there is a SunWatch dPDSI surface and a separate surface for Reinhardt for both periods (see Figures 5.4 and 5.5, respectively). Each of these surfaces is derived from the same data but illustrates different experiences of variation in the spatial distribution of relative moisture availability between the focal sites.

5.3. Results

Figures 5.4 through 5.8 present the results of our least cost modeling. For Figures 5.4 and 5.5, subfigures (A) are the LCPs to the faunal sites within the 40-km catchment over the actual topography and subfigures (B) are the LCPs to the 5-km cardinal catchment zones over the respective cumulative cost surfaces. Figures 5.6 through 5.8 represent the LCPs to all Late Prehistoric sites with fauna in the state. The background for these figures is the respective dPDSI surface for the focal site for the period of analysis.

5.3.1. Digital Elevation Model Least Cost Paths

5.3.1.1. *SunWatch*

Least cost paths from SunWatch to other Late Prehistoric sites with faunal remains are summarized in Figure 5.4a. There are 24 sites within 40 km of the SunWatch village. For most of these sites, the most cost-effective path is not drastically different from a straight-line path, as they tend to head more or less directly to the site. The exception is the sole site located to the west of SunWatch (Howard Thomas site, 33PR73). In this case it is cheaper to go southwest first and then cut due west. We also see that paths for most of the sites diverge from straight-line paths where the lowlands of major waterways ease travel costs. River travel may have affected actual intercommunity travel (see Surface-Evans this volume; Livingood this volume); however, this was beyond the scope of our current investigation. Notable exceptions are found to the east and southeast. As can be seen in Figure 5.4a, travel to these sites requires uphill, cross-drainage effort. There are no sites in the northwestern section of the catchment and almost no sites in the northeast section of the catchment.

Several interesting patterns come to light in the control LCPs from SunWatch, as shown in Figure 5.4b. First, there is significant overlap between the paths to the cardinal catchment zones and those of Figure 5.4a. Second, the cheapest way to move south or west, or northwest, is to first head southwest or north, respectively, due to the orientation of the Great Miami River and its major tributaries. Not surprisingly, the lowest cost is incurred if one moves along the paths of the extant waterways (the Great and Little Miami Rivers and their tributaries). When one is heading northwest, it is cheapest to follow tributary drainages of the Great Miami River north and then cut west toward the intended destination. Due to the orientation of drainages and the relatively flat character of the Wisconsin-aged till plains, it is best to go more or less straight to the destination when one is moving southeast or east.

5.3.1.2. *Reinhardt*

The LCPs generated for Reinhardt village show many similarities with those for SunWatch (Figure 5.5a). The sites and LCPs are mostly along the larger rivers and tributaries. Most of the OAI sites are accessible via the Scioto River, which runs north-south in the middle of the catchment just 150 m west of Reinhardt. However, the path to the Voss site (33FR52; westernmost site in the figure) is forced to go over land the majority of the way due to the meandering nature of the Big Darby Creek and this creek's joining the Scioto south of the focal site. It is cheaper to follow the Scioto River northward and then trek westward over the flat Wisconsinan glacial till. There are no OAI sites to the west or southwest of Reinhardt (Figure 5.5a).

There is much less overlap between the LCPs to the OAI fauna sites and the control LCPs to Reinhardt's cardinal catchment zones than there was for SunWatch (Figure 5.5b). This may be due to differences in the drainage structure of the landscape surrounding the Reinhardt and SunWatch sites. The absence of documented sites west and southwest of Reinhardt is especially strange given the low cost of traveling in those directions. From Reinhardt, the southeastern and northwestern sections are especially expensive to travel to, and efficient travel to these zones is confined to the trunk stream of the Scioto River followed by an eastern or western jog, respectively, to the closest point within each zone. Notice also that the cluster of sites in the dissected plateau in the western 35–40-km zone is nowhere near the LCPs for this zone. The cheapest avenue of travel, in general, is to follow the Scioto River south toward the cluster of sites in northern Ross County, including Gartner (33RO19; Mills 1904), Baum (33RO18; Mills 1906), and Kramer (33RO33; Ullman 1985).

We also note that the location of the sites to the east and east-northeast force the LCPs from Figure 5.5a through the sections of the cumulative cost surface (see Figure 5.5b) that are markedly more expensive to travel through. To illustrate, within the northern edge of the western catchment wedge we see a difference of approximately 5 km between areas that lie within the same cost zone. This difference corresponds roughly with the Wisconsinan glacial margin, leaving the unglaciated, dissected Appalachian Plateau with higher cost-distance ratios.

5.3.2. ad 1251–1300 Difference of PDSI

The original regional model (Nolan and Cook 2010a:Figure 9d–f) shows that AD 1251–1300 was a period of moderate year-to-year variability in PDSI (standard deviation of PDSI) for southwestern and central Ohio, with the most pronounced spatial variability for the region focused between the Miami and Scioto valleys. Both central and southwestern Ohio are predicted to fall near W-K case B (see Figure 5.2), and therefore communities in these two drainages become likely trading partners. Exchange between the Scioto and Miami valleys is especially likely at this time period as there is very little difference in PDSI to the east and west, respectively.

5.3.2.1. *SunWatch*

The large-scale pattern for the whole valley is still present in the SunWatch dPDSI surface (Figure 5.6). What becomes clear from this new surface is that the Reinhardt site and its neighbors in the upper Scioto Valley are the closest communities (especially Voss) with a relatively large difference (for the period) in mean precipitation. Any travel along the course of the Miami Rivers will provide little access to varied mean precipitation regimes and therefore highly correlated harvests. Even more interestingly, trade along the Ohio River would yield little variety. A detail that is not obvious from the original surface (Nolan and Cook 2010a:Figure 9d) is that the farther east one goes from SunWatch, the greater the difference obtained; however, near maximum difference is achieved just past Reinhardt in southern Franklin County and western Fairfield County (toward the same cluster of sites in the dissected plateau referenced in section 5.3.1.2). The path to the greatest variability goes nearly through the Reinhardt site in northern Pickaway County. A plateau of similar PDSI values extends west, north, south, and southeast. Moreover, no further difference in PDSI is obtained by going farther than central Ohio. One final detail of note is that, with few exceptions, the LCPs run nearly directly to the destination and there is no notable corridor for access to variable moisture regimes. This underscores the limited relative variability within Ohio during this time period. Based on these results, it is expected that communities in southwestern Ohio, including SunWatch, would seek to alleviate periodic shortages by trading with communities located to the east, perhaps as far as central Ohio (through either direct or indirect means).

5.3.2.2. *Reinhardt*

There is a slightly lower degree of difference in PDSI within the state from the perspective of the Reinhardt village (Figure 5.7; compare the range of 667–1000 for SunWatch with the 706–1000 for Reinhardt). Additionally, the model suggests that there is nothing to be gained by engaging in trade with the communities to the east, north, and south. A plateau of near unity in PDSI envelops the whole of central Ohio, and the only "downhill" run is located to the west. There is a relatively pronounced difference in the PDSI values along the western third of the state and notably a single corridor of ecological LCPs to the Miami Valley sites. This may account for the strong pattern of interaction between the Scioto and Miami valleys previously recognized by the distribution of *Anderson*-like ceramics (Brady-Rawlins 2007; Graybill 1981; see also Prufer and Shane's 1970 discussion of the relation between *Baum* and *Anderson* series ceramics). In contrast to the fairly even spread of LCPs from SunWatch (see Figure 5.7), the Reinhardt LCPs form several clusters, predicting a more constrained flow of exchange in each direction.

5.3.3. Difference of PDSI, ad 1301–1350

For the entire Ohio Valley, the period AD 1301–1350 represents a significant decrease in spatial variability (Nolan and Cook 2010a:Figure 10a). This would make it very difficult to access variable precipitation regimes through trade. Temporal variability also increased during this period.

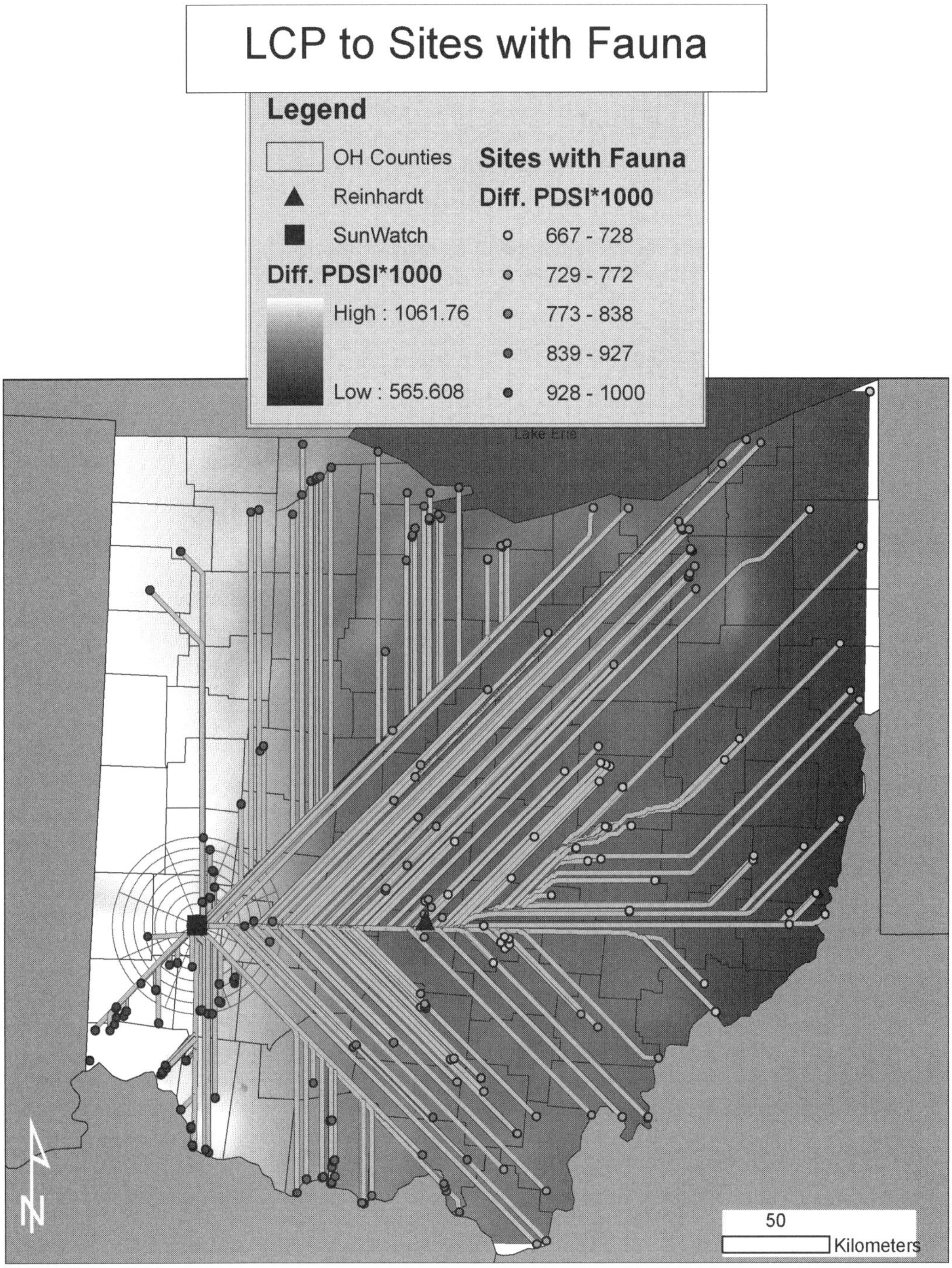

FIGURE 5.6. Least cost paths from SunWatch for difference of PDSI for the period ad 1251–1300. Cost surface is the absolute value of the difference in PDSI from the value for SunWatch subtracted from 1 and multiplied by 1000. The higher the value, the greater the similarity in PDSI values for the period.

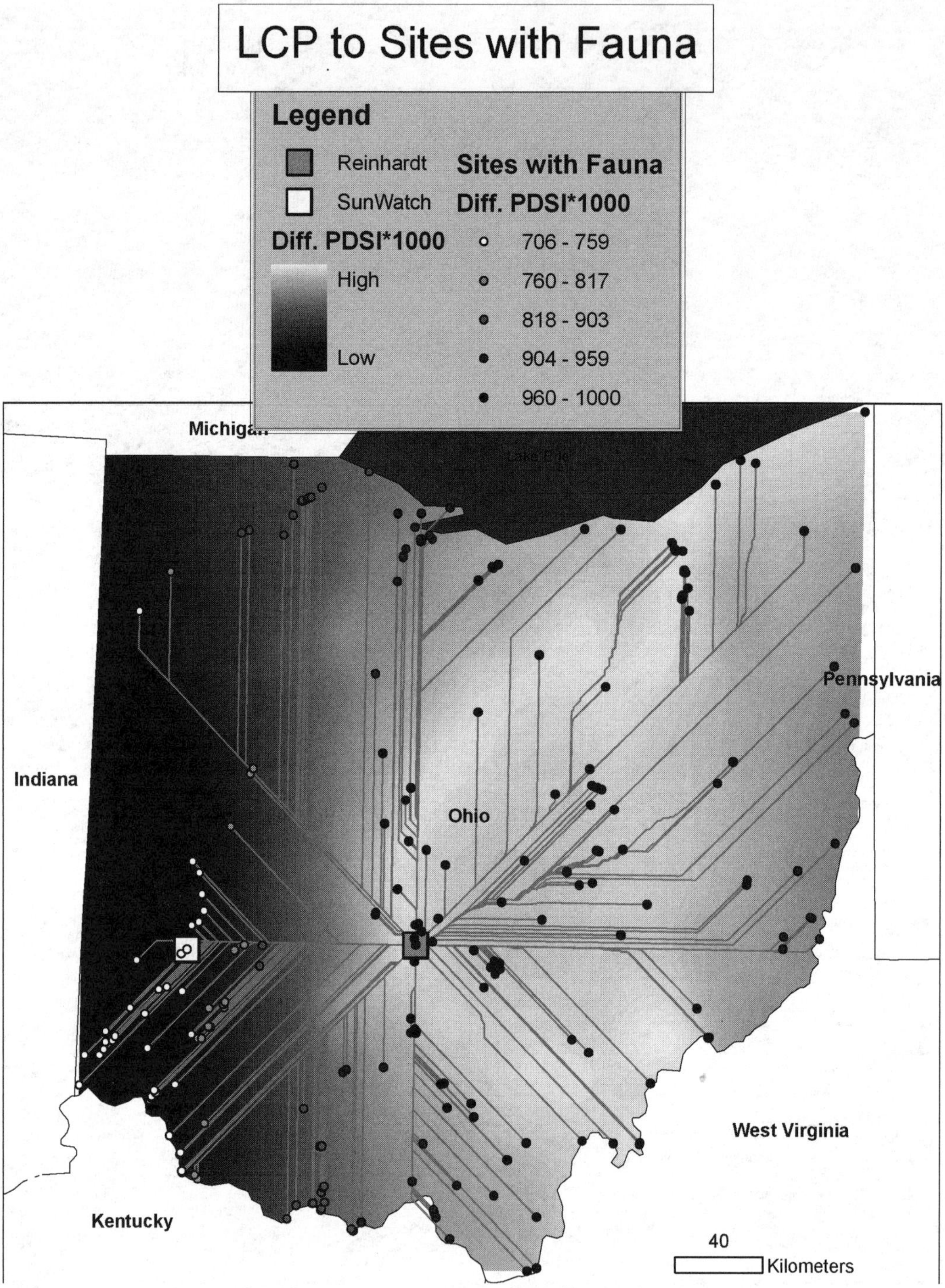

FIGURE 5.7. Least cost paths from Reinhardt for difference of PDSI for the period ad 1251–1300. Cost surface is the absolute value of the difference in PDSI from the value for SunWatch subtracted from 1 and multiplied by 1000. The higher the value, the greater the similarity in PDSI values for the period.

We predicted increasing size of east-west exchange networks, especially among more complex Mississippian communities and their neighbors (including southwestern Ohio Fort Ancient societies), but also an increased emphasis on territorial defense (W-K case A; see Figure 5.2) (Nolan and Cook 2010a).

5.3.3.1. *SunWatch*

The most obvious result for this surface is the constricted range in both interpolated (dPDSI surface) and site-specific values of difference of PDSI (Plate 5). Though the range of values is small, there is a noticeable "sink" in the dPDSI surface centered on Franklin and Licking Counties in central Ohio. That is, central Ohio represents the maximum difference of PDSI from the SunWatch perspective, and this difference is particularly pronounced for this period. The resulting LCPs for this period are more clustered than for the previous period; the majority of paths start out due east toward the sink. The LCP clustering results from a more uneven dPDSI surface than the gently sloping dPDSI surface of the previous period.

5.3.3.2. *Reinhardt*

The dPDSI LCPs from Reinhardt for the period AD 1301–1350 are nearly identical to the paths for the previous period (Figure 5.8; cf. Figure 5.7). There is very little access to different precipitation regimes from Reinhardt during this period. The greatest difference is found to the east, but sites with modest relative difference are found closer in the Miami drainage (within 100 km).

5.4. Discussion

The most interesting result of all sets of LCPs modeled here is the high likelihood of trade between communities in the Miami and Scioto valleys. Not only is this a moderate cost per unit of distance (Figures 5.4 and 5.5), it is the easiest way for communities in these two valleys to obtain access to variable patterns of moisture availability for the modeled periods. As a corollary, neither valley has access to variable precipitation via topographically inexpensive routes. If topography were the only concern in accessing trade partners, Figures 5.4 and 5.5 indicate that movement would be limited to a northeast-southwest and north-south orientation for SunWatch and Reinhardt, respectively, following existing major waterways. Both patterns would lead the traders eventually southward toward the Ohio River. Obviously, topography is just one factor (and according to our modeling, a less significant one) affecting choice of trading partners. In neither case would the topographically most efficient paths lead to the establishment of ecologically beneficial trade partnerships.

The prediction of trade interactions between the two valleys is pronounced for the AD 1301–1350 period. Though there is less overall variability, the middle and upper Scioto Valley (central Ohio) contains the maximum difference in mean PDSI from the SunWatch perspective. If we contextualize this expectation within the broader model (Nolan and Cook 2010a), we would expect the impetus for trade to be coming from the Miami Valley toward the Scioto. As previously predicted (Nolan and Cook 2010a) and supported by other recent analyses (Cook 2008; Cook and Schurr 2009), communities in southwestern Ohio would be precocious in developing complex sociopolitical organizations due to their prolonged interaction with more sociopolitically complex Mississippian communities to the west and possibly augmented by immigration to Fort Ancient communities by Mississippian individuals and families (Cook and Schurr 2009). It is significant to note that the predicted direction and extent of exchange by the original and especially the community-based model fit previously observed patterns of stylistic similarity between *Baum* (Scioto Valley) and *Anderson* (Miami Valley) phase ceramics (Graybill 1981; Prufer and Shane 1970). Braun (1991) argues that style of the emblemic (sensu Weissner 1983) variety will be more common the greater the frequency of extra-community interaction. Specifically, Braun argues that this fits the midwestern pattern of elaborate decoration during the Middle Woodland (Hopewell/Havanna; 250 BC/AD 0–AD 400/500) period and the near absence of stylistic elaboration during the Late Woodland (AD 400/500–1000).

In the initial definition of Fort Ancient subdivisions (Griffin 1966), the Scioto and Miami valleys were characterized by two separate foci: Baum and Anderson, respectively. However,

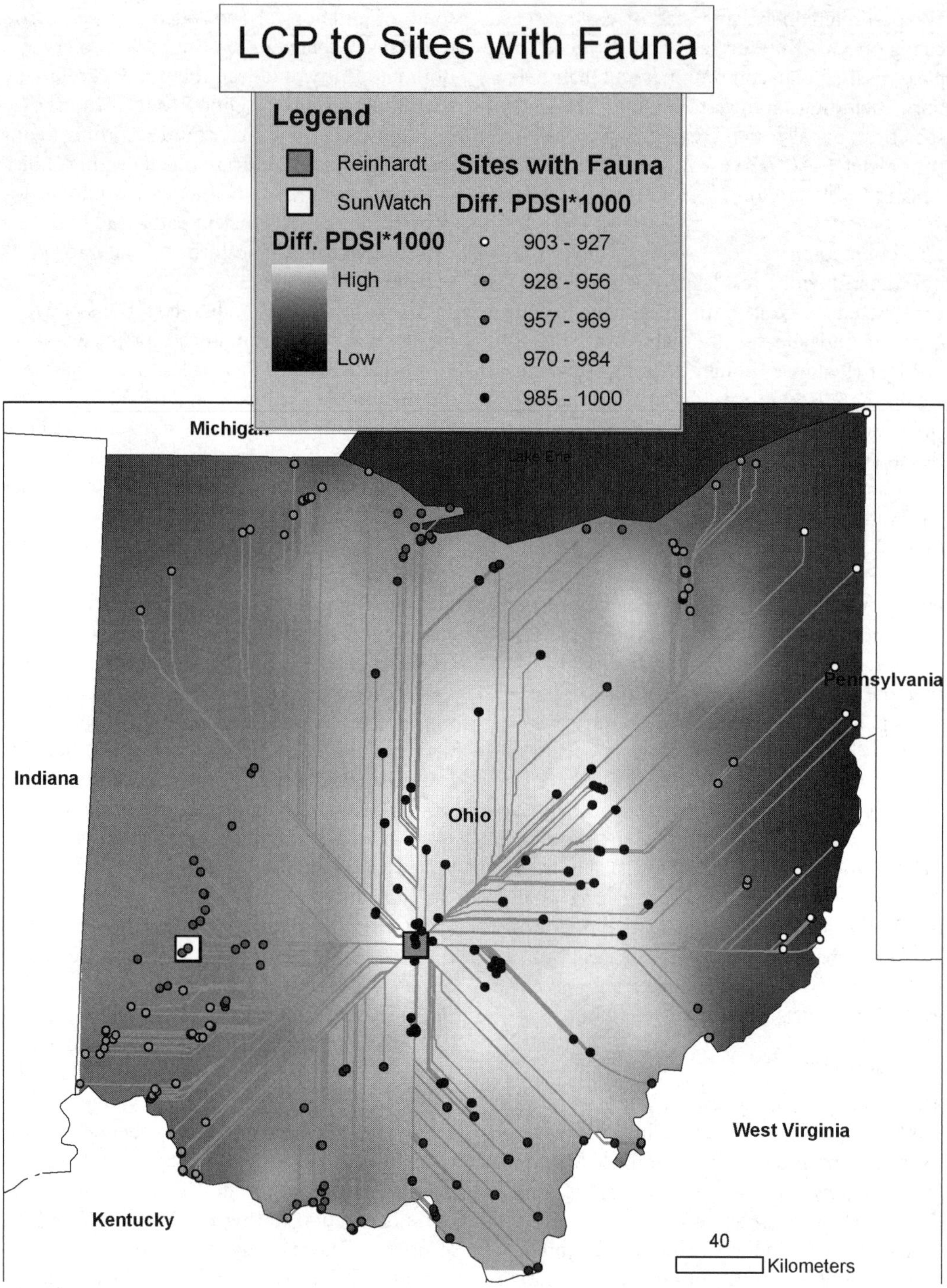

FIGURE 5.8. Least cost paths from Reinhardt for difference of PDSI for the period ad 1301–1350. Cost surface is the absolute value of the difference in PDSI from the value for SunWatch subtracted from 1 and multiplied by 1000. The higher the value, the greater the similarity in PDSI values for the period.

subsequent investigations have noted the significant similarity between the pottery of the middle and upper Scioto Valley and the Anderson wares of the Miami drainage (Brady-Rawlins 2007; Graybill 1981:Figure 1; Prufer and Shane 1970). If pottery decoration is conceived of as style (sensu Dunnell 1978; i.e., homologous similarity), then communities with shared symbols indicate shared, transmitted cultural ideas and constitute "populations" in the evolution of cultural traditions (Hart 2008; Hart and Brumbach 2009; Lipo et al. 1997). Following this logic, the classification of pottery from both drainages as *Anderson* connotes transmission of information between communities across drainage divides likely associated with exchange of nonperishable and, possibly, perishable goods.

This similarity in style of decoration can be seen in the extant collections from both the focal sites. A rimsherd with an incised curvilinear guilloche recovered from the Reinhardt site (Figure 5.9a) has a near twin in a sherd illustrated by Griffin (1966:Plate LII figure 2) from the Steel Dam site in Dayton (Miami drainage). The ceramics from SunWatch (Figure 5.9b) also exhibit a style of decoration similar to that found in the Reinhardt ceramics.

We do *not* mean to imply direct trade between these two communities, only that they were participating in similar interaction spheres that were navigated through manipulation of shared stylistic elements. Our model provides another level of explanation for the previous observations but also points out the need for more thorough analysis of shared attributes across sites in these two drainages and more rigorous dating of these interactions (preferably by directly dating the artifacts of interest [see Feathers 2009]). In order to fully understand the nature and extent of these hypothesized interactions, we need to move beyond a reliance on culture-historic phases and types (e.g., Anderson, Baum, and Fort Ancient). As pointed out by Essenpreis (1978), the construction of phases obscures real variation and creates artificial breaks between collections (see also Hart et al. 2005; Means 2007; Lyman and O'Brien 2006). Lipo et al. (1997) observe that culture-historic taxa often approximate real patterns of social interaction as indicated by shared stylistic attributes.

Another interesting result of our analysis is the greater agreement between the LCPs to documented OAI sites and the cardinal catchment control for SunWatch (Figure 5.4) than for Reinhardt (Figure 5.5). This may have implications for the interpretation of regional interaction networks or it could indicate problems with differential site survey or even differences in the structure of the landscape at the SunWatch and Reinhardt sites. One potential interpretation of this patterning is that access to SunWatch was considered important when site locations were selected in the Miami Valley, while the same cannot be said for Reinhardt. This provides another avenue of investigation into intradrainage interaction patterns, especially in combination with intervisibility analyses (i.e., viewshed; for a good example, see Jones 2006). To fully explore this type of spatial relationship, we would need to repeat the types of analyses performed here for each site in the catchment and/or contemporaneously occupied site. In addition, further systematic archaeological survey in the vicinity of either or both focal sites could shed light on this apparent difference in site interaction patterning. It is important to reemphasize here that the chronological control provided by the OAI database is not adequate for assessing contemporaneity between the sites used in this analysis. It is possible that the patterns of correspondence between the two sets of topographic LCPs is spurious. Again, our model requires more intensive and systematic survey if we are to evaluate the potential patterns indicated by our analysis.

The analyses presented herein have provided several intriguing glimpses into the explanations of patterns of interaction in southern Ohio. We can also offer suggestions for improvements of this type of analysis. The DEM cost-surface analysis is very beneficial and holds much potential alone for the investigation of settlement patterning and interaction patterns—as demonstrated by the other contributions to this book—or in combination with other methods. While the LCP derivation from the modeled PDSI surfaces sheds light on some aspects of interaction, this may not be the best use of these surfaces for the intended analyses. The way that the LCPs were calculated means that there is a lot of noise in the resulting images. Although examination of

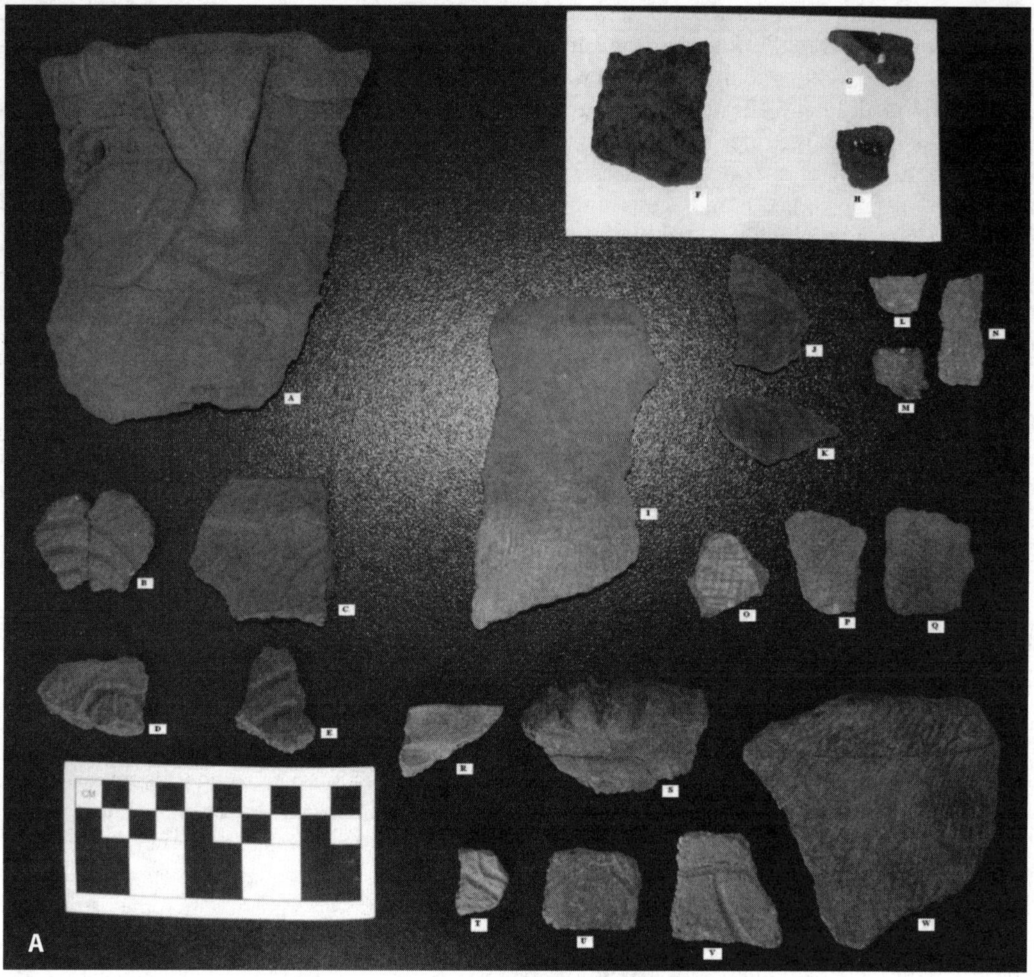

FIGURE 5.9. Diagnostic pottery from Reinhardt and SunWatch. (A) Decorated, shell-tempered, and miscellaneous rim sherds from 2008 Reinhardt excavations (see Nolan 2009). (B) Examples of decorated ceramics from SunWatch.

a dPDSI surface itself in tandem with the topographic LCPs can be informative, it may be more effective to analyze the dPDSI surfaces with hydrological modeling. This is best illustrated with the AD 1301–1350 SunWatch dPDSI surface. As noted in the Results section, there is a distinct depression, or sink, in central Ohio for the AD 1301–1350 period. If we were to model flow accumulation for this surface, we would eliminate the paths to sites of similar mean PDSI for the period and we would see that the modeled interaction flow from SunWatch would pool around Reinhardt. Due to restrictions of processing time, we have not yet performed this additional analysis.

To fully realize the potential of the W-K model in site-focused analyses would require analyses of the standard deviation surfaces and even the intraperiod fluctuation in difference of PDSI in an analog fashion similar to the analysis conducted by Benson et al. (2009) for Cahokia. Each of these analyses could be accomplished with methods similar to the ones employed here. However, it will be impossible to evaluate the model based on the current database. The extant OAI database is biased to an unknown degree by uneven survey coverage and uneven reporting of sites. A further hindrance to evaluation of the predictions for interaction is the paucity of well-dated sites. Nevertheless, we feel that our control LCPs have the potential to provide meaningful heuristic models even in situations of possible site identification bias.

5.5. Conclusion

The analyses presented herein have provided several intriguing glimpses into the explanations of patterns of interaction in southern Ohio during the Late Prehistoric period (AD 1000–1650). We have also presented the first least cost analysis for the southern Ohio Late Prehistoric period. This has been a cursory foray into the local application of our model of interaction for the whole Middle Ohio Valley. However, our main goal was to illustrate how to drill down into the original model, allowing us to perform analysis at multiple scales and to draw contrasts among competing alternative explanations. In doing so, we have identified several promising avenues for future exploration of the model's community-based predictions and identified shortcomings in both the approach taken here and the extant database. In addition to noting these deficiencies in the database, we have presented a method for controlling for bias inherent in the way the database was compiled. Our control cost paths (i.e., subcardinal catchment zones are targets) identify the most efficient paths to each zone of the catchment and can be used either to identify areas in need of survey coverage and/or to evaluate the nature of the economy of site choice in a regional interaction network. In our region, either application requires additional data collection. Intensified efforts to build up the extent and quality of the available database are necessary. We also need to shed old conceptual models that inhibit our ability to critically assess variability in the archaeological record.

Acknowledgments

We express our gratitude to Brent Eberhard and the Ohio Historic Preservation Office (OHPO) for access to the Ohio Archaeological Inventory data used in this chapter. Partial funding for data acquisition and the ESRI ArcInfo license used in this analysis was provided to the primary author by the OHPO Certified Local Government grant program (on behalf of the City of Columbus, Ohio, award number 39-08-21740) and the National Science Foundation Doctoral Dissertation Improvement Grant program (BCS0832272), respectively. We also are indebted to Devin White and Sarah Surface-Evans for helping us make our contribution stronger and our presentation more clear. Of course, any errors or omissions are our responsibility.

Notes

1. It is important to note that the PDSI does not measure precipitation; it is a measure of the moisture balance (which is the result of precipitation, temperature, and other environmental variables) and specifically that quantity available for plants. We injudiciously used precipitation and moisture availability interchangeably in our original article (Nolan and Cook 2010a) and would like to clarify this issue up front. This does not change the argument or the justification for the model, but in some areas the failure to distinguish the two variables could have consequences for their interpretation.
2. The Ohio Archaeological Inventory (OAI) is the statewide database of officially recorded

archaeological sites maintained by the Ohio Historic Preservation Office (OHPO) and available to researchers. The OAI database contains data on location, time period(s) of use, site type, and types of material present/observed at the site. The OHPO maintains this database in a GIS housed at the Ohio Historical Society, Columbus.

3. To populate these PDSI fields, we used the Extract Values to Points function of ArcToolbox. This function creates a new shapefile with the point values. The values of the newly created point shapefile needed to be copied into the respective column of the original for the analysis to proceed. This can be accomplished in ArcMap, but it is more easily accomplished in a spreadsheet or database program. Any version of Microsoft Excel prior to 2007 should be able to open the dbaseIV (.dbf extension) and save the changes into the original file. In order for these changes to be performed, all ArcInfo products must be shut down or the changes will not register with the program. Unfortunately, Excel 2007 does not retain the ability to save in dbaseIV format. To work around this difficulty, we saved the modified version of the original dbaseIV file in Excel format (.xls or .xlsx extensions) and then imported it into Microsoft Access (alternatively, the changes could be made in Access, but the Paste function in Excel is easier to work with). Once the table is open in Access, it can be exported into a dbaseIV file. After the changes were made, we returned to ArcMap for the rest of the analysis.
4. Note that this formula is presented here in generic form and that in application in the ArcMap field calculator the appropriate column headers or variable values would need to be inserted (e.g., P_1 = PDSI1251 where PDSI1251 is the column that was populated by the Extract Value to Point function; $F_{r1} = -0.0359$).
5. In cases where there are bigger differences, this step will need to be modified.
6. The settings used for the spline function were 12 points, weight 0.1, cell size 65.616 ft.

References

Benson, Larry V., Timothy R. Pauketat, and Edward R. Cook
2009 Cahokia's Boom and Bust in the Context of Climate Change. *American Antiquity* 74:467–483.

Benson, Larry, Kenneth Peterson, and John Stein
2007 Anasazi (Pre-Columbian Native-American) Migrations during the Middle-12th and Late-13th Centuries—Were They Drought Induced? *Climatic Change* 83:187–213.

Brady-Rawlins, Kathleen
2007 The O.C. Voss Site: Reassessing What We Know about the Fort Ancient Occupation of the Central Scioto Drainage and Its Tributaries. Unpublished Ph.D. dissertation, Ohio State University, Columbus.

Braun, David P.
1991 Why Decorate a Pot? Midwestern Household Pottery, 200 BC–AD 600. *Journal of Anthropological Archaeology* 10(4):360–397.

Bryson, R. A., and K. M. DeWall (editors)
2007 *A Paleoclimatology Workbook: High Resolution, Site-Specific, Macrophysical Climate Modeling.* The Mammoth Site, Hot Springs, South Dakota.

Carskadden, J., and J. Morton
2000 Fort Ancient in the Central Muskingum Valley of Eastern Ohio: A View from the Philo II Site. In *Cultures Before Contact: The Late Prehistory of Ohio and Surrounding Regions*, edited by R. A. Genheimer, pp. 158–193. Ohio Archaeological Council, Columbus.

Cobb, C. R., and B. M. Butler
2002 The Vacant Quarter Revisited: Late Mississippian Abandonment of the Lower Ohio Valley. *American Antiquity* 67:625–641.

Cook, E. R., D. M. Meko, D. W. Stahle, and M. K. Cleaveland
1999 Drought Reconstructions for the Continental United States. *Journal of Climate* 12:1145–1162.

Cook, Edward R., Connie A. Woodhouse, C. Mark Eakin, David M. Meko, and David W. Stahle
2004 Long-Term Aridity Changes in the Western United States. *Science* 306:1015–1018.

Cook, Robert A.
2007 Single Component Sites with Long Sequences of Radiocarbon Dates: The SunWatch Site and Middle Fort Ancient Village Growth. *American Antiquity* 72:439–460.
2008 *SunWatch: Fort Ancient Development in the Mississippian World.* University of Alabama Press, Tuscaloosa.

Cook, R. A., and J. Burks
2011 Determining Site Size and Structure: A Fort Ancient Example. *American Antiquity* 76(1): 145–162.

Cook, Robert A., and Mark Schurr
2009 Eating between the Lines: Mississippian Migration and Stable Carbon Isotope Variation in Fort Ancient Populations. *American Anthropologist* 111:344–359.

Cooper, Judith
2008 Bison Hunting and Late Prehistoric Human Subsistence Economies in the Great Plains.

Unpublished Ph.D. dissertation, Southern Methodist University, Dallas, Texas.

Cowan, C. Wesley

1986 *Fort Ancient Chronology and Settlement Evaluation in the Great Miami River Valley, Volume II: Excavation and Chronology*. Report Submitted to the Ohio Historic Preservation Office by the Cincinnati Museum of Natural History.

Drooker, P. B.

1997 *The View from Madisonville: Protohistoric Western Fort Ancient Interaction Patterns*. Memoirs No. 31. Museum of Anthropology, University of Michigan, Ann Arbor.

Dunnell, Robert C.

1961 A General Survey of Fort Ancient in the Kentucky–West Virginia Area. Manuscript on file, Department of Anthropology, University of Kentucky, Lexington.

1972 The Prehistory of Fishtrap, Kentucky. *Yale University Publications in Anthropology* No. 75.

1978 Style and Function: A Fundamental Dichotomy. *American Antiquity* 43:192–202.

Essenpreis, P. S.

1978 Fort Ancient Settlement: Differential Response at a Mississippian–Late Woodland Interface. In *Mississippian Settlement Patterns*, edited by Bruce D. Smith, pp. 141–167. Academic Press, New York.

1982 *The Anderson Village Site: Redefining the Anderson Phase of the Fort Ancient Tradition of the Middle Ohio Valley*. Ph.D. dissertation, Harvard University, Cambridge, Massachusetts. University Microfilms, Ann Arbor, Michigan.

Feathers, James K.

2009 Problems of Ceramic Chronology in the Southeast: Does Shell-Tempered Pottery Appear Earlier Than We Think? *American Antiquity* 74:113–142.

Feathers, James K., and Jacob E. Deppen

2009 Appendix C: Chronometric Dating Report for the Reinhardt Site (33PI880). In *Archaeological Survey of the Reinhardt Tract Property through a Certified Local Government (CLG) Grant on Behalf of the City of Columbus in Harrison Township, Pickaway County, Ohio, Volume I: Survey Results*, by K. C. Nolan, pp. 133–140. Report submitted to the Ohio Historic Preservation Office in compliance with CLG contract agreement.

Feinman, G., and J. Neitzel

1984 Too Many Types: An Overview of Sedentary Prestate Societies. *Advances in Archaeological Method and Theory* 7:39–102.

Graybill, J. R.

1981 The Eastern Periphery of Fort Ancient (AD 1050–1650): A Diachronic Approach to Settlement Variability. Unpublished Ph.D. dissertation, University of Washington, Seattle.

Greenlee, D. M.

2002 *Accounting for Subsistence Variation among Maize Farmers in Ohio Valley Prehistory*. Ph.D. dissertation, University of Washington, Seattle. University Microfilms, Ann Arbor, Michigan.

Griffin, J. B.

1966 [1943] *The Fort Ancient Aspect: Its Cultural and Chronological Position in Mississippi Valley Archaeology*. University of Michigan Press, Ann Arbor.

Halstead, Paul, and John O'Shea

1989 Introduction: Cultural Responses to Risk and Uncertainty. In *Bad Year Economics: Cultural Responses to Risk and Uncertainty*, edited by P. Halstead and J. O'Shea, pp. 1–7. New Directions in Archaeology. Cambridge University Press, New York.

Hart, John P.

2008 Pots, Maize, and Longhouses: Reflections on the Evolution of the Northern Iroquoians. Keynote address for the 54th Midwest Archaeological Conference, Milwaukee.

Hart, John P., and Hetty Jo Brumbach

2009 On Pottery Change and Northern Iroquoian Origins: An Assessment from the Finger Lakes Region of Central New York. *Journal of Anthropological Archaeology* 28:367–381.

Hart, John P., John P. Nass, and Bernard K. Means

2005 Monongahela Subsistence Settlement Change. *Midcontinental Journal of Archaeology* 30:327–365.

Heilman, James M., Malinda C. Lileas, and Christopher Turnbow (editors)

1988 *A History of 17 Years of Excavation and Reconstruction—A Chronicle of 12th Century Human Values and the Built Environment, Volume 1: Excavation*. Dayton Museum of Natural History, Dayton, Ohio.

Henderson, A. G.

1998 *Middle Fort Ancient Villages and Organizational Complexity in Kentucky*. Ph.D. dissertation, University of Kentucky, Lexington. University Microfilms, Ann Arbor, Michigan.

Henderson, A. G., and D. Pollack

2001 Fort Ancient. In *Encyclopedia of Prehistory*, vol. 6, edited by Peter N. Peregrine and Melvin Ember, pp. 174–194. Kluwer Academic/ Plenum, New York.

Jones, Eric E.
2006 Using Viewshed Analysis to Explore Settlement Choice: A Case Study of the Onondoga Iroquois. *American Antiquity* 71:523–538.

Kelly, Robert L.
1995 *The Foraging Spectrum: Diversity in Hunter-Gatherer Lifeways*. Smithsonian Institution Press, Washington, D.C.

Kennedy, W. E.
2000 Interpreting Fort Ancient Settlement Variability. Unpublished Master's thesis, Kent State University, Kent, Ohio.

Lipo, Carl P., Mark E. Madsen, Robert C. Dunnell, and Tim Hunt
1997 Population Structure, Cultural Transmission, and Frequency Seriation. *Journal of Anthropological Archaeology* 16:301–333.

Lyman, R. Lee, and Michael J. O'Brien
2006 *Measuring Time with Artifacts: A History of Methods in American Archaeology*. University of Nebraska Press, Lincoln.

Martin, Kristie Rae
2009 Eastern Agricultural Complex Traditions in Small Fort Ancient Communities: The Wildcat Example. Unpublished Master's thesis, Ohio State University, Columbus.

Mayer-Oakes, W. J.
1955 *Prehistory of the Upper Ohio Valley: An Introductory Archaeological Study*. Annals of Carnegie Museum Vol. 34. Anthropological Series No. 2. Carnegie Museum, Pittsburgh.

Means, Bernard K.
2007 *Circular Villages of the Monongahela Tradition*. University Alabama Press, Tuscaloosa.

Mills, William C.
1904 Explorations of the Gartner Mound and Village Site. *Ohio State Archaeological and Historical Society Publications* 13:129–191.
1906 Baum Prehistoric Village. *Ohio State Archaeological and Historical Society Publications* 15: 45–136.

Moorehead, W. K.
1892 *Primitive Man in Ohio*. G. P. Putnam's Sons, New York.

Nolan, Kevin C.
2009 *Archaeological Survey of the Reinhardt Tract Property through a Certified Local Government (CLG) Grant on Behalf of the City of Columbus in Harrison Township, Pickaway County, Ohio, Volume I: Survey Results*. Report submitted to the Ohio Historic Preservation Office in compliance with CLG contract agreement.
2010 Multi-staged Analysis of the Reinhardt Village Community: A Fourteenth Century Central Ohio Community in Context. Unpublished Ph.D. dissertation, Ohio State University, Columbus.
2011 Distributional Survey of the Reinhardt Site (33PI880), Pickaway County, Ohio: A Strategy for Deciphering the Community Structure of a Fort Ancient Village. *Midcontinental Journal of Archaeology* 36(1):105–130.

Nolan, Kevin C., Jarrod Burks, and William S. Dancey
2008 Discovering Ohio's Newest Earthwork: Geophysics and Distributional Survey of a Fort Ancient Village and a Hopewellian Enclosure at the Reinhart Site, Pickaway County, Ohio. Poster presented at the 73rd Meeting of the Society for American Archaeology, Vancouver, British Columbia. Electronic document, http://www.ohioarchaeology.org/joomla/index.php?option=com_content&task=view&id=236&Itemid=32.

Nolan, Kevin C., and Robert A. Cook
2008 Was Social Complexity Impossible during the Late Woodland but Mandatory during the Late Prehistoric? An Evolutionary Ecological Model of Cultural Change in the Ohio Valley. Paper presented at the 54th Midwest Archaeological Conference, Milwaukee, Wisconsin.
2010a An Evolutionary Model of Social Change in the Middle Ohio Valley: Was Social Complexity Impossible during the Late Woodland but Mandatory during the Late Prehistoric? *Journal of Anthropological Archaeology* 29:62–79.
2010b Volatile Climate Conditions Cahokia: Comment on Benson, Pauketat and Cook 2009. *American Antiquity* 75(4):978–983.

Ohio EPA/USGS
2004 Ohio 10m Resolution Digital Elevation Model. Ohio EPA Division of Emergency and Remedial Response, Columbus.

Pollack, David, and A. Gwynn Henderson
1992 Towards a Model of Fort Ancient Society. In *Fort Ancient Cultural Dynamics in the Middle Ohio River Valley*, edited by A. G. Henderson, pp. 281–294. Monographs in World Archaeology No. 8. Prehistory Press, Madison, Wisconsin.
2000 Insights into Fort Ancient Culture Change: A View from South of the Ohio River. In *Cultures before Contact: The Late Prehistory of Ohio and Surrounding Regions*, edited by R. A. Genheimer, pp. 194–227. Ohio Archaeological Council, Columbus.

Prufer, Olaf H., and Orin C. Shane III
1970 *Blain Village and the Fort Ancient Tradition in Ohio*. Kent State University Press, Kent, Ohio.

Putnam, F. W.
1886 Report of the Curator. *Peabody Museum Annual Report* 18:401–419, 477–502.

Rafferty, Janet E.
1974 *The Development of the Fort Ancient Tradition in Northern Kentucky*. Ph.D. dissertation, University of Washington, Seattle. U.M.I. Dissertation Information Service, Ann Arbor, Michigan.

Redmond, B., and R. McCullough
2000 The Late Woodland to Late Prehistoric Occupations of Central Indiana. In *Late Woodland Societies: Tradition and Transformation across the Midcontinent*, edited by Thomas E. Emerson, Dale L. McElrath, and Andrew C. Fortier, pp. 643–684. University of Nebraska Press, Lincoln.

Rossen, J.
1992 Botanical Remains. In *Fort Ancient Cultural Dynamics in the Middle Ohio Valley*, edited by A. G. Henderson, pp. 189–208. Prehistory Press, Madison, Wisconsin.

Smith, B.
1990 Introduction. In *The Mississippian Emergence*, edited by Bruce D. Smith, pp. 1–8. Smithsonian Institution Press, Washington, D.C.

Smith, H.
1910 *The Prehistoric Ethnology of a Kentucky Site*. Anthropological Papers Vol. 6, Pt. 2. American Museum of Natural History, New York.

Stahle, David W., Falko K. Fye, Edward R. Cook, and R. Daniel Griffin
2007 Tree-Ring Reconstructed Megadroughts over North America since AD 1300. *Climatic Change* 83:133–149.

Surface-Evans, Sarah
2007 "Where Many Paths and Errands Meet": Modeling Archaic Hunter-Gatherer Movement in the Ohio Falls Landscape. Poster presented at the 72nd Annual Meeting of the Society for American Archaeology, Austin, Texas.
2009 Hunter-Gatherer Cultural Landscapes: A Case Study for a GIS-Based Reconstruction of the Shell Mound Archaic in the Falls of the Ohio Region of Indiana and Kentucky. Unpublished Ph.D. dissertation, Michigan State University, East Lansing.

Ullman, Kyle L.
1985 *The Ceramics from the Kramer Village Site (33RO33), Ross County, Ohio*. Research Papers in Archaeology No. 5. Department of Sociology and Anthropology, Kent State University, Kent, Ohio.

Wagner, G. E.
1987 *Uses of Plants by the Fort Ancient Indians*. Ph.D. dissertation, Washington University, St. Louis, Missouri. University Microfilms, Ann Arbor, Michigan.

Weissner, Polly
1983 Style and Social Information in Kalahari San Projectile Points. *American Antiquity* 48:235–276.

Williams, S.
1990 The Vacant Quarter and Other Late Events in the Lower Valley. In *Towns and Temples along the Mississippi*, edited by David H. Dye and Cheryl A. Cox, pp. 170–180. University of Alabama Press, Tuscaloosa.

Winterhalder, Bruce
1986 Diet Choice, Risk, and Food Sharing in a Stochastic Environment. *Journal of Anthropological Archaeology* 5:369–392.

PART 2

Nontraditional Applications of Existing Methods

CHAPTER 6

Walking and Watching

New Approaches to Reconstructing Cultural Landscapes through Space Syntax Analysis

Erin J. Hudson

6.1. Introduction

Space syntax was originally developed in the field of architecture and planning (Hillier and Hanson 1984; see Cooper 1995). The common presence of architecture at archaeological sites has allowed the application of these methods throughout the U.S. Southwest. The majority of space syntax studies in this region have focused on sites in Chaco Canyon and Chacoan outliers (Bustard 2003; Cooper 1995; Van Dyke 1999), but the method has also been applied to other Ancestral Puebloan sites, such as Sand Canyon Pueblo (Bradley 1993), historic Zuni settlements (Ferguson 1996), and Casas Grandes (Paquimé) (Wilcox 1999). Space syntax analysis provides archaeologists one way to measure quantitatively the spatial relationships within architecture. These relationships are believed to provide insight into the social organization of the people who built and occupied a site.

Space syntax, as initially conceived by Hillier and Hanson, focuses exclusively on the built environment, ignoring the landscape on which structures are constructed. However, the landscape may play an important role in restricting or enabling access to sites in a way that is analogous to altering access to spaces within a site. There is little doubt that the location of a pueblo on a mesa top, surrounded by steep cliffs on all sides, tells us something about the interactions between the people in that pueblo and outsiders. Yet archaeologists usually do not go beyond these general observations to quantify the accessibility of sites on a landscape.

The research presented here builds on previous space syntax analyses in the Southwest. But instead of quantifying access within a site, it focuses on social integration and access to sites at a landscape scale. In addition, it tracks changes in settlement patterns during a time of social reorganization. Specifically, this chapter incorporates the concepts of symmetry/asymmetry and distributed/nondistributed into a landscape-scale study of social organization and settlement patterns by using Geographic Information Systems (GIS) technology. The overall goal of this research is to create a measure of the openness and accessibility of prehistoric pueblo habitation sites. Least cost pathways (LCP) are used here as merely a proxy for accessibility. The actual cost of traveling those paths is of secondary importance; rather, I consider the number of pathways to a site a measure of accessibility.

The study area and time periods selected for this research are the Pueblo II (AD 950–1150) through Pueblo IV (AD 1300–1600) occupations of the Gallinas and Bear Mountains, northwest of the modern town of Magdalena, New Mexico (Figure 6.1). In particular, this research focuses on a cluster of sites dating to the Pueblo II period, commonly referred to as the Lion Mountain community, and the late Pueblo III (AD 1150–1300) site of Gallinas Springs, a 500-room pueblo located east of Lion Mountain. This area was selected due to the presence of numerous well-documented Pueblo II sites. Also, the site of Gallinas Springs has been excavated, and unpublished excavation

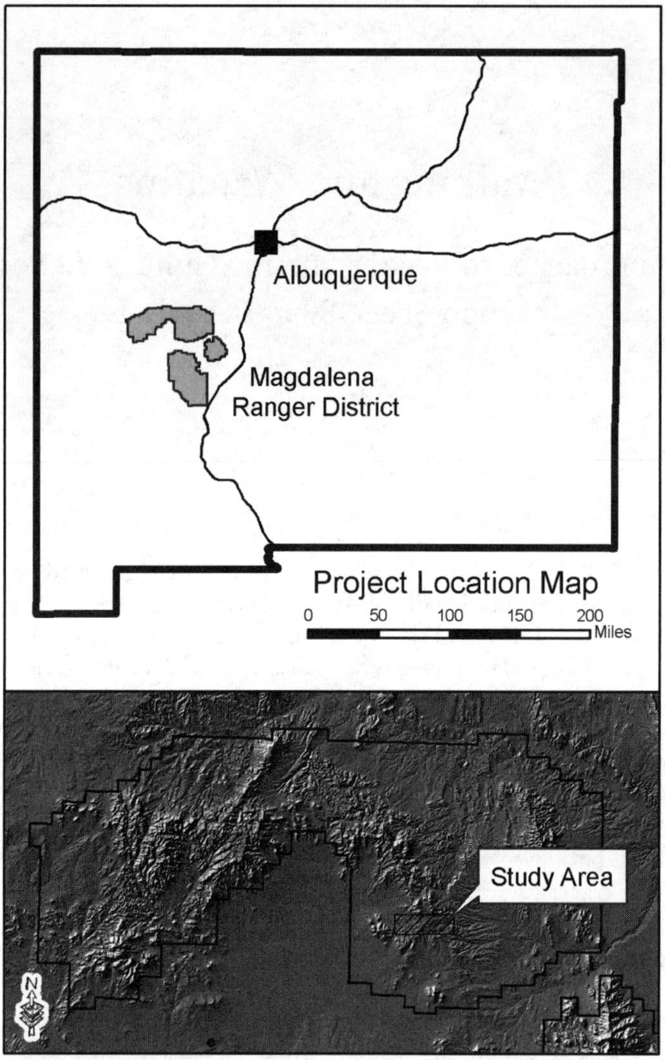

FIGURE 6.1. Spatial extent of the study area, northwest of Magdalena, New Mexico.

reports provide additional data. Most important, there is evidence for significant social reorganization in the Pueblo II/III periods as site settlement patterns shift from smaller, dispersed settlements to large, aggregated pueblos, which makes the study area an excellent candidate for a consideration of a dynamic space syntax analysis.

This chapter describes a preliminary attempt to apply space syntax theory to questions concerning landscape-scale use of space. The GIS methods employed were selected in an effort to create measures of accessibility that are similar to those employed in traditional space syntax studies (e.g., access graphs). This research was accomplished through the use of readily available tools in GIS to analyze these relationships. The methods employed here could be followed by a beginner GIS user for similar research projects in other archaeological contexts.

6.2. Background

In the Southwest, space syntax analyses have relied on the measurement of symmetry/asymmetry and distributed/nondistributed patterns (Cooper 1995; Van Dyke 1999; Wilcox 1999). Access graphs are used to measure these concepts. A structure is considered symmetric when all spaces are equally accessible from a given starting point, and this is an indication of accessibility and integration. Asymmetry occurs when a space is accessible only through another space; this pattern is an indication of isolation and segregation (Cooper 1995; Van Dyke 1999). *Distributed/nondistributed* refers to the number of access routes to individual rooms or other loci in a structure. Distributed configurations are those with multiple routes of access between spaces (Van Dyke 1999). Nondistributed patterns occur when there is only one access route to a space. In other words, distributed or nondistributed patterns provide insight into the overall accessibility of a site.

In general, spatial segregation and limited access are assumed to be associated with more complex social organization and inequalities (Bustard 2003; Van Dyke 1999) or religious use of a room or structure (Wilcox 1999). Van Dyke (1999) bases this assumption on the work of Foucault and others who have shown that architecture is an important aspect of social power. Other researchers are less specific about how they develop this assumption (Wilcox 1999). In general, it is assumed that there is a direct relationship between social organization and the built environment. Therefore, segregation in architecture is considered to reflect segregation in other aspects of social life.

Bustard (2003:81) used access graphs to assess the permeability—"the degree to which closed cells, entrances, and spatial configurations encourage or discourage access to and movement through structures"—of Pueblo Bonito, Chaco Canyon, in order to ascertain the function of the structure. She concluded that the function of the site changed through time. It may have begun as a habitation site, but the spatial segregation and lack of household features, such as grinding bins and fire pits, in the parts of the structure constructed and in use after AD 1075 indicated that it served another function. Bustard used this shift over time to argue for a change in social organization that resulted in increasing social segmentation beginning around AD 1050. This ultimately led to changes in the function of Pueblo Bonito.

The change in social organization observed at Chaco Canyon is not reflected as clearly in the analyses of Chacoan outliers. These sites are similar to Chaco Canyon great houses but are located outside the canyon. Van Dyke's (1999) analysis of the outlier of Guadalupe ruin, occupied from AD 960 to 1300, indicated that there was minimal asymmetry of the site during its occupation. Interestingly, the only period with an asymmetric distribution at the site occurred during the Early Chaco phase (AD 960–1050), before the changes in social organization argued for by Bustard at Pueblo Bonito. However, the asymmetrical nature of Guadalupe ruin was minimal even during this phase. Van Dyke argues that the symmetry of the site indicates that it was a domestic structure and that it did not function solely as an elite residence, administrative site, or ritual structure.

Wilcox (1999) constructed access graphs and determined that some areas of Casas Grandes (Paquimé) were difficult to access. Instead of proposing that these were elite residences, he argued that the most difficult to access, nonpermeable areas of the site served a ritual function. One of the interesting ideas he developed is that of inhabitant/stranger and inhabitant/visitor interactions. Wilcox suggested that architecture can be used to control the movement of people, especially those who do not reside in a site.

Overall, studies of space syntax in the Southwest indicate that it is a useful tool that may shed light on the relationship between the built environment and social organization, particularly at the site level. Although the exact nature of this relationship is unclear and may in fact vary in space and time, there is little doubt that restricted access to portions of a site tells us something about the use of those rooms. Generally, such restriction indicates that the people using them do not want them available to everyone—friend or other. At the very least, space syntax analyses have resulted in a more clear understanding of how site access changes over the course of occupancy. Since the usefulness of this method has been demonstrated for cultural features, it is likely that it can be expanded to larger-scale analysis, to tell us something about the location of sites on a landscape.

6.3. Study Area

The Gallinas and Bear Mountains are situated between the plains of San Augustine to the west and San Lorenzo to the east. The study area is enclosed within the boundaries of the Magdalena Ranger District, Cibola National Forest. These boundaries helped to limit the selected study area. The elevation of the area ranges from approximately 6,500 feet above sea level on the plains to 8,442 feet at the top of Gallinas Peak (Tainter n.d.). The area is characterized by a mixed conifer vegetation in the mountains and desert grassland vegetation in the lower elevations (Tainter n.d.). Current rainfall in the area is approximately 15 inches per year.

Compared with other parts of the Southwest, this area has received little archaeological attention and the culture history is still somewhat unclear (Bertram 1990a). In general, the west-central region of New Mexico is characterized by a combination of attributes from both the Mogollon culture, located in southeastern New Mexico and southwestern Arizona, and Ancestral Puebloan culture, located in the more northern Southwest (see Cordell 1997; Kantner 2004; Plog 1997). Ancestral Puebloan sites, formerly known as Anasazi, are characterized by compact masonry architecture, kivas (circular subsurface ceremonial structures), and a variety of black-on-white ceramic traditions (Cordell 1997). The Mogollon culture area is characterized by pithouse villages and the use of a brownware ceramic (Cordell 1997). This is a vast generalization of these culture groups and there is extensive variability within each group (Kantner 2004).

The study area exists in a cultural "gray area" of the Southwest on the borderlands of these two groups. In general, the area is characterized by a blending of attributes from both the Ancestral Pueblo and Mogollon and has led some to call this area the "Mogasazi Frontier" (Bertram 1990a). As is common in the region, sites in the study area display characteristics of both traditions, including ceramics imported from both areas (Gomolak and Knight 1990; Hudson 2008) and architecture styles.

The early occupation of the study area is not well documented. Settlements with substantial architecture (e.g., masonry construction) and pottery appear at approximately AD 800 (Bertram 1990a). The area appears to have been continuously occupied from that time, with increasing site numbers suggesting population growth. Between AD 1050 and 1100, the Mogasazi Frontier underwent major changes in social organization, settlement patterns, and ceramic styles (Bertram 1990a). These changes are evident in the Bear and Gallinas Mountains. Unlike other areas of the Mogasazi Frontier, however, the Gallinas and Bear Mountains received few trade goods from outside areas (Bertram 1990a). Additionally, the area is known for its "uniquely recognizable" carbon-painted whiteware, named Magdalena Black-on-white (Bertram 1990a; Lekson et al. 2002). This ceramic type appears at only a few sites outside the study area, which has resulted in significant debates surrounding its origin (Lekson et al. 2002). At the end of the Pueblo III period, settlement shifts to a few large pueblos. These settlements are abandoned in the mid-1400s.

Despite these drawbacks, there are several reasons this region was selected for the study. First, the land is managed by the Cibola National Forest, and numerous Pueblo II through IV sites have been recorded as part of Section 106 compliance. This provides a suitable dataset for landscape-scale analysis. Also, there is evidence for significant social reorganization beginning in the mid-1200s; the settlement patterns changed from numerous dispersed, small habitations to large, aggregated pueblos (Bertram 1990a). Finally, other than Forest Service surveys and the excavations conducted at the site of Gallinas Springs (Green 1974; Tainter n.d.; Bertram et al. 1990), this area has received little archaeological attention.

The majority of the archaeological information pertaining to this region comes from the site of Gallinas Springs. This pueblo consists of four room blocks, four kivas, and approximately 500 rooms (Green 1974). The room blocks span both sides of a drainage. The exact date of occupation for this site has proven difficult to determine, but it is believed to date to the late thirteenth and early fourteenth centuries (Tainter n.d.; Bertram et al. 1990).

The site was originally recorded in the 1920s, and research continued sporadically through the 1970s (Lekson et al. 2002). The most extensive excavations of the site were undertaken by the Western Michigan University field school in the summer of 1974 and the University of New Mexico field school in 1977, led by Joseph Tainter. Excavations occurred again in 1987 before stabilization to prevent further erosion of the site from the nearby drainage.

The site has been a point of contention for archaeologists interested in the area. It is believed to represent an anomaly in the region due to the presence of a carbon-painted black-on-white ceramic that is similar in style to the McElmo Black-on-white ceramics of the Mesa Verde region (Lekson et al. 2002). The black-on-white ceramics of Gallinas Springs were named Magdalena Black-on-white in 1981, after Knight and Gomolak (1981) completed an analysis of the materials recovered from the 1977 excavations and concluded that the ceramics were a local development. Interpretations about who lived at this site, based on these ceramics, have gone back and forth between migrants from Mesa Verde and in situ development (see Lekson et al. 2002). Magdalena Black-on-white occurs at other sites in the area, including Pinnacle Ruin, but its distribution is limited (Lekson et al. 2002).

6.4. Linking GIS Methods with Human Behavior

Geographic Information Systems (GIS) programs are an ideal tool with which to examine spatial relationships at the landscape scale and can contribute a rich approach to space syntax analysis. For example, the Spatial Analyst extension to ArcGIS allows the user to determine line of sight, viewshed, and LCP. These analyses can then be used to determine the spatial relationships between sites and the surrounding landscape. Line of sight and viewshed are visual analyses that can be used to infer openness, or the ability of a community to view neighbors or others approaching a site. LCP analysis can be used to calculate hypothetical routes to a given site from different points on the landscape.

In this research, I argue that line of sight or viewshed modeling can be a proxy for measuring symmetry/asymmetry. These large-scale spatial analyses can be conducted in the Southwest because the terrain and vegetation allow for long-distance visibility. I also propose that LCP be considered a proxy for distributed/nondistributed measures in space syntax because it can be used to determine the number of routes to a given point. These landscape-scale methods of spatial analysis provide a means for evaluating the concepts of symmetry/asymmetry and distributed/nondistributed at the regional level, which is a novel application of LCP, line-of-sight, and viewshed analyses. The following regional-scale spatial relationships can be expected for sites that are symmetrical/distributed and asymmetrical/nondistributed. For this research, sites considered symmetrical/distributed will appear integrated across a landscape via line of sight, by long-distance visibility of the landscape (i.e., large viewsheds), and by numerous routes of access (as measured using LCP). Previous site-level space syntax studies identify these patterns as being indicative of open access, with both neighboring communities and outside groups (i.e., strangers) that people would not have interacted with on a regular basis.

Sites that are characterized as asymmetrical/nondistributed will be less visible from other sites (limited line of sight with other contemporary sites), will have smaller viewsheds, and will be difficult to access (i.e., fewer routes across the landscape to the site). This pattern is considered a sign of restricted access and an indicator of social complexity and inequality (Bustard 2003; Van Dyke 1999) or ritual use of a space (Wilcox 1999).

It should be noted that although these proxies do not serve as exact replacements for the methods employed in traditional space syntax studies, they provide a replicable method for measuring how people use the landscape. The most fundamental difference between traditional space syntax and my methods is that the regional-level landscape is not entirely constructed as the built space of a settlement. However, this difference is minimal when we consider that many archaeologists have noted that people modify the landscape and utilize it to reify and reinforce their worldview (e.g., Ashmore and Knapp 1999; Basso 1996; Hicks et al. 2007; Munson 2002; Potter 2004; Snead 2008; Ucko and Layton 1999; Van Dyke 2004, 2008). Thus, it is likely that people chose locations on the landscape either to make themselves available to others or to make themselves hidden, depending on their relationships with surrounding communities. This chapter attempts to create a connection between the theoretical concepts associated with space syntax and the methodological concepts of GIS to better measure how people use the landscape. The hope is that with the basic steps outlined here, archaeologists will be able to move away from simple statements about a community's interactions and site defensibility based on its location to a more quantitative measure of how the people at a site used the landscape to facilitate social interaction.

6.5. Data Collection and Analysis

The data for this chapter come from the Cibola National Forest Heritage Resource GIS Database and site records on file at the Cibola National Forest Supervisor's Office in Albuquerque, New Mexico. The study area was defined by the location of the site of Gallinas Springs in the east and the Forest Service boundary to the west (Figure 6.1). Preliminary site information, such as site identification numbers, were obtained from the Cibola GIS database and used to locate site records for each site within the study area. Since this chapter focuses on habitations, only sites with architecture were included. It is worth mentioning that isolated artifact scatters are rare in the study area. Forty-eight architectural sites, dating from the Pueblo I/II period to the late Pueblo III period, are located in the study area. Site location data were entered into a GIS geodatabase that included information such as site identification number, site size, and the presence of kivas or enclosed plazas.

The size of the rubble mound was used to calculate the number of rooms present at each site. Fieldwork in the area indicates that room size varies both within and between sites. Rooms range from 2 × 2 m to 4 × 4 m. An average room size of 3 × 3 m was used to determine the number of rooms per rubble mound. Although there are problems with using rubble mound size to determine room number (for example, it does not account for multiple-storied rooms), it allows for the separation of small sites that are likely field houses from the larger habitation sites. Differentiating the sites by size was essential for comparing accessibility of large, multifamily habitations and smaller field houses that may have been occupied by single families.

All the digital maps necessary to complete the GIS analyses were downloaded from the New Mexico Resource Geographic Information System, operated by the University of New Mexico and the New Mexico Information Technology Commission (http://rgis.unm.edu/). The digital data include topographic USGS 7.5-minute quadrangles and 10-m-resolution digital elevation models (DEMs) for the entire study area. Any contour maps needed were made from the DEMs with the ArcGIS Spatial Analyst extension. The Surface Analysis Contour tool was selected from the Spatial Analyst toolbar, and an analysis was run with a contour interval of 5 m and the standard defaults.

Line-of-sight analysis was conducted for all the sites within the Lion Mountain community ($n = 46$) and between Gallinas Springs and all known habitation sites nearby to determine intervisibility between sites. Viewshed analyses were conducted on twelve sites in the Lion Mountain community plus the site of Gallinas Springs. This analysis calculates the total area visible from a specific point. A subset of the Pueblo II/III sites was selected due to the time involved in conducting this analysis.

Each site was exported into an individual shapefile as a single point representing the center of the site. A viewshed analysis was then conducted for each of the individual points with the Spatial Analyst extension, Surface Analysis Viewshed tool. The DEM of the study area was used as the input surface, and the point shapefile for the site as the observer points. The Z factor (distance off the ground) of 2 m was used to represent the approximate eye level of a person standing on the ground in the site or sitting on the roof of a structure. The output cell size remained at 10 to match the cell size of the DEM used as the input surface.

Finally, LCP analysis was conducted to determine the number of "easy" routes of access to individual sites and between known sites. This study assumes that a site located on relatively featureless and flat terrain would be accessible from all directions whereas a site located on variable terrain would have restricted access, and therefore fewer paths, directly to the site. The inhabitants of this area had the option of building their homes on the nearby plains, which had relatively easy accessibility, but in most cases chose not to. The goal of this analysis was to determine whether large habitation sites were any more accessible to outsiders (i.e., people traveling through the area or visiting) than small sites, and to measure the ease of access between contemporary sites. LCP is a proxy for accessibility, and "easy" routes are considered the fastest paths between two points, given terrain. How accurately any given route represents the ease of travel between two points is of secondary importance for this study because the number of our easy routes is used as a simple measure for ease of access. Therefore, a basic LCP equation (based on slope and distance) was used to calculate the cost of traveling across a landscape in terms of time.

To build LCPs, we selected four source points outside the Lion Mountain community and Gallinas Springs. These points represent an individual traversing the landscape from any direction. The source points also correspond to areas of gentle terrain, such as plains or natural passes, which would have facilitated travel across this area. The sites in the Lion Mountain community were divided into separate layers based on the overall size of the site (e.g., 20–50 m^2, 50–100 m^2, up to sites over 600 m^2). These layers served as the set of destinations for each LCP analysis.

Cost-distance and cost-direction rasters were created for sites with a slope raster and the Spatial Analyst extension, Distance Cost Weighted tool. The cost-distance and cost-direction rasters are used to determine the cost of traveling from each of the source points (north, south, east, and west) to the destination sites based on the slope of the landscape. The use of slope and distance is a simple way to measure LCP. Like any model, it simplifies reality and may not represent the actual easiest path between two points, particularly if we take into account other natural barriers (such as vegetation) or potential human barriers (such as other settlements). However, these paths serve as a heuristic device or baseline for approximating accessibility.

Once the cost-distance and cost-direction rasters were created, an LCP analysis was conducted from each source point to each destination layer with the Spatial Analyst extension, Distance Shortest Path tool. A "For Each Cell" path type was used to allow for multiple paths between the source point and the sites destination layer. A site is considered to have multiple access paths if travel from each of the source points results in a separate route to the site. If the paths converge away from the site, resulting in a single "easy" route of access, then it is counted as one path.

TABLE 6.1. Total Viewshed for Each of the 12 Sites Analyzed.

Site Number	Size of Viewshed (acres)
03-177	**1635**
03-180	7024
03-184	7474
03-190	**439**
03-191	665
03-192	**713**
03-692	512
03-693	318
03-704	2049
03-707	3412
03-708	3110
New Site	**4422**
Gallinas Springs	76

Note: The largest sites are in boldface.

6.6. Results

6.6.1. Line of Sight and Viewshed

These analyses identified significant differences in access between the Lion Mountain sites and Gallinas Springs. First, sites in the Lion Mountain community are extensively integrated into line-of-sight networks, with at least one other site visible from almost every site (Figure 6.2). Sites in the Lion Mountain community that are located along ridge tops are often able to see numerous other sites. Interestingly, these sites tend to be larger than those located near the valley floor. Gallinas Springs, on the other hand, is not connected via line of sight to any other known habitation sites. Habitation sites in the Lion Mountain area also have extensive viewsheds of hundreds to thousands of acres (Figure 6.3, Table 6.1). None of the sites located in the Lion Mountain community have a viewshed that is restricted by site location or terrain. Instead, visibility from each site would have been restricted by one's eyesight more than any other factor. There are, however, some interesting patterns present in the viewshed analysis.

There is considerable variability in the size of viewsheds among the largest sites in the Lion Mountain community (those with more than 40 rooms)—between 439 and 4,422 acres (see Table 6.1). These sites tend to be located on the more southern of the two ridges that make up the landscape of the community. Sites located on the more northern ridge tend to have larger viewsheds. These sites are also smaller, generally with fewer than 30 rooms.

Larger viewsheds are expected from sites on the northern ridge because it is adjacent to the

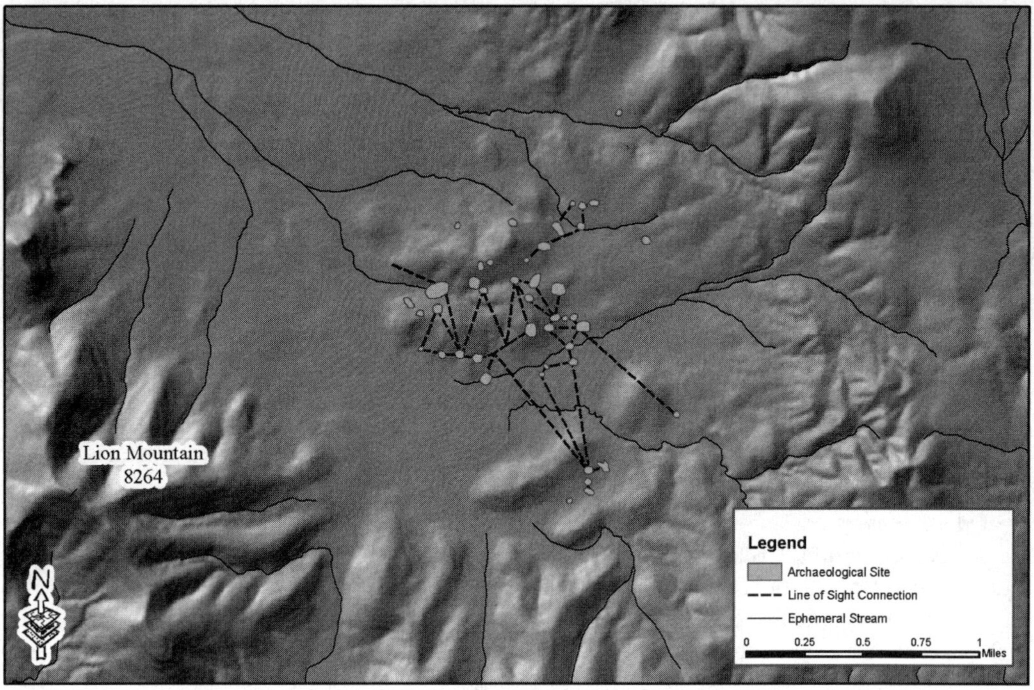

FIGURE 6.2. Sites connected via line of sight in the Lion Mountain community.

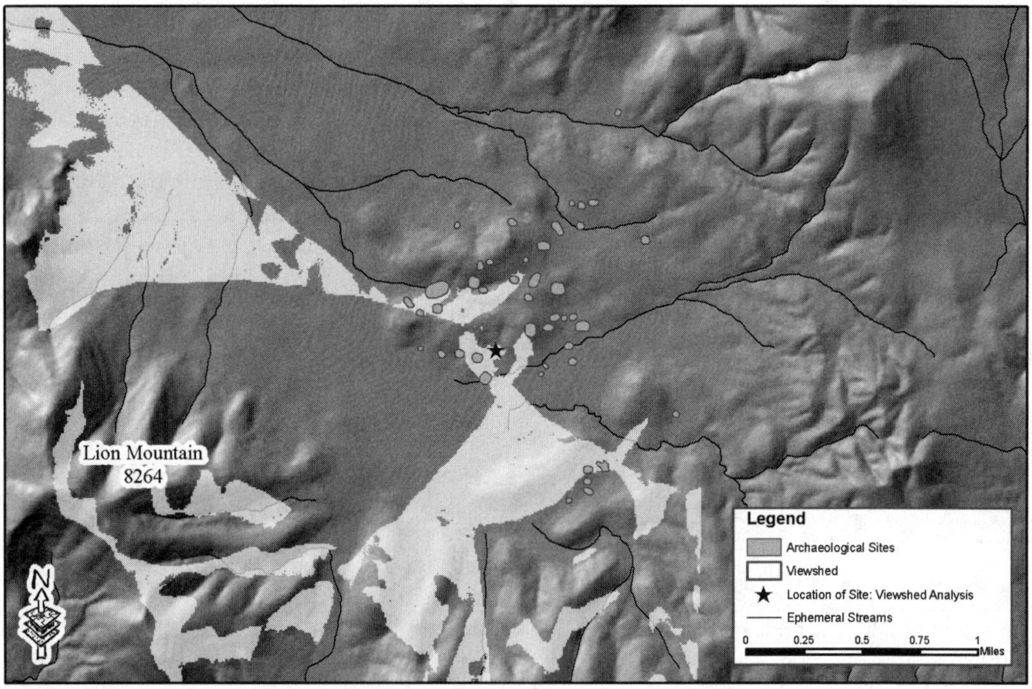

FIGURE 6.3. Viewshed analysis for the Lion Mountain community.

plains of San Augustine; however, it was assumed that larger sites would be associated with larger viewsheds in order to better control access. Instead, the opposite pattern appears to hold for the largest sites in the Lion Mountain community: the majority of these sites are located on the southern ridge, where viewshed would have been somewhat restricted by topography. It is possible that these communities wanted to see people coming but to avoid detection by people traveling through the San Augustine plains to the north. Further analysis of sites in the area is needed before this claim can be validated.

The people occupying Gallinas Springs would have been at a distinct disadvantage in this respect, with a viewshed of only 76 acres. It should also be noted that this viewshed includes several high peaks to the west that are visible from the site (including Gallinas Peak). These peaks could have served as lookouts, expanding the ability of people within the pueblo to monitor the landscape around them. Although there has been no formal survey of the areas around Gallinas Springs, preliminary reconnaissance indicates that site density may be higher than currently known. These sites may provide the needed line-of-sight connections to serve as lookouts for the Gallinas Springs community. Additional research is needed to verify this possibility.

The results of the viewshed analysis illustrate the relationship between the situation of sites on the landscape and their possible degree of connectedness. Sites in Lion Mountain are located on low ridges in the middle of a broad valley that is adjacent to the plains of San Augustine. Sites located on the more northern ridge of the community are more likely to have larger viewsheds because of the plains of San Augustine to the northwest. Also, the gently undulating topography in this area allows for line of sight with neighboring habitations. Therefore, the overall landscape around the Lion Mountain community allows the occupants to monitor the activities occurring in the valley and the habitations around them. Of course, it also allows approaching visitors to identify the location of the sites.

Gallinas Springs displays an opposite pattern that is innately tied to its location in a narrow drainage in rugged mountains. The landscape significantly reduces the ability to create line-of-sight networks with neighboring communities and to view large expanses of the surroundings. This may indicate a two-pronged attempt by people residing in the pueblo: first, to avoid detection by strangers by hiding in a place that cannot be seen from a distance, and second, to control access to the site by forcing people to enter it from the east or west along the drainage.

6.6.2. Least Cost Path Analysis

The LCP analysis provides additional interesting information concerning the accessibility of sites. First, the analysis of routes from outside the area to large habitation sites in Lion Mountain indicates that there are generally few relatively easy routes of access to any given large habitation site (Figure 6.4). No matter what direction people are coming from, the paths usually merge, resulting in only one or two ways to access a site. None of the sites have more than two LCPs leading to them. Additionally, the modeled paths often pass next to or through smaller sites. This pattern perhaps indicates a further restriction of access to the largest habitation sites since outsiders would be encountered in advance, providing time for warning the main community about the arrivals. These results are in line with our expectations about sites that are asymmetrical/nondistributed.

The opposite pattern is observed for the LCPs between sites at Lion Mountain. The relationship between large habitation sites at Lion Mountain indicates open access between sites. These results suggests a high degree of interconnectivity and accessibility within the region. Further, when the LCPs for all sites (regardless of size) are compared with the LCPs between large habitations, an interesting pattern emerges: all sites tend to fall along LCPs. This is the result of site location on the landscape, as most sites appear purposefully located on gentle ridges that facilitate movement between sites.

Access to Gallinas Springs is restricted to only two paths along the drainage. The reduced accessibility of Gallinas Springs is not unexpected given the steep slopes surrounding the drainage, so LCPs in this area do not provide much in the way of new information. The hypothetical paths pass by two known sites that may be contemporary, but their exact relationship is difficult to determine at this time. Future survey may locate

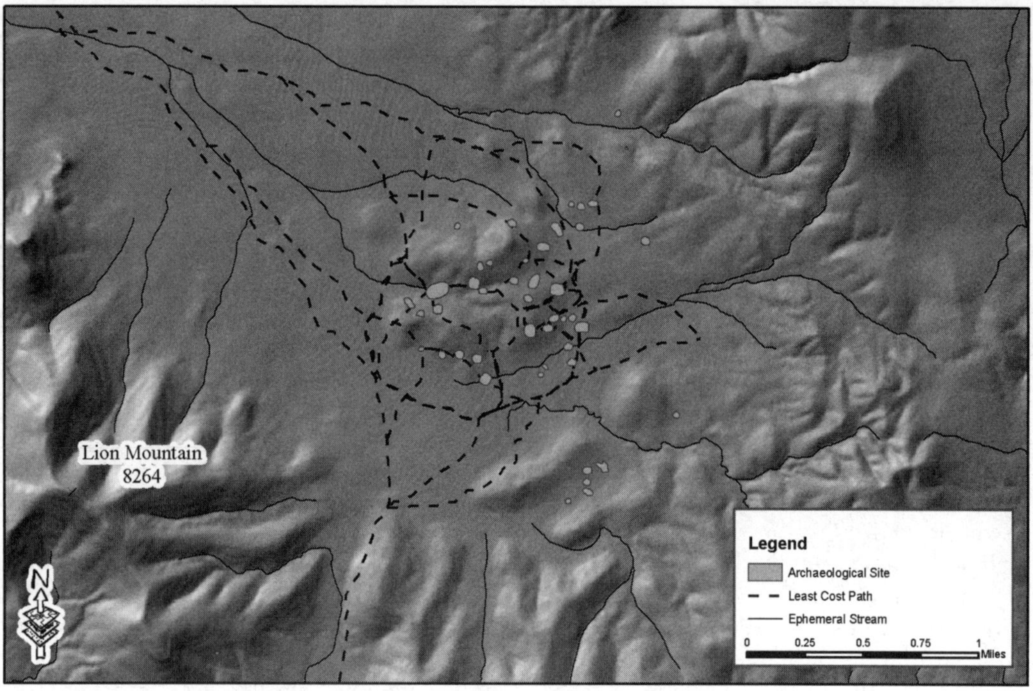

FIGURE 6.4. Least cost path analysis for the Lion Mountain community.

additional contemporary sites that expand our understanding of how Gallinas Springs is related to other sites via ease of access.

6.7. Discussion

These results provide some interesting and possibly conflicting data. The Lion Mountain community appears to be internally open, displaying a symmetrical/distributed pattern that is commonly indicative of social equalities. However, a second pattern appears when access by outsiders is considered—a pattern of symmetrical/nondistributed, indicating a general openness to visitors, but with conditions.

The people at Lion Mountain appear to have wanted to see and be seen, yet they restricted access to the larger residential sites by strangers. They accomplished this restriction by placing the largest habitation sites in areas that are more difficult to access, such as on top of ridges. This placement would have allowed the members of these communities to identify strangers as they approached and would have given them time to prepare should the strangers be unfriendly. Additionally, if people wanted to access these sites along the "easiest" paths, they would have had to pass near or through other, smaller habitation sites. The use of more difficult paths to reach the large Lion Mountain communities would have required more travel time, giving the community members additional time to identify people approaching.

Gallinas Springs is characterized by an asymmetrical/nondistributed pattern, which commonly indicates social inequalities and increasing sociopolitical complexity (see Bustard 2003; Van Dyke 1999). This interpretation may be strengthened by the size of the site (and possible monumental architecture). Unpublished excavation data also hint at long-distance trade and the possible presence of prestige goods. A preliminary analysis indicates that ceramic tradewares—painted ceramics from neighboring cultural groups such as the Mogollon and Rio Grande/Rio Puerco—occur throughout the site and may be related to social connections in these areas and/or prestige goods (Hudson 2008).

It is possible that the traditional interpretations employed by space syntax studies, such as that reduced access is a measure of social inequality, do not hold true for landscape-scale analysis. In fact, it is possible that the patterns revealed here are better explained by defensive strategies employed by the occupants of the sites. They may have used the landscape to stay hidden from strangers or to defend themselves. However, additional research is needed to confirm these interpretations. In the end, this is a preliminary study that will require refinement with additional site data and methodological improvements, such as more accurate measures of LCPs, so that we may better understand how space syntax can be used at a landscape scale.

6.8. Conclusion

This is a preliminary attempt to expand space syntax to the landscape level. Future research is needed to fully clarify whether this is a useful approach for understanding access across a landscape. First, high-quality temporal data are necessary for an accurate determination of the relationships between sites. These data are currently being generated for the Lion Mountain community.

Additionally, it may not be possible to interpret the results of landscape space syntax along the same lines as traditional space syntax. The above patterns may be more related to defensive strategies than to social organization. Thus, this preliminary study may have inadvertently developed a new method for evaluating relationships between communities and identifying defensive strategies. For example, the symmetrical/nondistributed pattern identified at Lion Mountain and the asymmetrical/nondistributed pattern at Gallinas Springs may represent two different defensive strategies. The first aimed at identifying strangers before they arrived and the other aimed at minimizing contact with outsiders, or at least restricting access to those who knew the location of the site. It would have been more difficult to "stumble" into the site of Gallinas Springs than it would have been to locate the Lion Mountain community. This method needs to be more fully articulated with archaeological theory on cultural landscapes, migrations, and defensive strategies before more in-depth interpretations can be developed.

To these ends, it would be useful to compare the results of this study with other areas in the Gallinas and throughout the Southwest to determine if similar patterns can be identified. The work reported in this chapter suggests that space syntax analyses can provide information about the social interactions within multisite communities as well as the relationships between those communities and outsiders.

Acknowledgments

I thank the staff at the Cibola National Forest, and in particular Forest Archaeologist Cynthia Benedict, for providing me with the GIS information needed to complete this study and for supporting this project. My husband, F. Scott Worman, deserves special thanks for letting me tap into his knowledge of landscape archaeology (and for his patience). I would not have been able to complete this project without my friend and colleague Natalie Heberling, who helped me work through many of my GIS issues. Thank you!

References

Ashmore, Wendy, and A. Bernard Knapp (editors)
1999 *Archaeologies of Landscape: Contemporary Perspectives.* Blackwell, Oxford.

Basso, Keith M.
1996 *Wisdom Sits in Places: Landscape and Language among the Western Apache.* University of New Mexico Press, Albuquerque.

Bertram, Jack B.
1990a Culture-Historical Overview. In Excavations in the South Block of Gallinas Springs Ruin (LA 1178), a Large Town of the Gallinas Mountains Phase (Late Pueblo III–Early Pueblo IV) on the Mogasazi Frontier. Manuscript on file, Cibola National Forest Supervisor's Office, Albuquerque, New Mexico.
1990b Introduction. In Excavations in the South Block of Gallinas Springs Ruin (LA 1178), a Large Town of the Gallinas Mountains Phase (Late Pueblo III–Early Pueblo IV) on the Mogasazi Frontier. Manuscript on file, Cibola National Forest Supervisor's Office, Albuquerque, New Mexico.

Bertram, Jack B., Andrew R. Gomolak, Steven J. Hogland, Terry L. Knight, Emily Garber, and Kenneth J. Lord
1990 Excavations in the South Block of Gallinas Springs Ruin (LA 1178), a Large Town of the Gallinas Mountains Phase (Late Pueblo III–Early Pueblo IV) on the Mogasazi Frontier. Manuscript on file, Cibola National Forest Supervisor's Office, Albuquerque, New Mexico.

Bradley, B.
1993 Planning, Growth, and Functional Differentiation at a Prehistoric Pueblo: A Case Study from SW Colorado. *Journal of Field Archaeology* 20:23–42.

Bustard, Wendy
2003 Pueblo Bonito—When a House Is Not a Home. In *Pueblo Bonito: Center of the Chacoan World*, edited by J. Neitzel, pp. 80–92. Smithsonian Books, Washington, D.C.

Cooper, Laurel M.
1995 *Space Syntax Analysis of Chacoan Great Houses*. Ph.D. dissertation, University of Arizona, Tucson. UMI Dissertation Services, Ann Arbor, Michigan.

Cordell, Linda
1997 *Archaeology of the Southwest*. 2nd ed. Academic Press, San Diego.

Ferguson, T. J.
1996 *Historic Zuni Architecture and Society: An Archaeological Application of Space Syntax*. Anthropological Papers of the University of Arizona 60. University of Arizona Press, Tucson.

Gomolak, Andrew R., and Terry L. Knight
1990 Ceramic Analysis. In Excavations in the South Block of Gallinas Springs Ruin (LA 1178), a Large Town of the Gallinas Mountains Phase (Late Pueblo III–Early Pueblo IV) on the Mogasazi Frontier. Manuscript on file, Cibola National Forest Supervisor's Office, Albuquerque, New Mexico.

Green, Ernestrene
1974 Excavations at Gallinas Springs Site, Magdalena District, Cibola National Forest, New Mexico. Report on file, Cibola National Forest Supervisor's Office, Albuquerque, New Mexico.

Hicks, D., L. McAtackney, and G. Fairclough (editors)
2007 *Envisioning Landscape: Situations and Standpoints in Archaeology and Heritage*. Left Coast Press, Walnut Creek, California.

Hillier, B., and J. Hanson
1984 *The Social Logic of Space*. Cambridge University Press, Cambridge.

Hudson, Erin J.
2008 People, Pots and Houses: What Do Ceramics Reveal about Migration at the Site of Gallinas Springs? Unpublished paper prepared for Ceramic Theory course, University of New Mexico.

Kantner, John
2004 *Ancient Puebloan Southwest*. Cambridge University Press, Cambridge.

Knight, Terry L., and Andrew R. Gomolak
1981 The Ceramics of LA 1178, Gallinas Springs, New Mexico. Report on file, Cibola National Forest Supervisor's Office, Albuquerque, New Mexico.

Lekson, Stephen H., Curtis P. Nepstad-Thornberry, Brian E. Yunker, Toni S. Laumbach, David P. Cain, and Karl W. Laumbach
2002 Migrations in the Southwest: Pinnacle Ruin, Southwestern New Mexico. *Kiva* 68(2):73–102.

Munson, Marit
2002 On Boundaries and Belief: Rock Art and Identity on the Pajarito Plateau. Unpublished Ph.D. dissertation, University of New Mexico, Albuquerque.

Plog, Stephen
1997 *Ancient Peoples of the American Southwest*. Thames and Hudson, London.

Potter, James M.
2004 The Creation of Person, the Creation of Place: Hunting Landscapes in the American Southwest. *American Antiquity* 69(2):322–338.

Sheldon, Charles M.
1977 A Report on the Human Osteological Remains from Gallinas Springs. Report on file, Cibola National Forest Supervisor's Office, Albuquerque, New Mexico.

Snead, James E.
2008 *Ancestral Landscapes of the Pueblo World*. University of Arizona Press, Tucson.

Tainter, Joseph A.
n.d. Social and Economic Organization of Gallinas Springs Pueblo. Manuscript on file, Maxwell Museum of Anthropology, Albuquerque, New Mexico.

Ucko, Peter J., and Robert Layton (editors)
1999 *The Archaeology and Anthropology of Landscape*. Routledge, London.

Van Dyke, Ruth M.
1999 Space Syntax Analysis at the Chacoan Outlier of Guadalupe. *American Antiquity* 64(3):461–473.
2004 Memory, Meaning and Masonry: The Late Bonito Chacoan Landscape. *American Antiquity* 69(3):413–431.
2008 Sacred Landscapes: The Chaco-Totah Connection. In *Chaco's Northern Prodigies: Salmon, Aztec, and the Ascendancy of the Middle San Juan Region After AD 1100*, edited by P. Reed, pp. 334–348. University of Utah Press, Salt Lake City.

Wilcox, David R.
1999 A Preliminary Graph—Theoretical Analysis of Relationships at Casas Grandes. In *The Casas Grandes World*, edited by C. F. Schaafsma and C. L. Riley, pp. 93–104. University of Utah Press, Salt Lake City.

CHAPTER 7

Social Interaction at the Maya Site of Copán, Honduras

A Least Cost Approach to Configurational Analysis

HEATHER RICHARDS-RISSETTO

Most archaeologists agree that the way in which ancient peoples arranged their physical surroundings, or in other words their built environment, provides a window to the past (e.g., Ashmore 1991, 1992; Ashmore and Sabloff 2002, 2003; Blanton 1989; DeMarrais et al. 1996; Lawrence and Low 1990; Moore 1996a, 1996b, 2005; Preziosi 1979a, 1979b; Reese-Taylor 2001; A. Smith 2003; M. Smith 2003, 2007). This is especially true for the ancient Maya, who scholars believe laid out their houses, monuments, and even roads to serve as a map of their worldview (Ashmore 1991; Ashmore and Sabloff 2002, 2003; Coggins 1980; Guillermin 1968; Houk 1996; Maca 2002). Most research of this nature tends to focus on cardinality, linking north, south, east, and west to representations of the heavens, earth, and underworld. Although such work is critical to our understanding of the ancient Maya, I believe that the advent of new technologies such as Geographic Information Systems (GIS) provides archaeologists with opportunities to begin to study Maya site configuration in new and more subtle ways.

In this chapter, I employ least cost paths to measure the relationship between site configuration and social connectivity at the ancient Maya site of Copán, Honduras. My research investigates two questions: First, did people of different social classes experience different degrees of social connectivity? And second, did people living in different parts of the city experience different degrees of social connectivity? From a theoretical perspective, my work is based on Charles Peirce's (1966) views of semiotics and regards site layout not simply as a reflection of ancient life but also as a mechanism that shaped ancient life (Giddens 1984; Jakobson 1980; King 1980; Moore 2005; Silverstein 1976). Along these lines, I view archaeological sites not just as anthropological features but as a combination of the built environment and the natural landscape. Ultimately, the goal of my work is to modify traditional configurational analysis using least cost methods to identify how social hierarchy was embedded in the landscape and how the ancient Maya may have strategically manipulated the landscape to structure social interaction and community organization at Copán.

At Classic period sites in the southern Maya lowlands, a social hierarchy existed that placed rulers at the top, members of the royal court just below, lesser nobles further down, and commoners at the bottom. As in many other ancient societies, cosmology provided the template and legitimization for this social structure. However, it was the daily routinization of these social categories that reinforced both the social and cosmic order (Joyce and Hendon 2000). This routinization was carried out, in part, through mechanisms such as access and visibility, which facilitated either social integration or segregation, depending on how societies employed them. The accessibility and visibility of buildings, roads, and other features serve as signs that influence how people move about landscapes, and people make use of this fact by organizing their surroundings to restrict access, channel movement, and display visual messages to elicit distinct responses from different groups of people (see, e.g., Fletcher 1981; Hudson this volume; Llobera 2006). Ultimately, the way in which people respond to the access and visibility of signs influences how different groups of people interact in the landscape. Although my research treats both accessibility and visibility, this chapter focuses explicitly on the role access may have played in establishing and maintaining sociopolitical relationships at Copán.

Other scholars have carried out accessibility studies in the Maya region (e.g., Sanchez 1997; Stuardo 2003; Yermakhanova 2005); however, my research differs from these studies in three important ways. First, instead of focusing on the internal spatial organization of a single architectural complex—one that is usually civic, ceremonial, or elite in nature (e.g., Ashmore 1991; Sanchez 1997; Stuardo 2003)—I examine a city's configuration as a whole, taking into account the spatial organization of architecture from all facets of society, including civic-ceremonial buildings, royal compounds, and elite and commoner residences as well as roads and reservoirs. I also incorporate natural features such as rivers, *quebradas* (stream cuts), hills, and mountains. Second, I introduce an innovative methodology that uses GIS to integrate the natural and built environments in the form of a raster dataset called the Urban Digital Elevation Model (DEM) (Ratti 2005; Richards-Rissetto 2007). This Urban DEM serves as the base dataset from which to create least cost paths, thereby allowing archaeologists to quantify accessibility for entire landscapes rather than simply within individual buildings or architectural complexes. Third, my research is multiscalar, studying access and visibility at four scales, from Copán's subcommunities to its physiographic zones to its urban core and hinterlands to the city as a whole.

7.1. Access among the Ancient Maya

Archaeological evidence suggests that accessibility and visibility served as mechanisms of social integration and/or social segregation in ancient Maya society (e.g., Hammond and Tourtellot 1999; Houston et al. 2006; Stuardo 2003). The Maya intentionally constructed their built environment to control access, manipulate the flow of movement, and send visual messages (Hammond and Tourtellot 1999; Keller 2001; Tourtellot et al. 2003; Tourtellot et al. 1999; Stuardo 2003). David Webster (1998:40) writes that Maya builders obviously intended to "to channel movement and create visual impressions of sanctity and power" through the organization of architecture. For example, at Copán the east and west *sacbeob* channeled people into the large, open Great Plaza, presumably for ritual events that brought together people from all walks of life (Baudez 1994; Sanchez 1997). It is likely that the accessibility of these plazas sent a message of unity—"we are one"—and created a sense of community and shared identity that helped to maintain social cohesion between commoners and elite.

In contrast, the highly restricted spaces of the East and West Courts of the Acropolis most likely sent different messages to different people (Figure 7.1). At most Maya sites, intimate access to the royal court was "restricted to the nobility and invited guests, spatial control being an integral part of the orchestration and wielding of regal power" (Reents-Budet 2001:225). On the one hand, it forged social bonds between the royal elite and other elite. On the other hand, it segregated the elite from the commoners by not permitting commoners access to certain spaces. This segregation helped to establish and maintain social inequalities. By making these royal spaces more inaccessible and separating the elite from the commoners, the ancient Maya were effectively replicating the order of the cosmos, in which supernatural beings and lords were separated from lesser or lower beings (Houston et al. 2006). Archaeologists have talked about the accessibility or inaccessibility of spaces within courtyard groups, but no one has empirically evaluated whether this same phenomenon is replicated for cities as a whole.

7.2. Configurational Analysis

Configurational analysis states that a city's configuration, or its morphological form, is a cultural product, and the way in which it is laid out influences how cultural information is transmitted (e.g., Hillier 1999; Hillier and Hanson 1984; Hillier et al. 1993; Marcus 1983; Preziosi 1979a, 1979b). Through mechanisms of accessibility

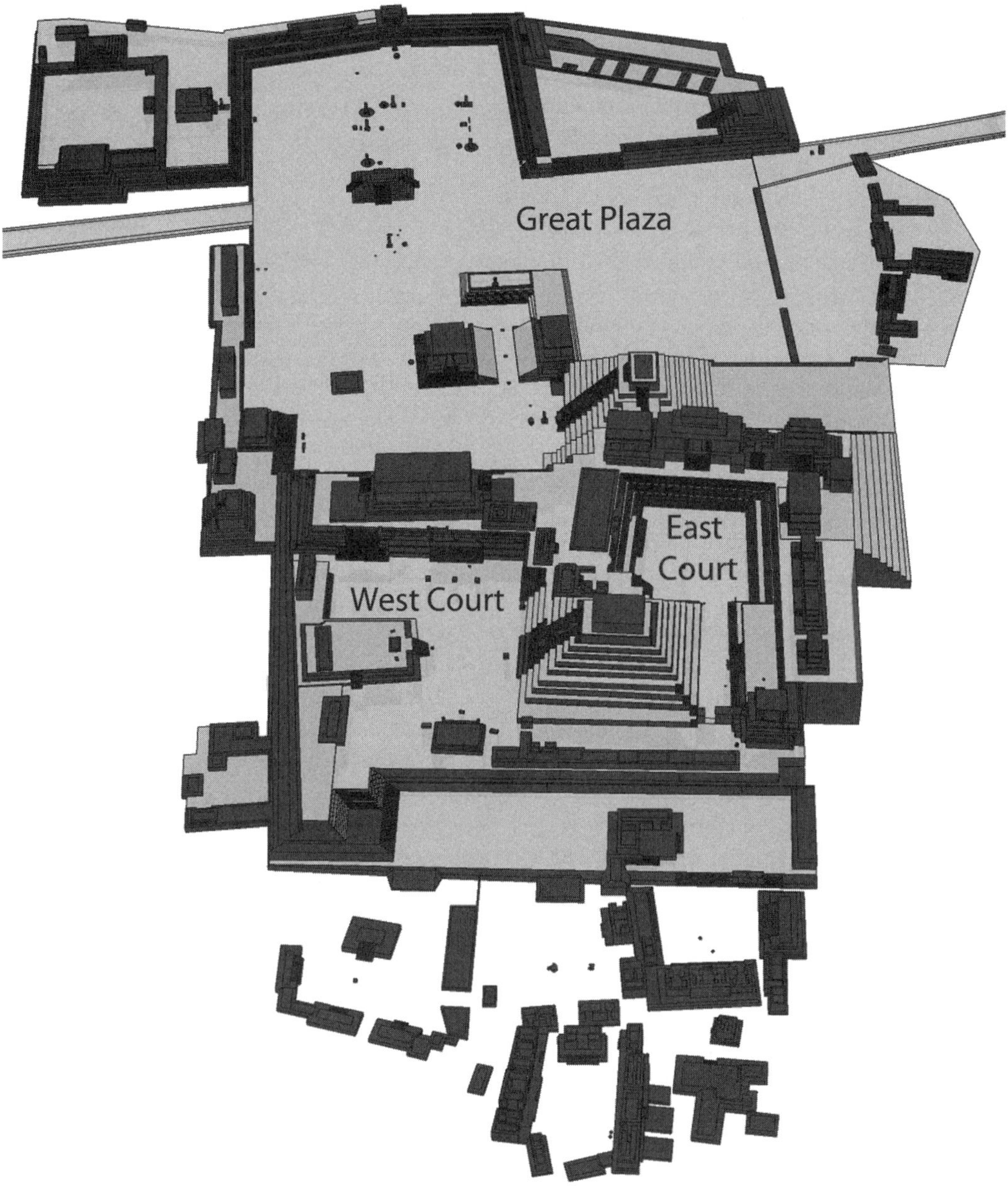

FIGURE 7.1. GIS-based Google SketchUp reconstruction of Copán's Principal Group.

and visibility, people send messages that help integrate some people while segregating others. Access structures social interaction by influencing pedestrian movement *to* and *through* space, and visibility does so by visually connecting certain groups to one another and not to others (e.g., Bustard 1996; Ferguson 1996; Hillier et al. 1993; Hillier and Hanson 1984; Ratti 2004, 2005; Shapiro 2005).

Most archaeological studies of accessibility use a form of configurational analysis called space syntax, which analyzes the structure of space to predict pedestrian movement (e.g., Bustard 1996; Ferguson 1996; Hillier and Hanson 1984; Hillier et al. 1993; Ratti 2005; Shapiro 2005; Stuardo 2003; see Hudson this volume for an alternative approach to space syntax). This work is based on studies indicating that spatial configurations are the primary generators of patterns of movement (Hillier et al. 1993). In other words, people are more likely to walk to or through certain spaces than others because of the way in which buildings and spaces are laid out. Spaces that people are more likely to walk to or through are considered to be more connected with the system as a whole, whereas those spaces that people are less likely to walk to or through are less connected. This degree of connectivity is measured as an integration value. Locations with low integration values are more accessible (connected) than those with higher integration values (Hillier 1999; Hillier and Hanson 1984; Hillier et al. 1993). These degrees of accessibility are related to variables such as political control and ritual exclusion (Ferguson 1996; Hillier and Hanson 1984; Smith 2007).

Many archaeologists have employed space syntax to study social interaction in ancient societies from across the world, including Medieval Europe, North America, Mesopotamia, and Mesoamerica (Dawson 2002; Craane 2009; During 2001; Stuardo 2003). In the U.S. Southwest, archaeologists used space syntax to study social interaction at several ancestral pueblos, including Arroyo Hondo, Pueblo Bonito at Chaco Canyon, and Zuni Pueblo (Bustard 1996; Shapiro 2005; Ferguson 1996). T. J. Ferguson's (1996) work at Zuni Pueblo, New Mexico, illustrated how changes through time in architectural configurations reflected wider sociopolitical changes. The space syntax results indicated that from AD 1400 to 1800 Zuni's inhabitants built structures that served to increasingly restrict accessibility to particular spaces within the community. These changes corresponded to ongoing threats of Apache and Navajo raids at the pueblo, suggesting that these relatively inaccessible areas may have been used to shelter women and children during raids.

In regard to Maya studies, space syntax has been applied to examine differences and similarities in access patterns within royal compounds across the Maya region. An example of such work is Rodrigo Liendo Stuardo's (2003) comparisons of access between Classic (AD 250–950) royal architecture at the sites of Palenque, Tikal, and Uaxactún in the southern lowlands and Early Postclassic (AD 950–1250) royal architecture at Uxmal,

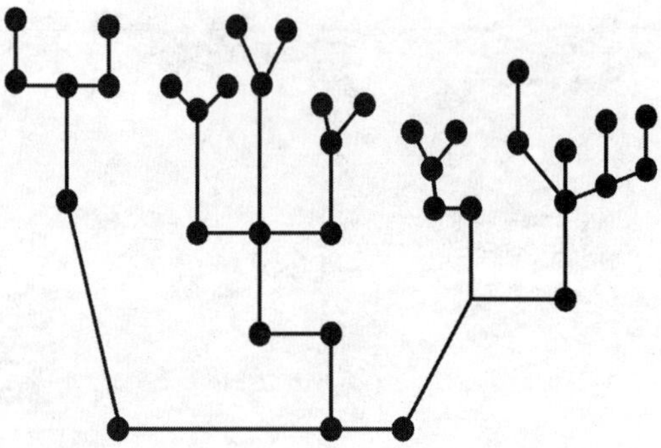

FIGURE 7.2. Hypothetical axial map of architectural complex.

Labna, Kabah, and Sayil in the northern Yucatan. His work demonstrates that simple access patterns existed in the elite architectural complexes of the northern Yucatan, while more complex patterns existed in southern lowland palaces. These differences suggest changes in political organization from the Classic to Early Postclassic periods, which Stuardo believes reflect a Postclassic departure from Classic forms of rulership to a more decentralized system of rulership under a council of nobles (Schele and Friedel 1990). Although space syntax has proven useful for providing insight into ancient social interaction within architectural compounds (e.g., Bustard 1996; Ferguson 1996; Shapiro 2005; Stuardo 2003), I believe that because of the way in which it measures integration, its utility for studying access in large Maya centers is limited (Cutting 2003).

7.3. Limitations of Space Syntax Methods

I contend that the limitations of space syntax are predominantly due to the traditional methodology of measuring integration with axial maps, which rely on simple longest-line-of-sight mapping derived from planimetric representations (Figure 7.2). Axial maps are problematic because measurements are two-dimensional rather than three-dimensional (Batty 2004; Hudson this volume; Ratti 2004, 2005). Such maps may be sufficient for measuring the accessibility of interior spaces for buildings or even architectural compounds; however, they cannot accurately measure accessibility across large Maya cities. This is because they do not take into account distance, topography, or the effects of barriers and facilitators on the *cost* of movement in the landscape.

Taking these factors into account is important because Maya communities comprise both the built environment and the natural landscape (Plate 6). Among contemporary Maya the term for large community is *kahkab*. *Kah* means "populated place," and *kab* means "land" or "earth"; in joining these words, the Maya essentially combine the built and natural environments (Marcus 2000:236). In other words, unlike the Western concept of city as human-made, the Maya view their communities as a construct of both the natural and built worlds. The ancient Maya seem to have had similar ideas. Along the Usumacinta River in Guatemala, they constructed temples atop caves that during the wet season were filled with fast-flowing water that sent a roaring sound up through these temples (Brady and Ashmore 1999). By fusing their built and natural surroundings, they were able to create an impressive auditory effect that produced a ritually charged atmosphere at specific times of the year. At the site of Copán in Honduras, the ancient Maya used the natural backdrop of the hillsides to heighten certain ceremonial and/or elite structures, making them appear larger than they truly were (Leventhal 1979, 1983).

These examples show that the ancient Maya integrated their built and natural surroundings in order to express ideas and structure events, and thus it follows that they would have also taken into consideration both built and natural features as they configured their surroundings to influence social connectivity within their cities. Given that spatial layout is a primary factor in facilitating and impeding movement and that pedestrian movement to or through particular spaces affects a location's degree of social connectivity, archaeologists must consider the *cost* of movement in configurational analyses. Unfortunately, axial maps are two-dimensional datasets and consequently cannot be used to measure the cost of movement, which is better approximated using three-dimensional data; however, by making use of the capabilities of GIS and least cost analysis (LCA), we can surmount such problems and measure the cost of movement, and ultimately social connectivity, across the Maya *kahkab*.

7.4. An Alternative Approach to Measuring Integration

Zipf's Principle of Least Effort states that interactions between places are inversely proportional to the cost of travel between them (Surface-Evans and White this volume; Zipf 1949). This means that people are more likely to travel to places that they can more easily reach or to which they will expend less energy traveling. Therefore, it follows that people are more likely to interact with people living at locations that are more easily reached than those living at hard-to-reach places. This often translates into greater interaction with one's neighbors, that is, those individuals who live close by rather than those who live far away. However, proximity, or distance, is only one variable affecting travel cost. Topography, hydrology, cultural features, and other factors also affect travel cost or the likelihood for interaction to occur (e.g., Kantner 2004; Llobera 2000; Miller 2006).

A GIS can simultaneously evaluate the effect these many variables have on the cost of pedestrian movement; thus, it is ideal for developing an alternative to axial graphs for measuring integration at ancient Maya sites. In GIS terms, axial graphs make measurements using a vector map; however, a much more powerful data type is available: the raster map. In a recent issue of *Environment and Planning B: Planning and Design* (2005), Carlo Ratti explores how the Urban Digital Elevation Model)—a raster map that stores elevations and building heights—can serve as a better alternative to the axial maps typically used in space syntax. Ratti suggests measuring the integration, or connectivity, of particular locations by using a cost-of-passage function, friction or impedance, to model travel costs across ancient landscapes. In the GIS, this translates to using the Urban DEM and employing algorithmic functions to merge several raster datasets into a single friction surface embedded with data on facilitators such as *sacbeob* (roads) and barriers such as structures, rivers, and *quebradas* to create least cost paths (Figure 7.3).

The friction surface for this study was created (1) using the union function to combine several shapefiles (representative of different archaeological features) into a single shapefile, (2) converting

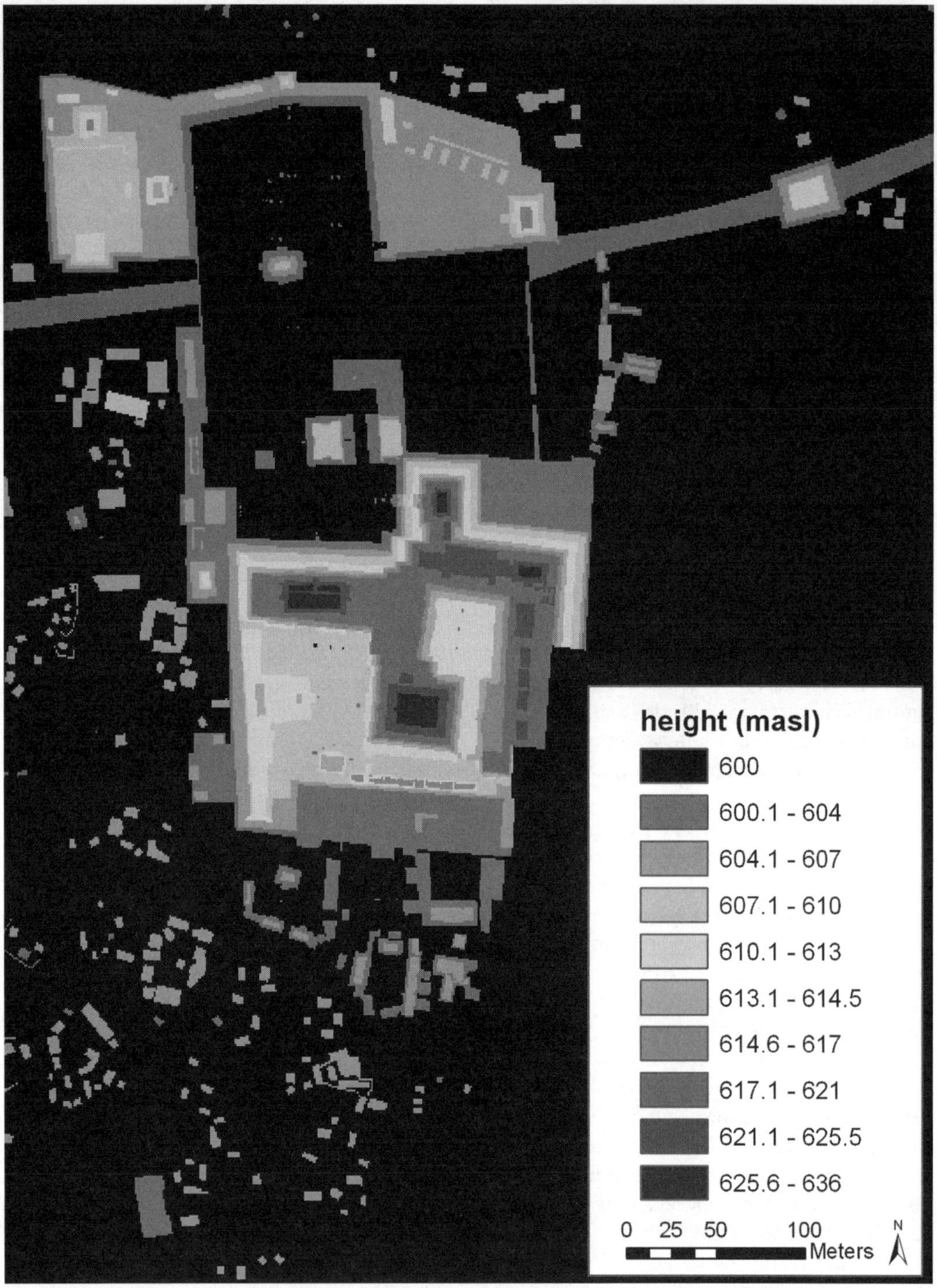

FIGURE 7.3. Urban DEM of Copán's Principal Group showing building heights in a raster (pixel-based) format that allows for LCA in a GIS.

this shapefile into a raster dataset, and (3) reclassifying the *sacbeob*, structures, and hydrology features into three classes: facilitators, barriers, and no change. The *sacbeob* were classified as facilitators and assigned a value = 0.9 because they attract and facilitate pedestrian movement. The Río Copán and *quebradas* were classified as barriers, or features that would increase the cost of movement, and were assigned a value of 3. The reservoirs and the structures were also classified as barriers and were assigned a value of 999—a high enough value to ensure that they would not be crossed. Spaces without archaeological features or hydrological features were classified as areas of no change and were assigned a value of 1. Ultimately, the reclassified dataset was integrated with the Urban DEM to take into account impedance (cost of movement) and generate least cost paths.

Least cost paths are not necessarily the shortest or quickest routes, but routes that involve the lowest travel costs (based on input criteria). In a GIS, a cost-of-passage function is employed to calculate the accumulated cost of moving from a source or set of sources to a destination or set of destinations. The path with the lowest value, or cost, is highlighted as the least cost travel route. The *average* values of these least cost paths indicate the likelihood that movement will occur to or through a particular space, that is, the likelihood that an individual will pass through that particular space. For example, people are more likely to walk to or through those sites with lower path costs than those with higher path costs (Hillier 1999; Hillier and Hanson 1984; Hillier et al. 1993). Significantly, this provides a method to quantify the degree of connectivity between spaces, which serves as a proxy for determining how integrated or segregated different groups of people (based on site type and site location) were in ancient landscapes. The ability to quantify connections using cost-based measures was lacking in previous space syntax studies.

Using the Urban DEM and friction surface, we can determine the average cost of travel using least cost paths from a certain point to all other relevant points. In this way, differences in cost to travel from one type of household to other types of households or to other points of interest such as stelae or monumental architecture can be measured. These measures are referred to as integration values (as in space syntax) and are used to provide information on interaction patterns between different social groups and different parts of cities, allowing archaeologists to tackle the question whether social integration and social segregation in prehistoric cities can be *quantitatively* addressed. The *average* value of the least cost paths from a source site to a subset of sites provides the integration values that indicate the degree of connectivity between people of different social groups or people living in different parts of Copán. The maps in Figures 7.4 and 7.5 compare least cost travel paths from Group 11L-13 in Copán's El Bosque suburb to two different site types (type 1 and type 2). The integration value for travel to type 2 sites from Group 11L-13 is 4068.67 (Figure 7.4). The integration value for travel to type 1 sites from Group 11L-13 is 5681.47 (Figure 7.5). The lower integration value (or lower cost of travel) for travel to type 2 sites indicates that people living in El Bosque were more socially connected with people living at type 2 sites than with people living at type 1 sites. These integration values allow archaeologists to assess the potential for interaction between different groups by quantifying the degree of connectivity between them.

Several steps were involved in the design and development of the GIS data that were used to create these least cost path maps and generate integration values:

1. Scan maps and architectural plans and drawings
2. Georeference scanned images
3. Digitize archaeological and natural features from these georeferenced images to create shapefiles
4. Attribute shapefiles (e.g., site types, group ID, structure ID, elevation)
5. Convert polylines shapefiles to polygon shapefiles
6. Convert shapefiles to raster files
7. Create digital terrain model (DTM)
8. Create Urban DEM from DTM and raster files
9. Create friction surface
10. Generate slope from the Urban DEM
11. Integrate slope and friction surface to derive cost-distance and cost-direction surfaces

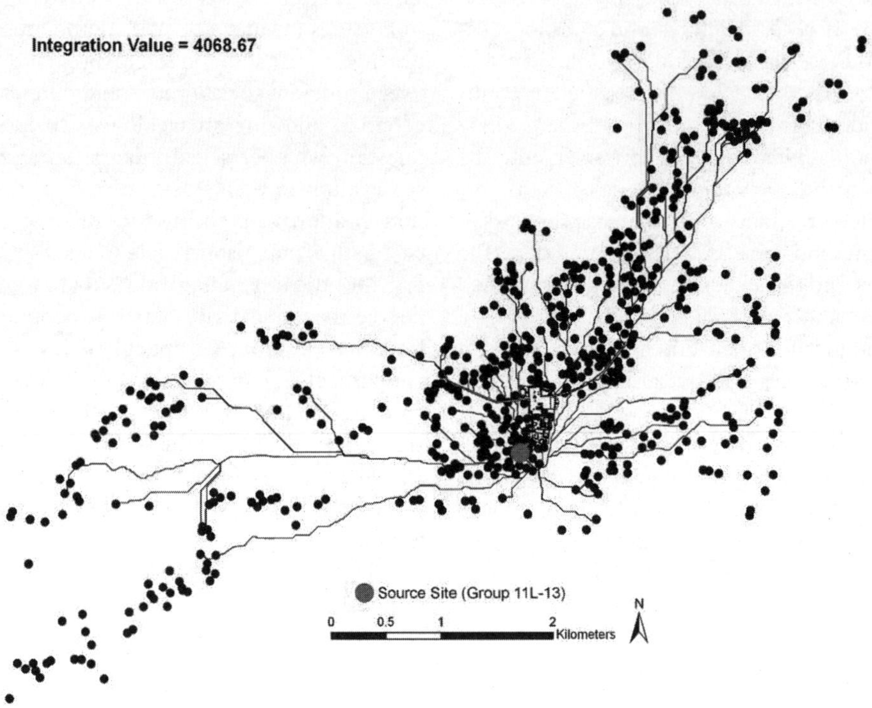

FIGURE 7.4. LCP map for travel to type 2 sites from Group 11L-13 in the suburb of El Bosque at Copán, Honduras.

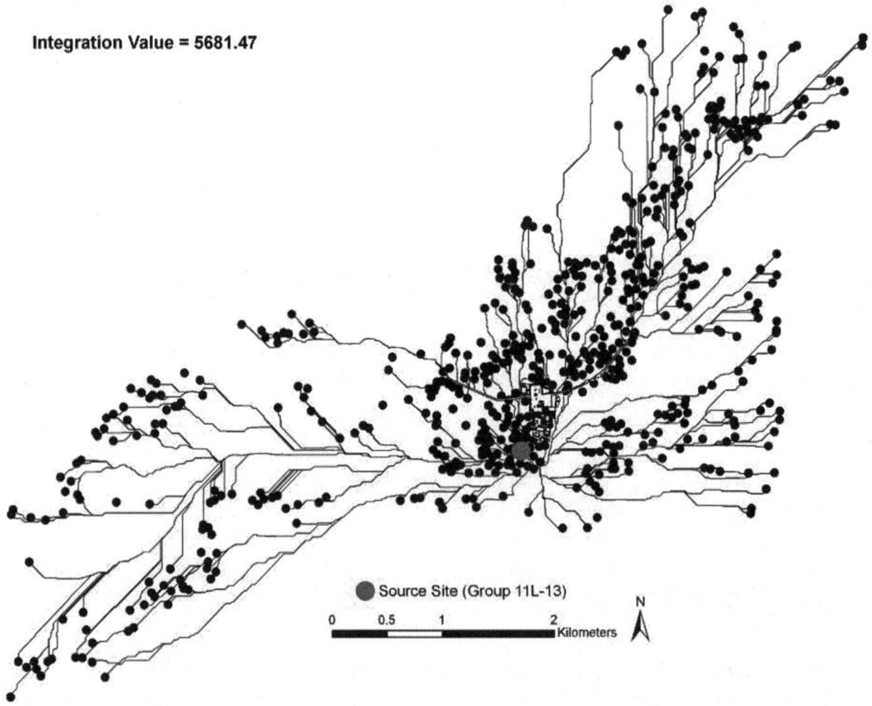

FIGURE 7.5. LCP map for travel to type 1 sites from Group 11L-13 in the suburb of El Bosque at Copán, Honduras.

FIGURE 7.6. Three-dimensional reconstruction of the Copán Valley in the late eighth and early ninth centuries (ad 763–820) showing settlement pattern in relation to natural topography and hydrology.

12. Generate least cost paths from 74 sample sites to all (594) sites at Copán
13. Classify least cost paths based on site types 1–5
14. Export classified path costs (integration values) to Excel
15. Import path costs (integration values) to Minitab 15 for statistical tests

The data were created using the Environmental Science Research Institute's (ESRI) ArcGIS 9.1, a GIS software package, and analyzed using ArcGIS 9.2. Minitab 15, a statistical software package, was used to evaluate the statistical significance of the integration and visibility analyses.

7.5. Case Study: Copán, Honduras

This research focuses on site organization in the late eighth and early ninth centuries (AD 763–820) at the ancient Maya site of Copán, Honduras (Figure 7.6). This site serves as an ideal case study, for two reasons. First, for practical reasons, its long history of research provides voluminous survey and excavation data. Archaeologists have carried out a full-coverage survey (100 percent) and instrument-mapped all visible archaeological features in the Copán Valley (27 km^2) at a scale of 1:2000 (Baudez 1983; Fash and Long 1983; Leventhal 1979; Willey et al. 1978), and architects have used photogrammetric studies to map Copán's Principal Group and several elite complexes at a scale of 1:200 (Hohmann 1995; Hohmann and Vogrin 1982). These analog data were scanned, georeferenced, and digitized to create the GIS data used in this research.

Second, the Harvard Site Typology—a five-part classification scheme at Copán—provides a means to analyze how people living at different site types organized themselves within the city because it correlates site types to socioeconomic status (Willey et al. 1978; Willey and Leventhal 1979). The typology comprises five site types (1–5); it is assumed that commoners lived at site types 1 and 2, and the elite occupied site types 3 and 4 (Plate 7). There is only one type 5 site at Copán—the city's major civic-ceremonial complex, the Principal Group (see Figure 7.1).

When speaking of restricted access at ancient Maya sites, scholars typically refer to access to the interior spaces of elite residences and civic-ceremonial complexes. They assume that access is limited to other elites (unless the person is a member of the household or serves as some sort of laborer); this is most likely a valid assumption, but it does not address how integrated the elite living at a specific compound may be with respect to society as a whole and to people from different social groups. My research differs from other accessibility studies because it explicitly measures social connectivity across the city as a whole and addresses such questions as, Were elites living at type 4 sites more or less integrated with society as a whole, with people living at specific site types (1–4), with the ruler, or with the site's civic-ceremonial center? I asked the same questions of elites and commoners living at site types 1–3 at Copán.

7.6. Results

The way ancient people constructed and organized their physical surroundings and spaces provides information about their past lives; physical configurations are created that both mirror and shape social interaction among different social groups (Cutting 2003; Moore 2003, 2005). These configurations occur on many scales, from single households to multifamily architectural complexes to neighborhoods to cities. A benefit of multiscalar studies is that they bridge household studies and settlement pattern studies, often focused on commoners and everyday life, with those oriented toward the elite (Marcus 2000; Yaegar and Canuto 2000).

Given that the goal of my research is to investigate social connectivity across social groups, I employed a multiscalar analysis to examine accessibility at four different scales in the Copán Valley: (1) valley-wide, (2) physiographic zone, (3) urban core–hinterland, and (4) subcommunity. The objectives were to study social connectivity:

1. between (rather than within) Copán's twenty-one subcommunities to study differences and similarities across subcommunities
2. between the urban core and its hinterland to study core-periphery relationships
3. between the valley's five physiographic zones to understand the potential influence of ecological variables on structuring pedestrian movement
4. for the city as a whole to understand how Copán's different social groups may have interacted across the valley.

A nonparametric statistical test, the Kruskal-Wallis test, was used to evaluate whether the differences in integration values between site types were statistically significant. (This statistical test offers an alternative to the ANOVA—one-way analysis of variance—for nonparametric data.) Summary tables and results for each analytical scale are presented below.

7.6.1. Valley-Wide

The valley-wide accessibility results identify a pattern that supports the assumption that the ancient Maya of Copán used accessibility to differentially channel pedestrians throughout the valley. The city's layout seems to have served as a guide to daily interactions, facilitating pedestrian movement from across the valley toward the highly accessible main civic-ceremonial complex, the Principal Group. The integration values presented in Table 7.1 indicate that people were channeled toward elite sites, suggesting that the elite were more socially connected to society as a whole than were commoners. The data show that the cost for people living at any site type to travel to the large, open Great Plaza is lower than the cost to travel to any other site type in the valley. Elites living at type 4 sites were situated at strategic locations, affording them the greatest access, excepting the king, to all the city's residents.

The data also indicate that people living at type 3 sites were more integrated with society as a whole than people living at type 1 and type 2 sites. (Interestingly, however, the values for type 2 [commoner households] and type 3 sites are more similar than the values for type 3 and type 4 sites, presumably both elite complexes—an important point that I revisit below.) The valley-wide access pattern indicates that as social status increased, accessibility increased. Previous studies show that people living at highly integrated locations can more easily exercise their authority as a result of their greater accessibility to both people and resources (Hillier 1999; Hillier and Hanson 1984), and thus the results suggest that the elites living at type 3 and type 4 sites positioned themselves at locations on the landscape that would help centralize their power.

TABLE 7.1. Valley-Wide Integration Values, Copán.

Site Type	*N* (paths)	Integration Value
1	25890	7246
2	3465	6297
3	1469	5842
4	925	5136
Great Plaza	586	3412
Acropolis	587	4130
Royal Courtyard	583	4061

p-value = <0.0001

Residents of type 1 sites were the least integrated, which suggests that the roles they played in society did not necessitate a high degree of connectivity with people across the valley. Residents of type 1 sites were less likely to interact with Copanecos of any social status, and thus they are considered more socially isolated. Perhaps the roles they played in society did not require routine communication with occupants living in many other parts of the valley. Their relative social isolation may have meant that they were a lesser target of social control, and their lower status may have provided them with more autonomy than other social groups at Copán.

Tables 7.2–7.4 illustrate the degrees of integration (connectivity) between Copán's Principal Group areas and the city's residential site types. They show that the city's configuration facilitated interaction between the type 4 occupants and the city's major civic-ceremonial complex, providing them with the greatest access to the ruler, other members of the royal court, and the ritual ceremonies performed in the city center (Inomata and Houston 2001). These tables also illustrate another interesting observation. According to the Harvard Site Typology, elites lived at type 3 and 4 sites and commoners lived at type 2 sites; therefore, I expected that the integration values for type 3 and 4 sites would be more similar than the integration values for type 2 and 3 sites. However, as the tables show, this is not the case: the integration values for type 2 and 3 sites are more similar. These results may indicate that the occupants of type 2 and type 3 sites played similar economic, social, and/or political roles at Copán, suggesting that the Harvard Typology requires some refinement.

TABLE 7.2. Integration Values for the Great Plaza.

Site Type	*N* (paths)	Integration Value
1	434	3987
2	107	2366
3	25	2080
4	16	1569

p-value = <0.0001

TABLE 7.3. Integration Values for the Acropolis.

Site Type	*N* (paths)	Integration Value
1	434	4853
2	107	3003
3	25	2970
4	16	2553

p-value = <0.0001

TABLE 7.4. Integration Values for the Royal Courtyard.

Site Type	*N* (paths)	Integration Value
1	433	4599
2	107	2917
3	25	2672
4	16	2128

p-value = <0.0001

In sum, the valley-wide results indicate that as accessibility increased, social status increased, and therefore the findings support the assumption that accessibility served as a mechanism to help create and maintain distinct social categories and ostensibly reinforce the cosmic order at Copán.

7.6.2. Physiographic Zones

The results of the physiographic zone portion of this study indicate that ecological variables alone are not responsible for differences in social connectivity across the valley (Table 7.5). In some cases, zones with very different landforms and topography have similar integration values, while zones with similar landforms and topography often have very different integration values. An interesting observation is that zone 5, which has the shortest occupation history (archaeological remains date only to the Middle Preclassic and

TABLE 7.5. Integration Values for the Physiographic Zones.

Zone	*N*	Integration Value
2	5136	4425
3	14759	5951
4	6817	5535
5	9781	11126

p-value = <0.0001

TABLE 7.6. Integration Values for Site Types by Physiographic Zone.

Site Type	Zone 2	Zone 3	Zone 4	Zone 5
Type 1	5746	6389	6049	11346
Type 2	3771	5245	5146	11224
Type 3	3541	5312	4838	10361
Type 4	2675	4464	4435	9344
Great Plaza	1783	3372	3723	9886
Acropolis	2515	4308	4509	10916
Royal Courtyard	2083	4469	3757	10839

p-value = <0.0001

TABLE 7.7. Integration Values for the Urban Core vs.the Hinterlands.

Area	*N* (paths)	Integration Value
Urban Core	11294	4387
Hinterland	25199	7970

p-value = <0.0001

Late Classic periods with no intermittent occupation), was the least integrated area of the site. The cost to travel from sites in zone 5 to other parts of the valley was two to three times more than it was for any other physiographic zone.

One explanation is that zone 5's high integration values may reflect less sociopolitical control on the part of Copán's ruling class. As Copán's ruler and elite experienced more difficulties in the later years of the Late Classic, they may have had less control over where people lived. Instead of being restricted to living in more highly centralized areas of the city, people may have enjoyed a new freedom to build in more distant locations. These decentralizing tendencies may reflect a loss of power for Copán's ruling authority (Fash 2001). An alternate explanation is that these low values may simply be due to rapid Late Classic population growth that would have required the occupation of new and more distant lands, presumably in great part for agricultural purposes. Regardless of the causes, these high integration values indicate that people living in zone 5 were more socially isolated from society as a whole than were people living in other zones.

Like the valley-wide results, the data indicate that a marked difference exists in the integration values of type 3 and 4 sites (Table 7.6). In all zones (except zone 5), the integration values of site types 2 and 3 are closer than the integration values of site types 3 and 4. These results provide another line of evidence suggesting that the Harvard Typology may need to be refined.

7.6.3. Urban Core–Hinterlands

The integration values in Table 7.7 show that people living in the urban core exhibited a higher degree of social connectivity than did people living in the hinterlands. These differences are most likely due in part to variability in settlement density and site type distribution between the urban core and the hinterlands. However, they may also arise from temporal variation. Archaeological excavation and test units indicate that the urban core is the oldest and most continuously occupied part of the valley (Fash 1983; Hall and Viel 2004; Sanders 1986). The accessibility patterns indicate greater social control within the urban core and less social control within newer areas, that is, the hinterlands. These findings may reflect a pattern of growth and development that is difficult to test without more excavation data, although it is still worth noting.

The values in Table 7.8 indicate that pedestrian movement was channeled toward elite compounds (site types 3 and 4) in both the urban core and the hinterlands. By channeling people past their elaborate architecture, the elite were able to remind others of their power and wealth on a daily basis as well as facilitate social interaction between themselves and other members of society. While people were channeled to both urban and hinterland elite sites, the larger differences in the integration values of urban core sites suggest greater social control in this part of the Copán. In contrast, smaller differences in integration values between sites in the hinterlands suggest less sociopolitical control outside the city's center. Before the Late Classic, Copán's different site types may have been more equally dispersed across the landscape. As the years passed, Copán's rulers and other elite appeared to have intentionally aggregated their residences in the urban core, close to one another and to the royal precinct (Fash 2001). This aggregation suggests the development of greater social inequalities between the elite and the commoners, reflected in the older areas of the site but less evident in the younger parts of the valley.

Like the results for the valley-wide and physiographic zones, the differences in integration values for sites in the urban core versus those in the hinterlands point to additional problems with the Harvard Site Typology. For example, although the difference in type 1 and type 2 integration values in the urban core is striking, the difference is much less marked in the hinterlands. The results suggest that people living at type 2 urban sites were relatively integrated with society as a whole, and yet people living at type 1 urban sites were not. In contrast, people living at type 2 sites in the hinterlands (similar to those living at type 1 hinterland sites) were relatively segregated. Such differences suggest that people living at type 1 and type 2 sites in the urban core played distinct and different roles in Copaneco society, while individuals living at the same site types in the hinterlands did not. Consequently, they may point to spatial variation and social distinctions (e.g., farmers vs. servants) that are not accounted for in the Harvard Typology.

7.6.4. Subcommunities

The subcommunity results help to refine the urban core–hinterland results by subdividing the valley into three intermediate-level interaction spheres based on varying degrees of social connectivity. The integration values indicate a high degree of channeling to elite sites in the urban core, a moderate degree of channeling to elite sites in eastern subcommunities, and a low degree of channeling to elite sites in western and far eastern subcommunities. Large differences in integration values reflect high degrees of sociopolitical control and social segregation because they reflect greater control over pedestrian movement, and small differences reflect low degrees of sociopolitical control and social integration because they reflect less control over pedestrian movement (Ferguson 1996; Hillier 1999; Hiller and Hanson 1984; Hillier et al. 1993). Figure 7.7 illustrates the subcommunity findings, which provide the following information about sociopolitical control and class segregation at Copán.

1. People living in the urban core had the highest degree of social connectivity. However, because urban core residents were highly channeled toward elite sites, they experienced the greatest degree of sociopolitical control and class segregation.
2. People living in eastern subcommunities had a moderate degree of social connectivity and experienced moderate degrees of sociopolitical control and class segregation.
3. People living in western and far eastern subcommunities experienced the highest degree of segregation because the cost to travel to these areas was very high. However, there may have been an advantage to such circumstances—a lesser degree of sociopolitical control and less class segregation.

TABLE 7.8. Integration Values for Site Types in the Urban Core vs. the Hinterlands.

Site Type	Urban Core	Hinterlands
Type 1	5293	8212
Type 2	3572	7792
Type 3	3312	7294
Type 4	2623	6967
Great Plaza	1847	5727
Acropolis	2604	6637
Royal Courtyard	2476	6616

p-value = <0.0001

The subcommunity accessibility results provide additional data that highlight unexpected similarities between type 2 sites and type 3 sites. Like the findings for the other scales of analysis, the integration values for these two site types often lack statistically significant differences, again suggesting that the Harvard Site Typology needs to be reevaluated.

7.7. Reevaluating the Harvard Site Typology

The Harvard Site Typology focuses explicitly on material remains; in contrast, my research on accessibility investigates more subtle nuances that

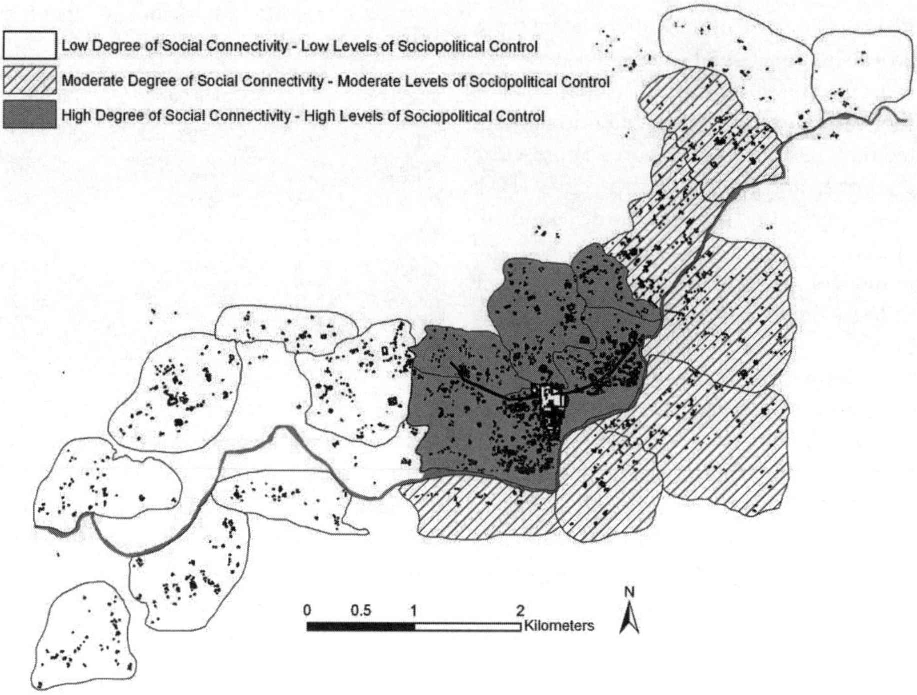

FIGURE 7.7. Three spheres of social connectivity and sociopolitical control, as defined by the LCA, in the late eighth and early ninth centuries at Copán.

reflect and influence sociopolitical organization. The accessibility results highlight similarities between site types 2 and 3 that have not been observed in previous studies, suggesting that perhaps the occupants of site types 2 and 3 played similar economic, social, and/or political roles at Copán or that the Harvard Typology has misclassified some of these sites.

The possible conflation of type 2 and type 3 sites provides evidence supporting the need to more clearly define the differences between site types. Type 2 sites are designated commoner households whereas type 3 sites are designated elite. The expectation is that there would be more similarities between type 1 and 2 sites, both considered commoner, and more similarities between type 3 and 4 sites, both considered elite; however, the access (and visibility) results do not meet this expectation. In many instances, the access (and visibility) values of type 2 and type 3 sites are more similar (see Richards-Rissetto 2010 for visibility analysis). Furthermore, test excavations reveal that some type 2 sites unexpectedly have dressed stone, vaulted roofs, and large numbers of fine wares—traits that are indicative of type 3 sites at Copán (Fash 1983; Webster et al. 2000). These findings lead me to ask, (1) Should some type 2 sites be reclassified as type 3 sites? (2) Should subtypes be created? (3) Is an entirely new classification system needed? and (4) Do we need a typology at all? The answers to these questions require future research.

7.8. Conclusion

Archaeologists traditionally rely on patterns in material culture to understand past societies and their dynamics through time. More recently, archaeologists have focused on the cultural use of space to enrich our understanding of ancient societies (e.g., Hendon 1991; Kent 1984; Lawrence and Low 1990), including investigating the way in which ancient peoples organized themselves in the landscape in order to reconstruct sociopolitical relations. While studies of site organization have been an important part of archaeology for the last half century, it was not until recently that scholars began to regard site layout not simply as a reflection of ancient life but also as a mechanism that shaped it. Seeking to contribute to the growing body of knowledge about the role that site organization played in influencing ancient sociopolitical relations, I developed a least cost approach to investigate the connections between site organization and sociopolitical relations in the late eighth and early ninth centuries at the ancient Maya site of Copán, Honduras.

The results indicate that, in general, Copán's layout replicated and reinforced society's hierarchical class structure—with higher accessibility correlated to higher social status—but that this pattern was not replicated at all societal levels. The access data reveal underlying complexities and sociopolitical relationships that point to the presence of varying degrees of sociopolitical control within the city. Returning to this research's two main questions—(1) Did people of different social classes experience different degrees of social connectivity? and (2) Did people living in different parts of the city experience different degrees of social connectivity?—we see from the study that the answers to these two questions are inextricably linked. People of different social classes experienced different degrees of social connectivity; however, the degree to which specific social groups were either integrated or segregated depended on where they lived. These differences in social connectivity translate into differential degrees of sociopolitical control at Copán. People living in the urban core and eastern subcommunities were channeled toward elite sites and thus experienced greater sociopolitical control than people living in western and far eastern subcommunities, who were not channeled toward elite sites.

In addition to providing information on the relationships among site organization, pedestrian movement, and social structure, the least cost approach to configurational analysis presented in this chapter indicates a need to modify the Harvard Site Typology. The access data suggest that, at least with respect to social connectivity, significant differences did not exist between type 2 and type 3 sites. Given that the Harvard Typology categorizes type 2 sites as commoner households and type 3 sites as elite complexes, these results were unexpected. The implication of the similarities in the integration values of type 2 and type 3 sites is that the typology needs to be refined, most likely in a way that incorporates more social or indigenously derived variables. Such an alteration is important because many interpretations of Copán's sociopolitical organization are heavily influenced by the current site typology.

To conclude, using a modified form of configurational analysis based on least cost paths to measure the relationship between site organization and social connectivity proved fruitful. The LCA identified patterns indicating that Copán's social hierarchy was embedded in the landscape. These patterns, in turn, provided information on how the ancient Maya may have strategically manipulated the landscape to structure social interaction and maintain sociopolitical control.

Acknowledgments

I thank the following organizations for their generous funding support for this project: the American Society of Photogrammetry and Remote Sensing (ASPRS); the Latin American and Iberian Institute, University of New Mexico (UNM); the Office of Graduate Studies, UNM; the Graduate and Professional Student Association, UNM; and the New Mexico Geographic Information Council. I am indebted to Jennifer von Schwerin for guidance and support. I am grateful as well to Jane Buikstra, David Dinwoodie, James Boone, and Richard Watson for comments and suggestions on my research. I thank Susan Sehi-Smith for her tutelage and assistance with the Google SketchUp models. Finally, I extend my gratitude to the Instituto Hondureño de Antropología e Historia for the opportunity to work at Copán.

References

Andrews, Wyllys E., and Cassandra Bill
2005 A Late Classic Royal Residence at Copán. In *Copán: The History of an Ancient Kingdom*, edited by E. Wyllys Andrews and W. Fash, pp. 239–314. School of American Research Advanced Seminar Series. School of American Research Press, Santa Fe, New Mexico.

Ashmore, Wendy
1991 Site-Planning Principles and Concepts Of Directionality among the Ancient Maya. *Latin American Antiquity* 2(3):199–226.
1992 Deciphering Maya Architectural Plans. In *New Theories on the Ancient Maya*, edited by Elin C. Danien and Robert J. Sharer, pp. 173–184. University Museum, University of Pennsylvania, Philadelphia.

Ashmore, Wendy, and Jeremy Sabloff
2002 Spatial Orders in Maya Civic Plans. *Latin American Antiquity* 13(2):201–215.
2003 Interpreting Ancient Maya Civic Plans: Reply to Smith. *Latin American Antiquity* 14(2):229–236.
Batty, Michael
2004 A New Theory of Space Syntax. Center for Advanced Spatial Analysis, London. Electronic document, http://www.casa.ucl.ac.uk/working_papers/paper75.pdf.
Baudez, Claude-Francois
1983 *Introducción a la arqueología de Copán, Honduras.* Proyecto Arqueológico Copán, Secretaria de Estado en el Despacho de Cultura y Turismo, Tegucigalpa, Honduras.
1994 *Maya Sculpture of Copán: The Iconography.* University of Oklahoma Press, Norman.
Blanton, Richards E.
1989 Continuity and Change in Public Architecture: Periods I through V of the Valley of Oaxaca, Mexico. In *Monte Alban's Hinterland, Part II: Prehispanic Settlement Patterns in Tlacolula, Etla, and Ocotlan, the Valley of Oaxaca, Mexico*, edited by S. Kowalewski, G. Feinman, L. Finsten, R. Blanton, and L. Nicholas, pp. 409–447. Memoirs of the Museum of Anthropology No. 23. University of Michigan, Ann Arbor.
Brady, James E., and Wendy Ashmore
1999 Mountains, Caves, Water: Ideational Landscapes of the Ancient Maya. In *Archaeologies of Landscape: Contemporary Perspectives,* edited by Wendy Ashmore and A. Bernard Knapp, pp. 124–145. Blackwell, Oxford.
Bustard, W. J.
1996 Space as Place: Small and Great House Spatial Organization in Chaco Canyon, New Mexico, AD 1000–1150. Unpublished Ph.D. dissertation, University of New Mexico, Albuquerque.
Coggins, Clemency C.
1980 The Shape of Time: Some Political Implications of a Four-Part Figure. *American Antiquity* 45:727–739.
Craane, Marlous L.
2009 The Medieval Urban "Movement Economy": Using Space Syntax in the Study of Medieval Towns as Exemplified by the Town of Hertogenbosch, the Netherlands. In *Seventh International Space Syntax Symposium Proceedings*, edited by D. Koch, L. Marcus, and J. Steen, pp. 01901–01914. KTH, Stockholm.
Cutting, Marion
2003 The Use of Spatial Analysis to Study Prehistoric Settlement Architecture. *Oxford Journal of Archaeology* 22(1):1–21.
Dawson, P. C.
2002 Space Syntax Analysis of Central Inuit Snow Houses. *Journal of Anthropological Archaeology* 21(4):464–480.
DeMarrais, Elizabeth, Luis Jaime Castillo, and Timothy Earle
1996 Ideology, Materialization, and Power Strategies. *Current Anthropology* 37(1):15–31.
During, B.
2001 Spatial Dimensions in the Architecture of Neolithic Catalhoyuk. *Anatolian Studies* 51:1–18.
Fash, William L.
1983 Maya State Formation: A Case Study and Its Implications. Unpublished Ph.D. dissertation, Harvard University, Cambridge, Massachusetts.
2001 *Scribes, Warriors, and Kings: The City of Copan and the Ancient Maya.* Thames and Hudson, London.
Fash, William L., and Kurt Z. Long
1983 Mapa arqueológico del valle de Copán. In *Introducción a la arqueología de Copán, Honduras, Tomo III.* Proyecto Arqueologico Copán, Secretaria de Estado en el Despacho de Cultura y Turismo, Tegucigalpa, Honduras.
Ferguson, T. J.
1996 *Historic Zuni Architecture and Society: An Archaeological Application of Space Syntax.* University of Arizona Press, Tucson.
Fletcher, Roland
1981 People and Space: A Case Study on Material Behavior. In *Pattern of the Past: Studies in Honour of David Clarke*, edited by I. Hodder, G. Issac, and N. Hammond, pp. 97–128. Cambridge University Press, Cambridge.
Freter, AnnCorinne
1994 The Classic Maya Collapse at Copan, Honduras: An Analysis of Maya Rural Settlement Trends. In *Archaeological Views from the Countryside*, edited by Glenn M. Schwartz and Steven E. Falconer, pp. 160–176. Smithsonian Institution Press, Washington, D.C.
Gerstle, Andrea I.
1987 Ethnic Diversity and Interaction at Copan, Honduras. In *Interaction on the Southeast Mesoamerican Frontier*, 2 vols., edited by E. Robinson, 2:328–356. BAR International Series 327. British Archaeological Reports, Oxford.
Giddens, Anthony
1984 *The Constitution of Society: Outline of the Theory of Structuration.* University of California Press, Berkeley.

Guillermin, G. F.
1968 Development and Function of the Tikal Ceremonial Center. *Ethnos* 33:1–35.
Hall, Jay, and René Viel
2004 The Early Copan Classic Landscape: A View from the Preclassic. In *Understanding Early Classic Copan*, pp. 17–28. University of Pennsylvania Museum of Archaeology and Anthropology, Philadelphia.
Hammond, Norman, and Gair Tourtellot
1999 Shifting Axes: Spatial Expressions of Power at La Milpa. Paper presented at the 64th Annual Meeting of the Society for American Archaeology, Chicago.
Hendon, Julia
1987 The Uses of Maya Structures: A Study of Architecture and Artifact Distribution at Sepulturas, Copán, Honduras. Unpublished Ph.D. dissertation, Harvard University, Cambridge, Massachusetts.
1991 Status and Power in Classic Maya Society: An Archaeological Study. *American Anthropologist*, new series, 93(4):894–918.
Hillier, Bill
1999 *Space Is the Machine: A Configurational Theory of Architecture*. Cambridge University Press, Cambridge.
Hillier, Bill, and Julienne Hanson
1984 *The Social Logic of Space*. Cambridge University Press, Cambridge.
Hillier, Bill, A. Penn, J. Hanson, T. Grajewski, and J. Xu
1993 Natural Movement; or, Configuration and Attraction in Urban Pedestrian Movement. *Environment and Planning B: Planning and Design* 20:29–66.
Hohmann, Hasso
1995 *Die Architektur der Sepulturas—Region von Copán in Honduras*. Academic Publishers, Graz, Austria.
Hohmann, Hasso, and Annegrete Vogrin
1982 *Die Architektur von Copan*. Academic Publishers, Graz, Austria.
Houk, Brett
1996 The Archaeology of Site Planning: An Example from the Maya site of Dos Hombres, Belize. Unpublished Ph.D. dissertation, University of Texas, Austin.
Houston, Stephen, David Stuart, and Karl Taube
2006 *The Memory of Bones: Body, Being, and Experience among the Classic Maya*. University of Texas Press, Austin.
Inomata, Takeshi, and Stephen Houston
2001 *Royal Courts of the Ancient Maya*. Volumes 1 and 2. Westview Press, Boulder, Colorado.
Jakobson, Roman
1980 *The Framework of Language*. University of Michigan, Ann Arbor.
Joyce, Rosemary A., and Julia A. Hendon
2000 Heterarchy, History, and Material Reality: "Communities" in Late Classic Honduras. In *The Archaeology of Communities: A New World Perspective*, edited by M. Canuto and J. Yaeger, pp. 143–160. Routledge, London.
Kantner, John
2004 Geographical Approaches for Reconstructing Past Human Behavior from Prehistoric Roadways. In *Spatially Integrated Social Science: Examples in Best Practice*, edited by M. F. Goodchild and D. G. Janelle, pp. 323–344. Oxford University Press, Oxford.
Keller, Angela
2001 Road Types and Political Statements at the Classic Period Maya Center of Xunantunich. Paper presented at the 2001 Chacmool Conference, Calgary, Alberta.
Kent, Susan
1984 *Analyzing Activity Areas: An Ethnoarchaeological Study of the Use of Space*. University of New Mexico Press, Albuquerque.
King, Anthony D.
1980 A Time for Space and a Space for Time: The Social Production of the Vacation House. In *Buildings and Society: Essays on the Social Development of the Built Environment*, edited by A. King, pp. 193–227. Routledge and Kegan Paul, London.
Lawrence, Denise L., and Setha M. Low
1990 The Built Environment and Spatial Form. *Annual Review Of Anthropology* 19:453–505.
Leventhal, Richard
1979 Settlement Patterns at Copan, Honduras. Unpublished Ph.D. dissertation, Harvard University, Cambridge, Massachusetts.
1983 Household Groups and Classic Maya Religion. In *Prehistoric Settlement Patterns: Essays in Honor of Gordon R. Willey*, edited by E. Vogt and R. Leventhal, pp. 55–76. University of New Mexico Press, Albuquerque; Peabody Museum of Archaeology and Ethnology, Harvard University, Cambridge, Massachusetts.
Llobera, Marcos
2000 Understanding Movement: A Pilot Model towards the Sociology of Movement. In *Beyond the Map: Archaeology and Spatial Technologies*, vol. 321, edited by G. R. Lock, pp. 65–84. IOS Press, Amsterdam.
2006 What You See Is What You Get? Visualscapes, Visual Genesis and Hierarchy. In *Digital*

Archaeology: Bridging Method and Theory, edited by P. Daly and T. Evans, pp. 148–167. Routledge, New York.

Lucero, Lisa
2007 Classic Maya Temples, Politics, and the Voice of the People. *Latin American Antiquity* 18(4): 407–427.

Maca, Allan Leigh, Jr.
2002 Spatio-Temporal Boundaries in Classic Maya Settlement Systems: Copan's Urban Foothills and the Excavations at Group 9J-5. Unpublished Ph.D. dissertation, Harvard University, Cambridge, Massachusetts.
2009 Ethnographic Analogy and the Archaeological Construction of Maya Identity at Copan, Honduras. In *The Ch'orti' Maya Area: Past and Present*, edited by B. Mentz, C. McNeil, and K. Hull. University Press of Florida, Gainesville.

Marcus, Joyce
1983 On the Nature of the Mesoamerican City. In *Prehistoric Settlement Patterns: Essays in Honor of Gordon R. Willey*, edited by E. Vogt and R. Leventhal, pp. 195–242. University of New Mexico Press, Albuquerque; Peabody Museum of Archaeology and Ethnology, Harvard University, Cambridge, Massachusetts.
2000 Toward an Archaeology of Communities. In *The Archaeology of Communities: A New World Perspective*, edited by C. Marcello and J. Yaeger, pp. 231–242. Routledge, London.

Miller, Heather M.
2006 Comparing Landscapes of Transportation: Riverine-Oriented and Land-Oriented Systems in the Indus Civilization and the Mughal Empire. In *Space and Spatial Analysis in Archaeology*, edited by Elizabeth C. Robertson, Jeffrey D. Seibert, Deepika C. Fernandez, and Marc U. Zender, pp. 281–291. University of Calgary Press, Calgary, Alberta.

Moore, Jerry D.
1996a *Architecture and Power in the Ancient Andes: The Archaeology of Public Buildings*. Cambridge University Press, Cambridge.
1996b The Archaeology of Plazas and the Proxemics of Ritual: Three Andean Traditions. *American Anthropologist* 98(4):789–802.
2003 Life behind Walls: Patterns in the Urban Landscape on the Prehistoric North Coast of Peru. In *The Social Construction of Ancient Cities*, edited by M. Smith, pp. 81–102. Smithsonian Books, Washington, D.C.
2005 *Cultural Landscapes in the Ancient Andes: Archaeologies of Place*. University Press of Florida, Gainesville.

Peirce, Charles Sanders
1966 *Collected Papers, Volumes 1–8*. Edited by Charles Hartshorne and Paul Weiss. Belknap Press of Harvard University Press, Cambridge, Massachusetts.

Preziosi, Donald
1979a *The Semiotics of the Built Environment: An Introduction to Architectonic Analysis*. Indiana University Press, Bloomington.
1979b Architecture, Language, and Meaning: The Origins of the Built World and Its Semiotic Organization. In *Approaches to Semiotics Vol. 49*. Mouton, New York.

Price, T. Douglas, J. Burton, P. Fullagar, L. Wright, J. Buikstra, and V. Tiesler
2008 87Sr/86Sr Ratios and the Study of Human Mobility in Ancient Mesoamerica. *Latin American Antiquity* 19(2):167–180.

Ratti, Carlo
2004 Space Syntax: Some Inconsistencies. *Environment and Planning B: Planning and Design* 31: 487–499.
2005 The Lineage of the Line: Space Syntax Parameters from the Analysis of Urban DEMs. *Environment and Planning B: Planning and Design* 32:547–566.

Reents-Budet, Dorie
2001 Classic Maya Concepts of the Royal Court: An Analysis of Renderings on Pictorial Ceramics. In *Royal Courts of the Ancient Maya, Vol. 1*, edited by T. Inomata and S. Houston, pp. 195–233. Westview Press, Boulder, Colorado.

Reese-Taylor, Kathyrn
2001 Ritual Circuits as Key Elements in Maya Civic Center Designs. In *Heart of Creation: The Mesoamerican World and the Legacy of Linda Schele*, edited by A. Stone, pp. 143–165. University of Alabama Press, Tuscaloosa.

Richards-Rissetto, Heather
2007 Exploring Social Dynamics at the Ancient Site of Copán, Honduras. *ESRI Professional Papers* (June), Paper 1475, San Diego.
2010 Exploring Social Interaction at the Ancient Maya City of Copán, Honduras: A Multi-Scalar Geographic Information Systems (GIS) Analysis of Access and Visibility. Unpublished Ph.D. dissertation, University of New Mexico, Albuquerque.

Sanchez, Julia Lynn
1997 Royal Strategies and Audience: An Analysis of Classic Maya Monumental Art. Unpublished Ph.D. dissertation, University of California, Los Angeles.

Sanders, William T.
1986 *Excavaciones en el área urbana de Copán, tomo*

I. Instituto Hondureño de Antropología e Historia, Tegucigalpa, Honduras.
1989 Household, Lineage, and State at Eighth-Century Copan, Honduras. In *The House of the Bacabs*, edited by David Webster, pp. 89–105. Studies in Precolumbian Art and Archaeology No. 29. Dumbarton Oaks, Washington, D.C.
Schele, Linda, and David A. Freidel
1990 *A Forest of Kings: The Untold Story of the Ancient Maya*. William Morrow, New York.
Shapiro, J. S.
2005 *A Space Syntax Analysis of Arroyo Hondo Pueblo, New Mexico: Community Formation in the Northern Rio Grande*. School of American Research Press, Santa Fe, New Mexico.
Silverstein, Michael
1976 Shifters, Linguistic Categories, and Cultural Description. In *Meaning in Anthropology*, edited by K. Basso and H. Selby, pp. 11–56. School of American Research Advanced Seminar Series. University of New Mexico Press, Albuquerque.
Smith, Adam T.
2003 *The Political Landscape: Constellations of Authority in Early Complex Polities*. University of California Press, Berkeley.
Smith, Michael E.
2003 Can We Read Cosmology in Ancient Maya City Plans? Comment on Ashmore and Sabloff. *Latin American Antiquity* 14(2):221–228.
2007 Form and Meaning in the Earliest Cities: A New Approach to Ancient Urban Planning. *Journal of Planning History* 6(1):3–47.
Stuardo, Rodrigo Liendo
2003 Access Patterns in Maya Royal Precincts: In *Maya Palaces and Elite Residences: An Interdisciplinary Approach*, edited by J. Christie, pp. 184–203. University of Texas Press, Austin.
Tourtellot, Gair, Gloria Everson, and Norman Hammond
2003 Suburban Organization: Minor Centers at La Milpa, Belize. In *Perspectives on Ancient Maya Rural Complexity*, edited by G. Iannone and S. Connell, pp. 93–107. Monograph 49. Cotsen Institute of Archaeology, UCLA, Los Angeles.
Tourtellot, Gair, Jason Gonzalez, and Francisco Estrada Belli
1999 Land and People at La Milpa, Belize. Paper presented at the 64th Annual Meeting of the Society for American Archaeology, Chicago.
Vogt, Evon Z.
1969 *Zinacantan: A Maya Community in the Highlands of Chiapas*. Belknap Press of Harvard University Press, Cambridge, Massachusetts.
Webster, David L.
1998 Classic Maya Architecture: Implications and Comparisons. In *Function and Meaning in Classic Maya Architecture*, edited by S. Houston, pp. 5–41. Dumbarton Oaks, Washington, D.C.
Webster, David, AnnCorinne Freter, and Nancy Gonlin
2000 *Copán: The Rise and Fall of an Ancient Maya Kingdom*. Harcourt College Publishers, Fort Worth, Texas.
Webster, David L., and Nancy Gonlin
1988 Household Remains of the Humblest Maya. In *Journal of Field Archaeology* 15(2):169–190.
Willey, Gordon R., and Richard M. Leventhal
1979 Prehistoric Settlement at Copan. In *Maya Archaeology and Ethnohistory*, edited by N. Hammond and G. Willey, pp. 75–102. University of Texas Press, Austin.
Willey, Gordon R., Richard M. Leventhal, and William L. Fash
1978 Maya Settlement in the Copan Valley. *Archaeology* 31:32–43.
Wisdom, Charles
1940 *The Chorti Indians of Guatemala*. University of Chicago Press, Chicago.
Yaeger, Jason, and Marcello Canuto
2000 Introducing an Archaeology of Communities. In *The Archaeology of Communities: A New World Perspective*, edited by M. Canuto and J. Yaeger, pp. 1–15. Routledge, London.
Yermakhanova, Olga
2005 Kingship, Structures and Access Patterns on the Royal Plaza at the Ancient Maya City of Altun Ha, Belize: The Construction of a Maya GIS. Master's thesis, Louisiana State University, Baton Rouge.
Zipf, George K.
1949 *Human Behavior and the Principle of Least Effort: An Introduction to Human Ecology*. Addison-Wesley, Cambridge, Massachusetts.

CHAPTER 8

Cost Catchments

A Least Cost Application for Modeling Hunter-Gatherer Land Use

Sarah L. Surface-Evans

8.1. Introduction

Site-catchment analysis has a long history in archaeological research (Foley 1985; Vita-Finzi and Higgs 1970). It is based on the observation that land use behavior and economy are spatially patterned in relationship to settlements; humans tend to utilize the area immediately surrounding a settlement most frequently. Thus, site-catchment analysis is used to characterize economic resources surrounding agricultural (Hunt 1992), pastoral (Ullah 2011), or hunter-gatherer settlements (Sanger 1996). However, site-catchment analysis is not without problems. Traditional site-catchment analysis assumes that resources are uniform throughout a landscape and does not account for potential restrictions to human movement. In highly variable or geographically constrained landscapes, it is not valid to assume that all areas surrounding a settlement are equally desirable or even equally accessible (Wheatley and Gillings 2002).

Computer-aided modeling techniques, such as least cost analysis (LCA), and the widespread availability of digital datasets now offer a new strategy for characterizing zones of land use surrounding settlements. LCA can overcome the aforementioned limitations of site-catchment studies by introducing additional detail to the modeling universe. Rather than assume that all portions of the landscape are accessible, with LCA we can quantify the area of land that may be accessed from a settlement. For example, environmental factors such as terrain, vegetation, and hydrology can be included in models of land use that employ LCA (see Livingood, Phillips and Leckman, Rademaker et. al, and Rissetto in this volume for additional examples). We can use these variables to understand how the movement of past peoples was influenced by physical characteristics of the landscape.

The research described here is one application of least cost techniques to derive cost-based catchments for Middle to Late Archaic period (ca. 7500–3600 rcy BP) hunter-gatherers (Stafford and Cantin 2005) known as the Shell Mound Archaic and is part of a larger study investigating the spatial characteristics of these sites in the Ohio River valley (Surface-Evans 2009b). Cost catchments can be considered an alternative to simple site catchments for viewing resource accessibility. Sites, resource patches, and other features of the landscape that lie within the cost catchment are more accessible than landscape features outside it. The principal difference between cost catchments and site catchments is that cost catchments take into account the cost of moving through the landscape whereas simple circular site catchments do not. This chapter describes the case study, details the methods used to create cost catchments, and then compares how site catchments and cost catchments differ in terms of characterizing hunter-gatherer land use patterns. The results of this research suggest that the methodology employed here has wide applicability for archaeological landscape studies.

8.2. Case Study: Shell Mound Archaic

Throughout the American midcontinent, intensification of long-distance trade networks, mortuary elaboration, subsistence specialization, and increasing sedentism can be recognized as early as the Middle Archaic period (Crothers 1999; Phillips and Brown 1983; Price and Brown 1985; Sassaman 2008). These trends are particularly notable in some portions of the rich riverine environments of the Midwest and Southeast, where archaeologists find a variety of Archaic period sites and artifacts that they define as the Shell Mound Archaic (SMA) tradition. Given the nature of SMA deposits, archaeologists infer that these sites represent the beginning of a shift in hunter-gatherer behavior toward practices that include subsistence intensification, increased sedentism, exchange, and mortuary elaboration (Marquardt and Watson 2005).

These changes are considered part of a larger, long-term trend toward increasing social and economic complexity in the Eastern Woodlands (Sassaman 2008; Marquardt and Watson 2005). Furthermore, these adaptations appear related to a shift toward a strategy that includes logistical mobility in a collector strategy (see Binford 1980). The underlying assumption is that SMA sites are positioned on the landscape in order to maximize hunter-gatherer access to seasonally abundant and diverse resources (Granger 1976; Janzen 1977). My research examines how SMA hunter-gatherers utilize the landscape as part of their economic and social strategies. This foray into computer-aided modeling and LCA was a deliberate attempt to get at questions of human decision making and how the decisions of the SMA people brought about patterning in their use of the landscape.

My research explicitly tests the model proposed by Janzen (1977) that SMA sites represent a shift from a foraging strategy to a collector strategy of resource procurement in which Archaic hunter-gatherers begin to focus their attention on particularly rich resource niches. If Archaic shell mounds primarily functioned as base camps or social centers where people gather to coordinate their activities, obtain and process a variety of resources, or even monitor and control access to key resource zones (see Binford 1980, 2001; Bettinger 1991; Kelly 1995), then we expect that their landscape contexts should include access to multiple resource zones or points of interest within a daily foraging area and high resource abundance and/or diversity. The catchment study described in this chapter is one method that was employed to characterize the nature of SMA settlement patterns and compare them with these expectations.

The Shell Mound Archaic of the central Ohio River valley is the particular case study for this research (Surface-Evans 2009b). The Falls of the Ohio River region is situated approximately 120 miles upriver from the confluence of the Green and Ohio Rivers (Figure 8.1). The Green River valley is considered the so-called heartland of the SMA cultural tradition (Marquardt and Watson 2005). The Falls of the Ohio region is a geographically defined region that is also archaeologically significant (Granger 1976; Janzen 1971, 1977, 2008). This region extends approximately 50 km from the Falls near present-day Louisville, Kentucky, and includes seven distinct physiological provinces. Consequently, this region is extremely diverse in terms of ecological habitats and resources. Not surprisingly, there is a high concentration of archaeological sites in this portion of the Ohio River valley, with more than 800 sites recorded for the Archaic period alone (Burdin 1998). Of these Archaic sites, a total of 29 are identified as shell mounds (Figure 8.2). The content and temporal range of Falls SMA sites is roughly comparable with the Green River valley. Falls region shell mounds date between 3800 and 5800 rcy BP (Bader and Granger 1989; Bellis 1981; Burdin 2002, 2008; Granger 1988; Janzen 1977; Surface-Evans 2006b), spanning the late Middle Archaic and early Late Archaic periods.

Although shell mound sites in the study region span roughly two thousand years, we currently do not have the data to determine which sites were inhabited contemporaneously in this region. The Ohio Falls SMA sites have not been investigated as heavily as those of the neighboring Green River valley, and many of the shell mounds remain unexcavated. Even fewer Falls SMA sites have been radiocarbon dated. Most sites have simply been assigned temporal ranges based on the minimum and maximum antiquity of diagnostic artifacts recovered during surveys. Given this lack of temporal control, for the purposes of this study no temporal distinction was made in

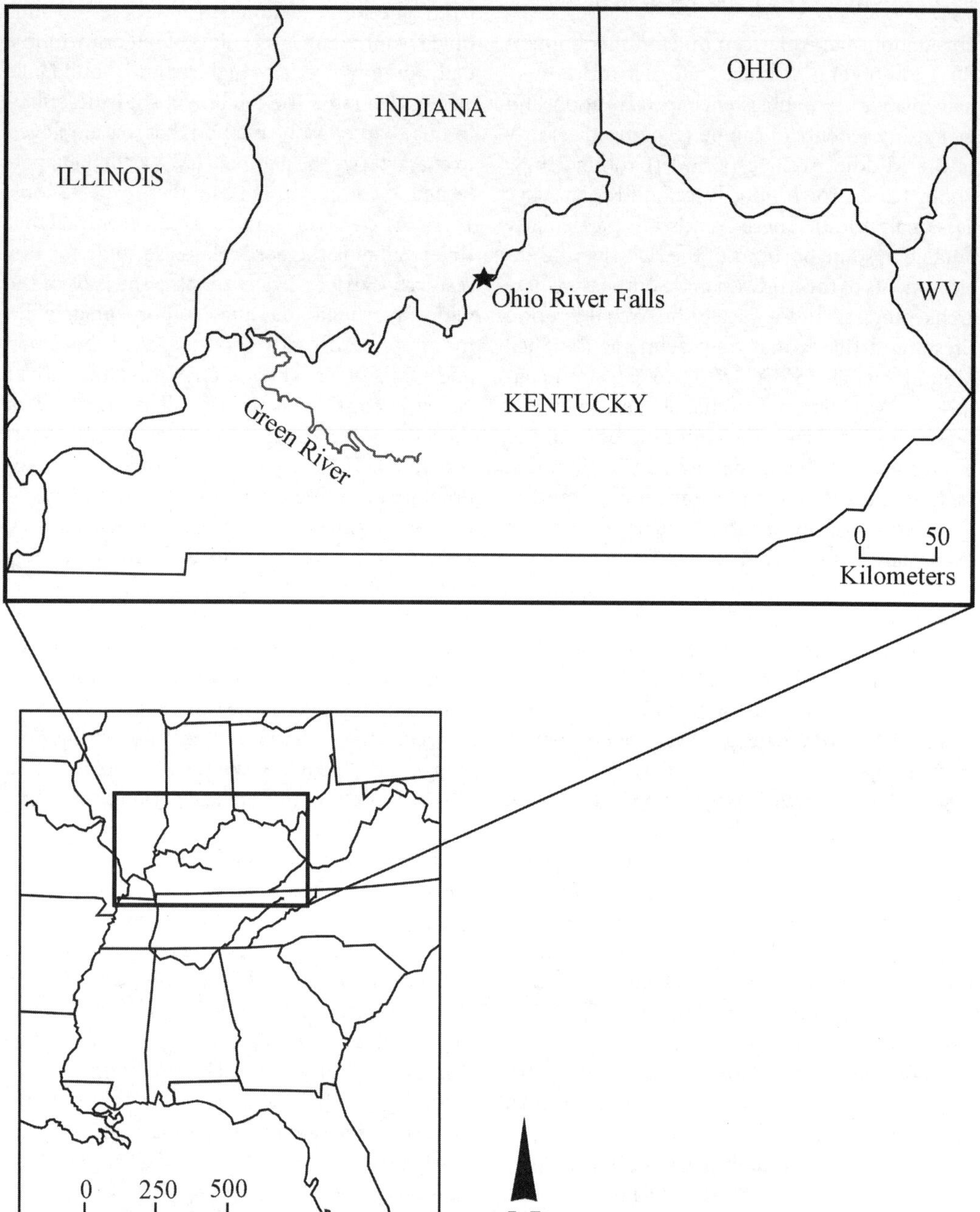

FIGURE 8.1. Map of the midcontinent of the United States showing the proximity of the Ohio Falls and Green River Shell Mound Archaic regions.

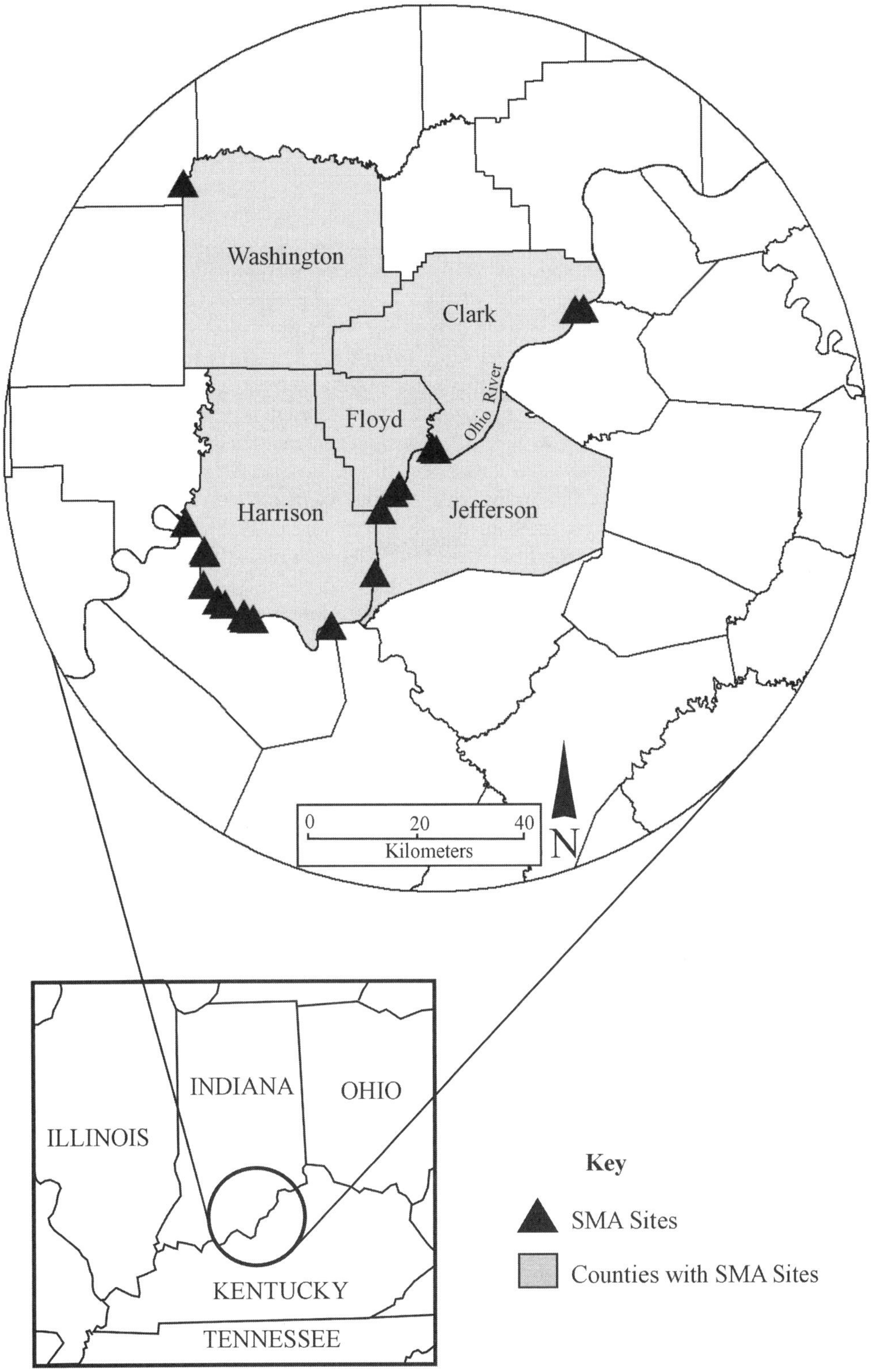

FIGURE 8.2. Map of the Falls Region counties showing the locations of Shell Mound Archaic sites (counties containing SMA sites are shown in gray).

the modeling universe. Given the regional-level scale of the model, the lack of temporal data does not detract from the findings.

Moreover, there is some evidence to suggest contemporaneity between site clusters. Spatial analysis of the scale of SMA site clusters indicates that the sites in this study occur in clusters of 2.5 km along the Ohio River (Surface-Evans 2009b). Figure 8.3 shows two examples from (A) the Maukport bottom and (B) the Middle Creek bottom where SMA sites occur in clusters. These examples are typical of the spatial clustering of sites within flood bottoms throughout the region. The limited temporal data available indicate that while the individual sites in these clusters may not all be contemporary with each other, if taken as a unit of settlement the clusters span the entire duration of the SMA. These small-scale clusters likely represent locales of preferential settlement that were used for generations but where the particular location of settlement in a flood basin varied over time (Surface-Evans 2009b). Therefore, even if the individual sites are not contemporary, important locales were clearly utilized repeatedly throughout the Middle–Late Archaic.

8.3. Research Methods

The focus of my overall research project was to characterize the economic landscape of SMA sites and test whether SMA sites are in fact "positioned" within the landscape to access and control key resources (Surface-Evans 2009b). I tested this expectation using three different measures of accessibility: traditional site catchments, LCP, and a cost-based catchment strategy. The aim of this research was to go beyond reconstructing static environmental contexts to modeling how people may have moved and interacted within their landscape. While numerous spatial studies were conducted to these ends (Surface-Evans 2006a, 2007, 2009a, 2009b), this chapter focuses on the application of LCA to derive cost catchments.

8.3.1. What Is Cost Distance and How Is It Modeled?

In simplest terms, cost-distance models are based on least cost assumptions — that humans will attempt to limit the energy expended while traveling in rugged terrain (see Surface-Evans and White this volume). As others in this book have pointed out, there are many reasons why humans may not always choose to limit costs. However, this does not negate the potential of LCA as a heuristic device for modeling idealized expectations for regional patterns of land use, against which behavioral hypotheses can be tested.

Early applications of LCA to archaeological problems focused on the structure, optimality, and accessibility of existing networks, such as roadways (e.g., Bell and Lock 2000; Kantner 1996; White 2007). Simulations of ancient networks for which there is no existing record are less common in the literature (e.g., Parslow 2006) but are well addressed in this book (see Livingood, Phillips and Leckman, Rademaker et. al, and Rissetto in this volume). LCA applications have generally focused on movement along pathways. Least cost pathways (LCPs) are defined here as a network between sites that is a *single raster cell* wide (Figure 8.4; see also Figure 1.2 this volume). LCP models represent a rather constrained channel of hypothetical movement. While they are useful for developing hypothetical models of movement between known sites, they do little to contextualize the cultural and natural resources that may also be accessible between sites.

Hunter-gatherers often deviate slightly from an idealized path in order to access resources, take advantage of scenic views, or visit economically or socially (or ritually) important locales (e.g., Binford 1980; Kelly 1995; Tacon 1999). While hypothetical LCPs were developed in this research (the results of which are detailed in Surface-Evans 2009b), I was also interested in developing an alternate method that could model a broad area of potential accessibility and travel as an alternative to traditional site-catchment studies based on Euclidean distance. This led me to *cost corridors*, which allow for wandering or deviation from the "ideal" pathway while still incorporating LCA modeling techniques and general assumptions.

Cost corridors are defined here as wide areas of potential travel. One can generate cost corridors by adding together the cost surface of an origin and the cost surface for a destination (Figure 8.5). The value of each cell in the resulting raster reflects the cumulative cost to reach that cell from both origin and destination. This creates a surface that indicates the costs of moving anywhere

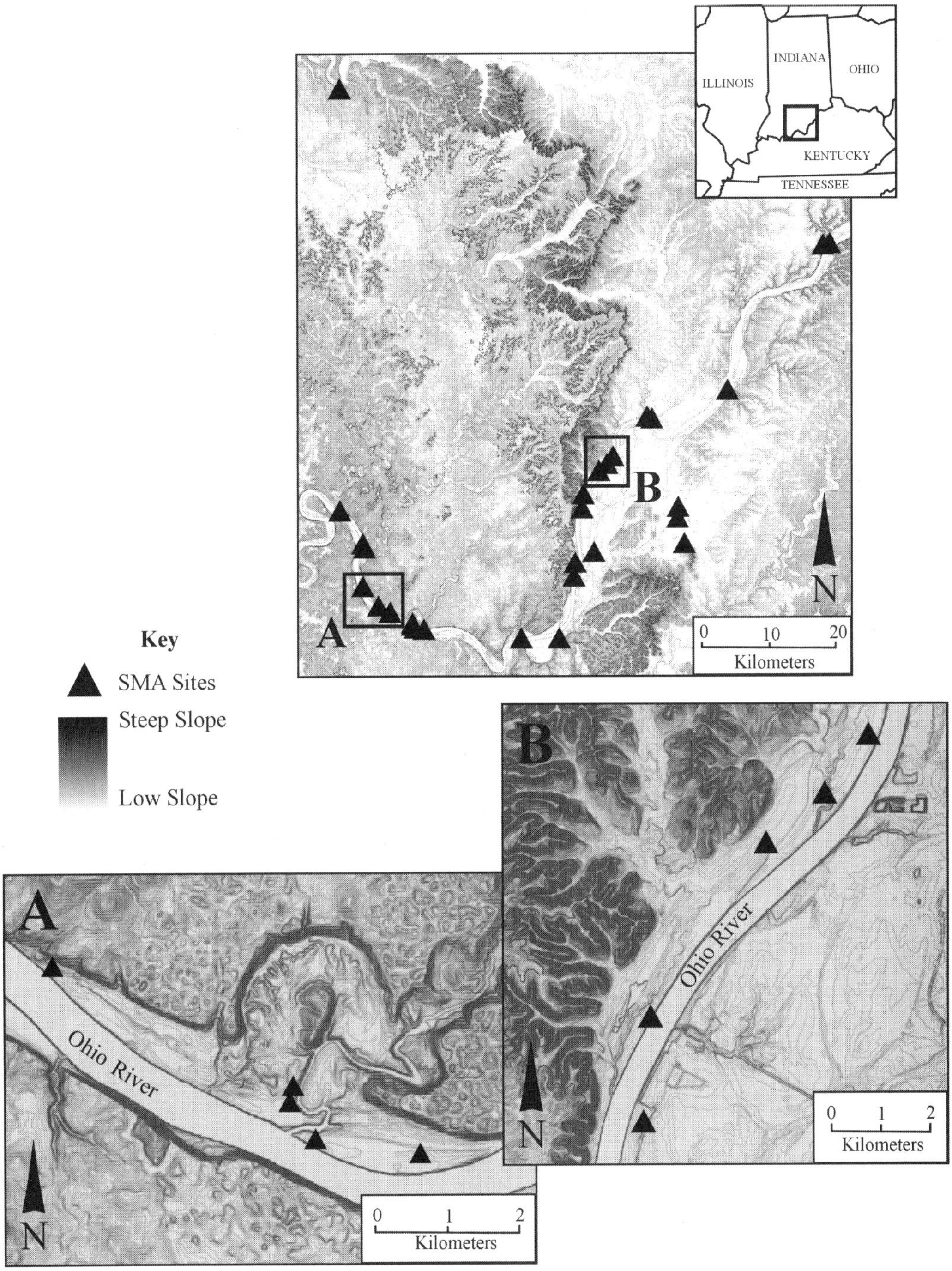

FIGURE 8.3. Examples of spatial clustering of SMA sites within two different flood basins in the study area: (A) Maukport bottom and (B) Middle Creek bottom.

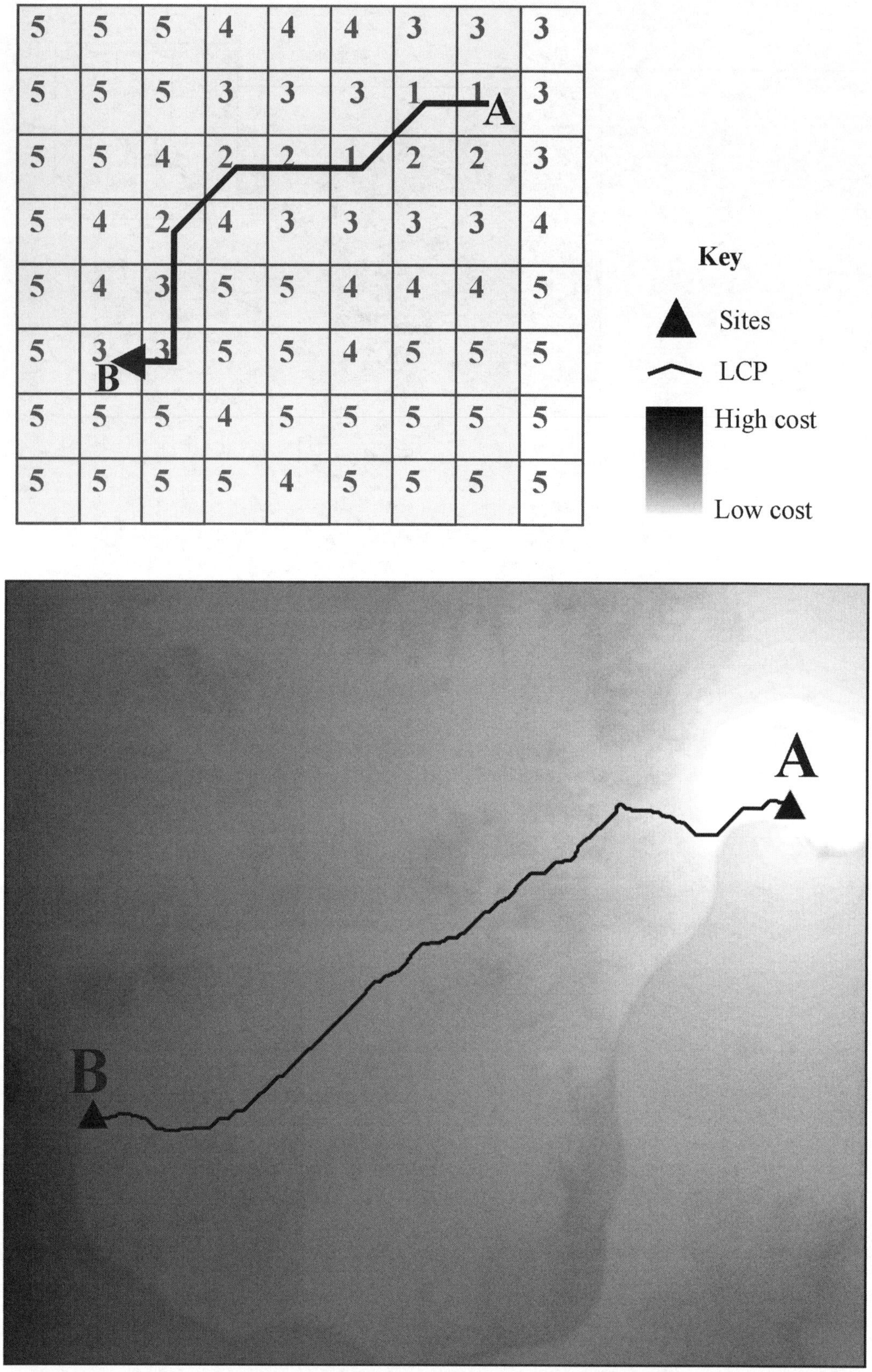

FIGURE 8.4. Example of an idealized LCP and actual LCP between sites A and B, as shown on the cost surface for site A.

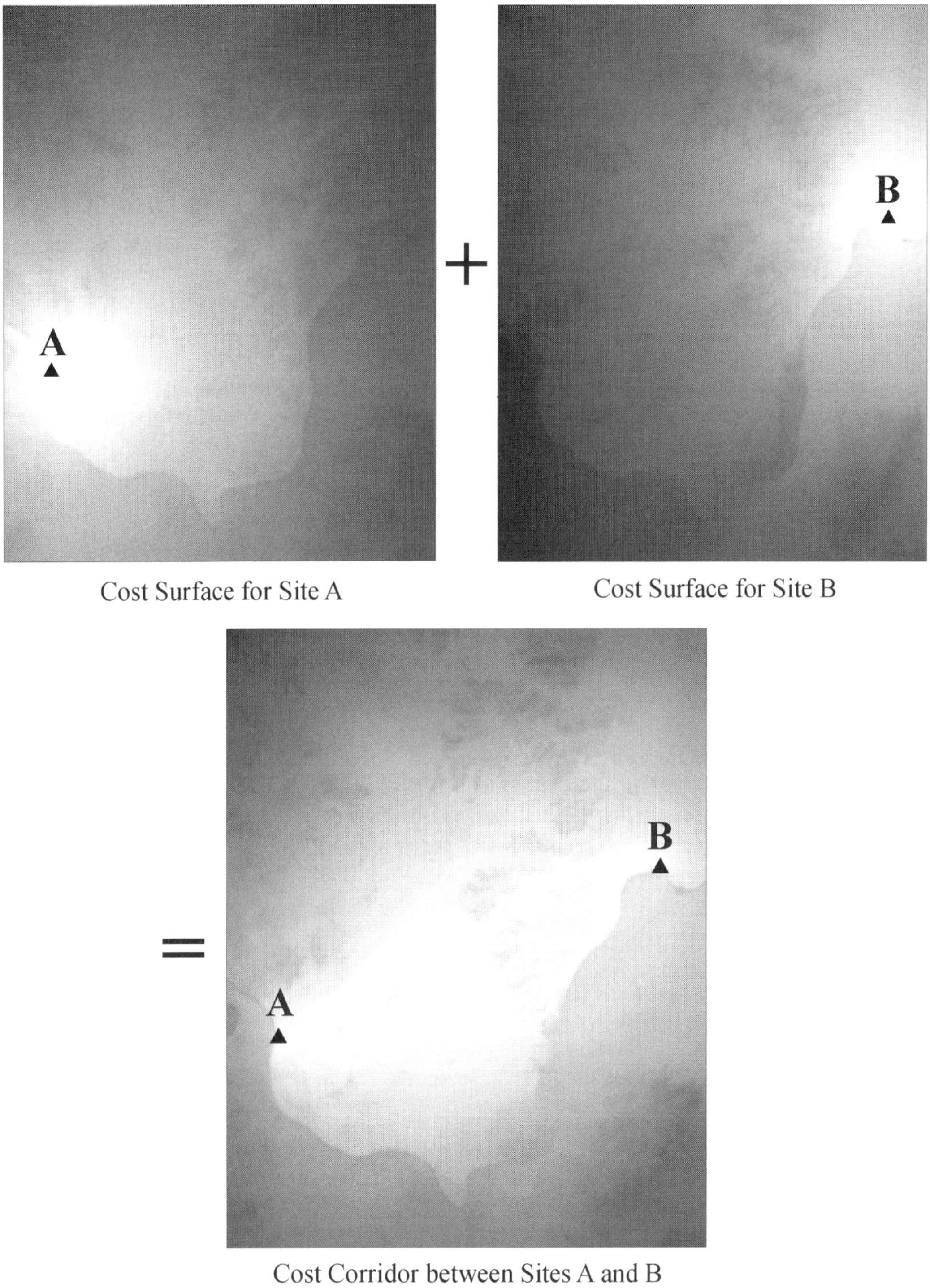

FIGURE 8.5. How the cost corridors are developed from the addition of cost surfaces for site A and site B.

between two (or more) points on the landscape. The resulting cost catchment is a more sophisticated way to view accessibility than either a circular site catchment or even a catchment produced by buffering a LCP because it produces a model of accessibility that is based on the constraints of the landscape. Areas of preferential movement or greater accessibility occur where costs are lower within the corridor. By themselves, cost corridors are not particularly useful for characterizing land use in a quantitative fashion. However, they can be the starting point for developing cost-based site catchments, which allow one to quantify the types and quantities of resources accessible for a site or between sites. Cost catchments are created by selecting a portion of the cost-corridor raster for further study. The specific methodologies employed in this research to develop cost surfaces, cost corridors, and ultimately cost-based catchments are described below.

8.3.2. Data Layers

Modeling hunter-gatherer land use during the Shell Mound Archaic requires the evaluation of the natural, as well as social, features of the landscape. The variables of site location, vis-à-vis other sites and natural features, are therefore of paramount importance. These variables allow us to model land use and accessibility, the contexts of SMA sites and their surroundings, and the ease with which people can move through a given terrain. Therefore, this research utilizes both archaeological site data and environmental data to construct the modeling universe.

I created the archaeological data layer used in this study after compiling spatial data for previously recorded SMA sites in the study area. This process required considerable data mining of reports and site records held at state offices in Indiana and Kentucky. A total of 29 previously documented SMA sites were identified in the Ohio Falls region and were included in this study. As a cautionary note, researchers utilizing previously recorded archaeological datasets should realize that the spatial data for site locations may not be accurate or even consistent (Surface-Evans 2005, 2009b). For example, the location of SMA sites was reported in latitude/longitude and Universal Transverse Mercator (UTM) Datums NAD83 and NAD27. I converted all these locations to the same projection and spatial coordinate system before analysis. Another way to offset this problem is to ground-truth site locations in the field with a GPS unit, which was necessary in some circumstances for this project (Surface-Evans 2005).

The environmental data selected for this research were in large part dictated by what data layers were available for the states of Indiana and Kentucky, since the study area includes portions of both states. At the time that this research was conducted, both states were developing digital repositories of environmental data and few datasets were available for the entire study area. Environmental data layers utilized for this portion of the study include:

National-level datasets:

- Terrain (10-m digital elevation models)
- Hydrology

State-level datasets:

- Karst spring features (as a proxy for wetland resources)
- Bedrock geology (as a proxy for chert outcrops)
- Physiographic zones (as a proxy for environmental diversity)

Topography and hydrology layers form the basis of the cost surface used to model potential constraints on human movement. Topography plays an important role in accessibility in the Falls region because it has fairly significant local relief for the American midcontinent. The eastern portion of this region is glaciated and has rolling to undulating topography in the uplands, dissected by numerous ravines formed by the downcutting of tributary streams (Campbell 1997; Gray 1997, 2000). The western portion of the region is a driftless landscape with karstic features such as springs, sinkholes, caves, and underground rivers (Gray 2000; Homoya and Huffman 1997). In both the glacial and karst portions of the region, the lowlands of the Ohio River are difficult to access from the uplands because of the steep bluffs overlooking the river. A large natural escarpment known as the "Knobs" also separates the glacial and karst landscapes and is considerably difficult to traverse on foot. Each of these features of the terrain can be characterized in terms of slope.

The Ohio River is a substantial barrier to travel

between north and south. In the central Ohio valley, the river is 1 to 2 miles wide. The only "natural" location for a portage occurs at the Falls itself, which is the only location in the course of the Ohio valley where bedrock is exposed across the entire riverbed (Powell 1999). The modern hydrology of the central Ohio valley is largely controlled by an extensive system of locks and dams, which has submerged five-sixths of the former area of rapids (Powell 1999). Before damming and canalization, the Falls was composed of a large expanse of bedrock interrupted by several chutes and rapids. When the first Europeans came to the Falls, they were astonished by the series of rapids that stretched across the width of the river (Powell 1999). The most notable characteristic of the pre-canalized Ohio River was seasonally shallow water conditions. The exposed bedrock during periods of seasonal low water may have allowed for easier access between the north and south banks of the river for short periods each year. Therefore, it is not surprising that several large shell middens are found adjacent to the Falls.

The other three environmental variables examined in this study include the locations of karst springs, bedrock geology, and physiography of the region. These variables were used to characterize the variety and types of resources accessible from SMA sites. Karst springs are found throughout the region (Campbell 1997; Homoya and Huffman 1997). These springs create natural wetlands in both upland and lowland settings and attract a variety of mammals and waterfowl, as well as numerous reptiles and amphibians. Consequently, the locations of springs on the landscape can be used as a proxy for access to wetland resources, which were heavily exploited by the SMA peoples (Dye 1996; Granger 1976; Styles 1986).

The bedrock geology of the Falls region is extremely significant, as particular formations are known to include highly sought-after Wyandotte (also called Indiana Hornstone) and Muldraugh cherts (Cantin 1994). Both cherts have excellent knapping characteristics and were used by native peoples since the Paleoindian period. Several significant Archaic period mines and workshops have been documented in the Falls region (e.g., Mocas and Smith 1995; Stafford and Cantin 1996). Because of the formations in which these and other chert sources have been documented (Cantin 1994), bedrock outcrops are a proxy for access to chert resources.

The Falls region has exceptional environmental diversity for a region of its size (Granger 1976; Gray 2000; Janzen 1977, 2008). One of the ways this diversity can be characterized is by physiography. There are seven different physiographic zones in the Falls region (Campbell 1997; Gray 2000; Homoya and Huffman 1997). The locations of SMA sites with respect to these zones and where they converge can be proxy for the variety of environmental niches and resources accessible by SMA peoples (Granger 1976; Janzen 1977).

8.3.3. Creating Cost Surfaces and Least Cost Corridors

Cost distance is calculated using the variables that constrain or place costs on movement. Terrain is probably the most commonly used variable for estimating cost distance (see Bell and Lock 2000; Gorenflo and Bell 1991; Tobler 1979, 1993). To calculate LCPs and cost corridors using terrain, I converted elevation data (in the form of digital elevation models, or DEMs) mathematically to slope. I then used the slope raster to create a cost surface, sometimes called a "friction" raster. The cost surface is the basis for modeling hypothetical LCPs and corridors between various destinations—in this case, shell mound sites. A conscious decision was made to use terrain and the presence of large hydrological features (such as the Ohio River) to create the cost surface for this study and to omit vegetation as a variable, because detailed reconstructions of plant communities are not currently available for the Middle to Late Archaic period in the region (Surface-Evans 2006b).

In this research, I developed two different models of movement based on the ability or inability to travel on the Ohio River: river-based and land-based. First, I created cost surfaces where river travel was permitted. Figure 8.6A illustrates an example of the river-based cost surface and resulting pathways from site 12-Cl-1 to all other SMA sites in the study area. In this model, the river is the preferential path of travel, given the comparatively high costs of moving into the uplands from the river floodplain. The river-based model assumes that prehistoric

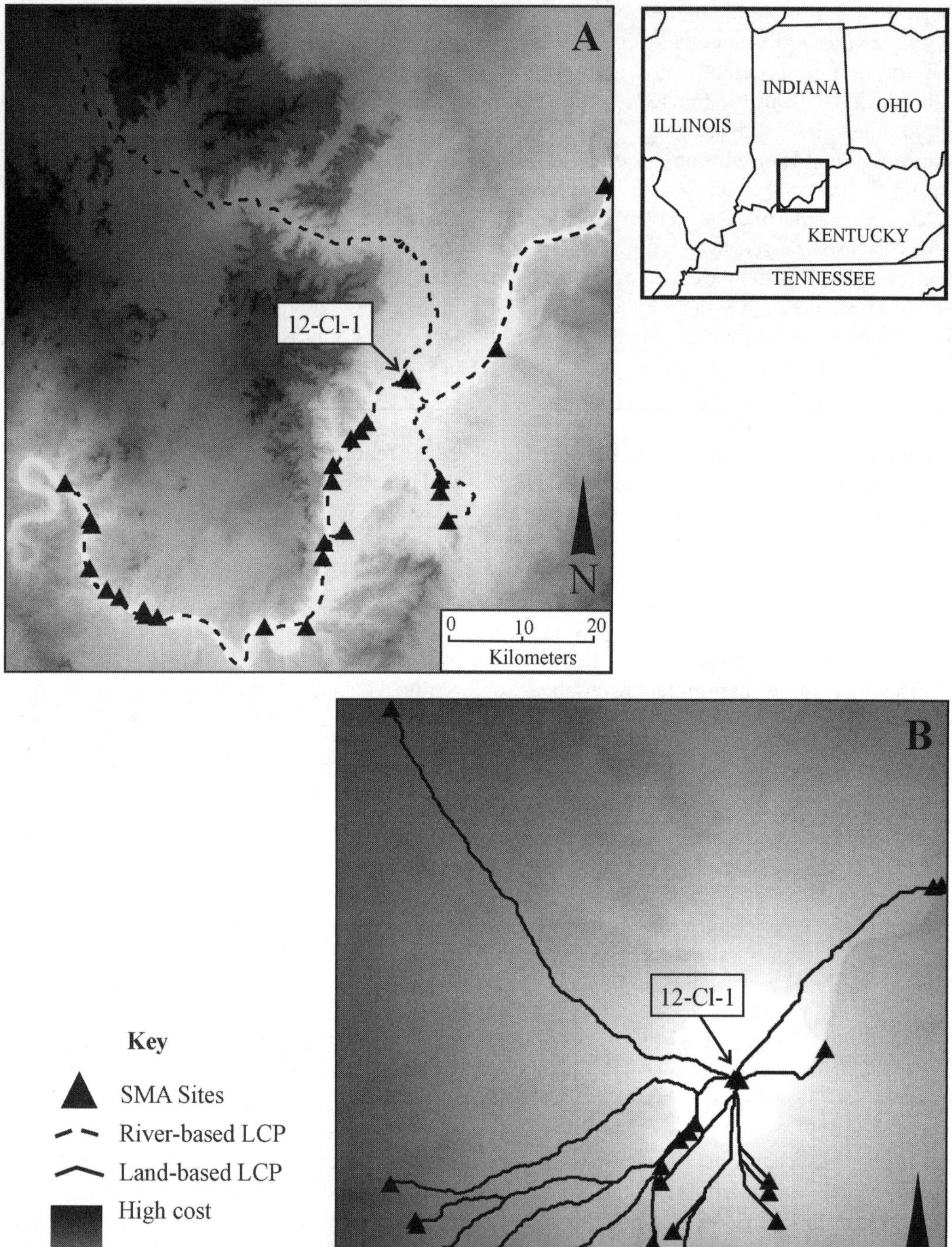

FIGURE 8.6. (A) River-based and (B) land-based LCPs and cost corridors developed for site 12-Cl-1.

peoples had access to canoes. Livingood (this volume) also considers riverine travel and presents a more sophisticated approach that includes factors such as river flow rates and paddling speed, which were not included in my model. While the Shell Mound Archaic peoples certainly traveled in canoes on the Ohio River, it was important also to consider travel exclusively by foot because this was most likely the primary method of travel within the region. I developed land-based cost surfaces by "weighting" the Ohio River, thereby making it extremely costly to follow the river and forcing movement into the uplands. Figure 8.6B is once again site 12-Cl-1 but with land-based pathways and cost surface—note how movement between sites avoids the Ohio River. Tributary streams were not excluded from the walking models because most of the streams in the study area can be forded on foot, particularly during the drier months of late summer and fall.

A series of hypothetical pathways and cost corridors were generated for each site in the study area, using both the river-based and land-based cost surfaces (see Surface-Evans 2009b for full discussion of the results of this analysis). The land-based models ultimately proved to be the most useful, particularly for the cost-catchment portion of this study. This was due to the fact that the river and bottomlands were so low cost compared with the uplands that the river was *always* the preferential mode of travel, particularly for SMA sites located immediately adjacent to it. Consequently, all the cost catchments in this study were developed from land-based cost surfaces, such as the one illustrated in Figure 8.6B.

As the studies by Rademaker et al. and Kantner in this volume illustrate, there are different algorithms for calculating cost and creating cost surfaces. In my research, cost surfaces used to model cost corridors and cost catchments were created via two methods. In one method, I used the built-in Cost Weighted Distance algorithm in ArcGIS Spatial Analyst. This algorithm simply considers cost in terms of slope. The Cost Weighted Distance function uses an origin point (shell mound site) and a slope surface to create two layers: cost direction and cost distance. The cost-distance layer contains cells with the cumulative cost to the closest source cell. Alternatively, a cost surface was also generated using the hiking function developed by Tobler (1993). The hiking function has been applied in numerous archaeological contexts (e.g., Gorenflo and Bell 1991; Kantner 1996; and Phillips and Leckman, Livingood, White, and Kantner in this volume). The hiking function converts slope to walking velocity. Both methods are fairly simplistic compared with energetic models such as the ones considered by Rademaker et al. and White in this volume, but they are reasonably straightforward for the GIS novice.

Cost corridors were created from both the built-in and hiking cost surfaces. The Corridor function in the Distance tools in the Spatial Analyst toolbox of ArcGIS essentially adds the cost surfaces of two different sites to create a combined cost surface. The output raster shows the total sum of all costs between two or more locations. In the case of the cost catchments, this was the cost from an origin to all other SMA sites in the study area, rather than a single destination. The resulting cost corridor calculates the cost of moving in any direction to other SMA sites in the entire region, rather than a single pathway.

8.3.4. Creating Cost Catchments

As stated previously, a cost catchment is created when a portion of a cost corridor is selected. This step converts cost corridors from a continuous data layer or raster into a polygon. Figure 8.7 shows cost catchments created for site 12-Cl-1 from the (A) built-in algorithm and (B) hiking function. As a polygon, with a discrete boundary and area, cost corridors can be examined statistically and it is possible quantify the cultural and natural features accessible within that portion of the cost corridor. However, before creating a cost catchment, one must consider *how* the resulting polygon is created.

With traditional site-catchment studies, the diameter of the catchment is specified based on some attributes of human behavior. For example, ethnographic research of contemporary hunter-gatherers suggests that a daily foraging radius of 5 km is a reasonable distance to walk round-trip (Kelly 1995). I used a foraging radius of 5 km to develop circular site catchments for each SMA site in the study area, as a control against which cost-based catchments could be tested. In the simple 5-km site catchment of land use and resource accessibility, Euclidean distance is the unit

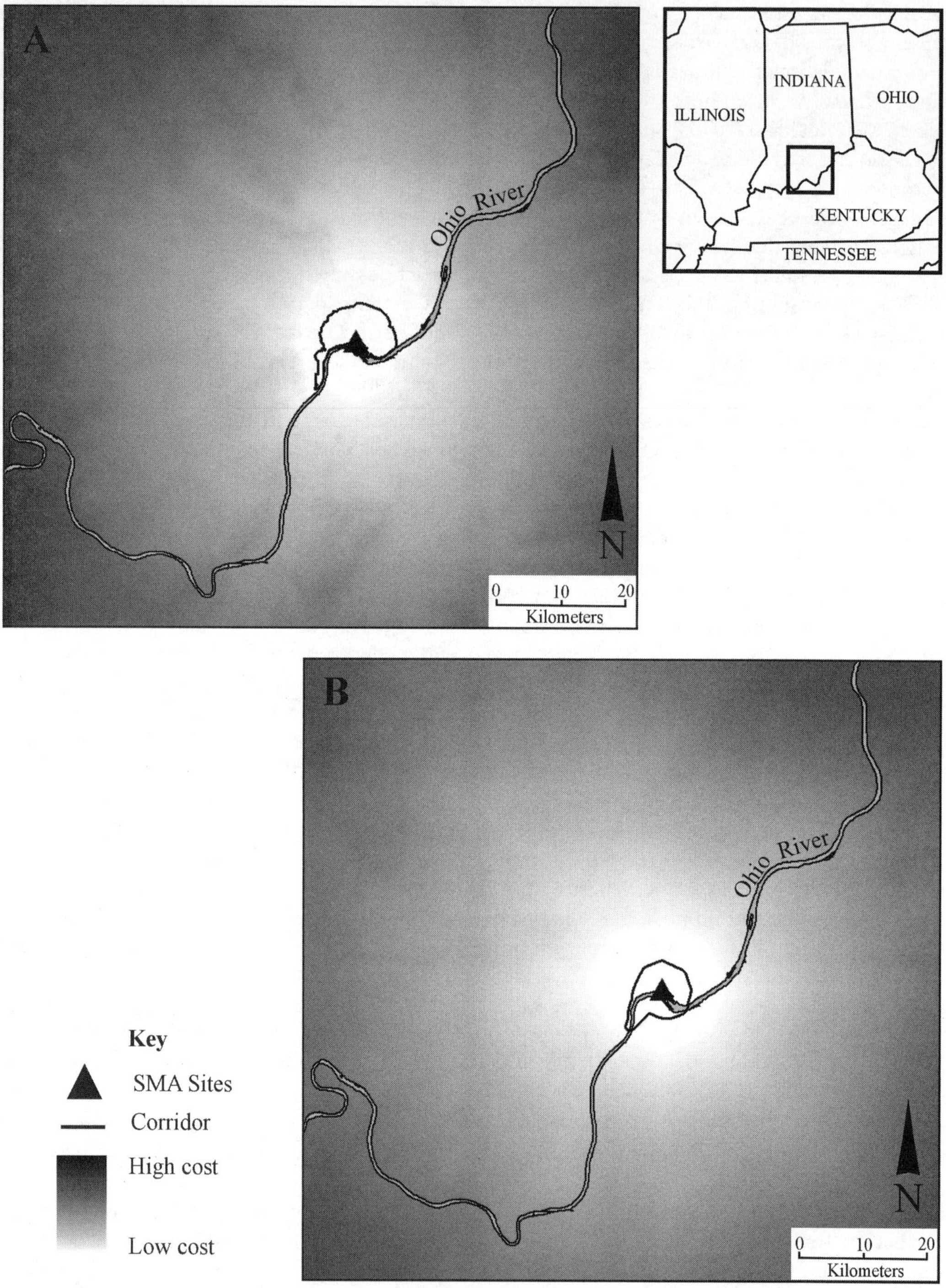

FIGURE 8.7. Examples of cost catchments created for site 12-Cl-1 from (A) the built-in algorithm and (B) the hiking function.

of cost and there is no consideration of the difficulty of terrain or other possible factors of the landscape. The use of cost-based catchments can overcome this limitation, but how does one determine the boundaries or threshold of the cost catchment?

Specifying the unit of cost that defines a cost catchment depends on (1) the measure of cost established by the cost surface or friction raster and (2) the selection of an appropriate cost value based in ethnographic or ethnoarchaeological data. In this study, cost was slope for the cost corridors derived from the built-in algorithm, and walking velocity for the cost corridors derived from the hiking function. Unfortunately, ethnographic and ethnoarchaeological studies have not quantified land use or movement based on either of these measures but instead tend to focus on linear distance (as above) or energetic costs (Kelly 1995; see also Rademaker et al. and White in this volume). Because the 5-km site catchments and cost catchments were not directly comparable, I focused on creating a method for delineating cost catchment that would result in a similar mean *area* between the methods. The rationale for approximating the same area was to allow for statistical comparisons between traditional site catchments and cost catchments.

The lowest 10 percent of cost was selected from each cost-corridor raster to produce a 10 percent threshold polygon: this was the basis for the cost catchments. The 10 percent cost raster was calculated using the Raster Calculator in the ArcGIS Spatial Analyst extension. The new cost raster was then converted to a feature layer using the Raster to Polygon function in the Conversion Tools of ArcGIS. Once converted to a polygon feature, the area of the corridor polygon can be calculated with the Calculate Areas tool in the Spatial Statistics Tools, Utilities. This procedure was repeated for both the built-in and hiking models for each site in the study area.

Why the lowest 10 percent? The 10 percent value was selected through exploratory data analysis. I tested several different values before settling on the value of 10 percent, which most closely resembled the 5-km site catchments in terms of area (Table 8.1). For consistency, the 10 percent threshold was used for both cost-catchment methods. However, the lowest 10 percent more closely approximates the 5-km site-catchment area in the built-in cost catchments than in the hiking cost catchments. There are several reasons why the catchment area produced by these two cost-based models would differ. The built-in algorithm produces a catchment model that provides a cost estimate in terms of slope. Tobler's hiking algorithm, however, produces a catchment model that is a measure of time, or walking velocity. Therefore, the lowest 10 percent of the hiking cost catchment does not measure the same type of cost as the lowest 10 percent of the built-in cost catchment, and consequently it is not surprising that the areas differ. In addition, it is worth noting that the hiking algorithm requires that the slope surface used to create the cost surface is in *percent slope rather than degrees slope*. This methodological difference between the models is also responsible for producing different outcomes for the cost-catchment models. It should be noted that the 10 percent threshold value was a compromise in an effort to find a value that could be statistically compared between catchment methods.

TABLE 8.1. Comparison of Mean Catchment Areas.

5-km Site Catchments	Built-in Cost Catchments	Hiking Cost Catchments
79 km^2	71 km^2	107 km^2

8.4. Research Results

The intent of my research is to evaluate the position of shell mound sites in the Ohio Falls landscape. Specifically, I investigated the types and quantity of resource zones accessible from SMA sites to determine whether they conform to expectations for residential base camps (i.e., access to multiple resource zones within a daily foraging radius). Resource accessibility was modeled through (1) traditional site-catchment analysis and (2) cost catchments derived from LCA. A catchment radius of 5 km was selected because it approximates a hunter-gatherer's daily foraging radius (Kelly 1995). Additional catchment radii of 1 km and 25 km were also examined in this study but are presented elsewhere (Surface-Evans 2006a, 2007, 2009a, 2009b). I compared the 5-km site catchments with the cost catchments derived from LCA to examine how these models of land use differ from each other.

8.4.1. Site Catchments vs. Cost Catchments

While the 5-km site catchments (Figure 8.8A) and cost catchments (Figure 8.8B) were roughly similar in terms of area (see Table 8.1), the shapes of the catchments differ considerably. The boundaries of both versions of the cost-corridor models roughly follow the constraints of the landscape, such as the Ohio River and the edges of the bluffs, as shown in Figure 8.8B. Although the Ohio River was weighted the equivalent of the steepest slope in the project area to prevent a model that would allow people to "walk on water," this technique was not entirely successful. Both cost catchments permit some access to the opposite side of the Ohio River—particularly hiking cost catchments. Perhaps future models should use a weighted value for the river that is exponentially higher than the greatest slope value so it is an effective barrier to foot traffic. This problem demonstrates how computer-aided modeling is an iterative process that requires frequent adjustments to improve the model in order to make it a better representation of human behavior.

To compare the resources accessible in the 5-km site catchments and the cost catchments, I counted and tabulated variables such as access to physiographic zones, bedrock geology, karst springs, and other SMA sites for each catchment model for all the SMA sites (Table 8.2). The data in Table 8.2 were compared statistically using a T-test to establish whether these methods differed significantly in their characterization of resource accessibility. The results of the T-test are shown in Table 8.3. The following hypotheses for the 5-km site catchments and the cost catchments were evaluated:

H_1 = There is a difference in the mean number of physiographic provinces (geological units, springs, or other SMA sites) within the 5-km site catchments and the built-in cost catchments.

H_2 = There is a difference in the mean number of physiographic provinces (geological units, springs, or other SMA sites) within the 5-km site catchments and the hiking cost catchments.

H_0 = There is no difference in the mean number of physiographic provinces (geological units, springs, or other SMA sites) within the cost catchments and 5-km catchment areas.

8.4.1.1. *Accessibility to SMA Sites*

Traditional site catchments and cost catchments differ significantly in accessibility to other SMA sites (see Figure 8.8). The T-test determined that there is a statistically significant difference in the number of shell mound sites in the both the built-in and hiking cost catchments vs. the site catchments (results in boldface in Table 8.3). The built-in cost catchments consistently contain more SMA sites on average than the 5-km catchments, even though the 5-km site catchments cover an additional area of 8,000 m^2, on average. The hiking catchments, with a slightly larger average area, produced an even greater accessibility to SMA sites.

The greater accessibility to other SMA sites in the cost-catchment models is perhaps due to the fact that SMA sites are not randomly distributed on the landscape. SMA sites are both clustered and preferentially located within lowland settings. Cost catchments model greater accessibility to lowland floodplains, where most SMA sites are located, compared with the 5-km site catchments. In other words, SMA sites are highly accessible to one another in terms of terrain, and the cost catchments capture this relationship. These results demonstrate that simple site catchments based on Euclidean distance do not accurately summarize all aspects of a settlement landscape, particularly with regard to the relationship between sites. Cost catchments provide a more nuanced picture of accessibility to aspects of a social landscape—in this case, other shell mound sites.

8.4.1.2. *Accessibility to Environmental Variables*

The environmental resources accessible within all the various catchments demonstrate that SMA peoples clearly had a preference for access to aquatic resources. This result is not unexpected given the nature of these sites. The largest multicomponent shell mound sites have access to highly productive aquatic settings and to high levels of resource diversity. However, there was

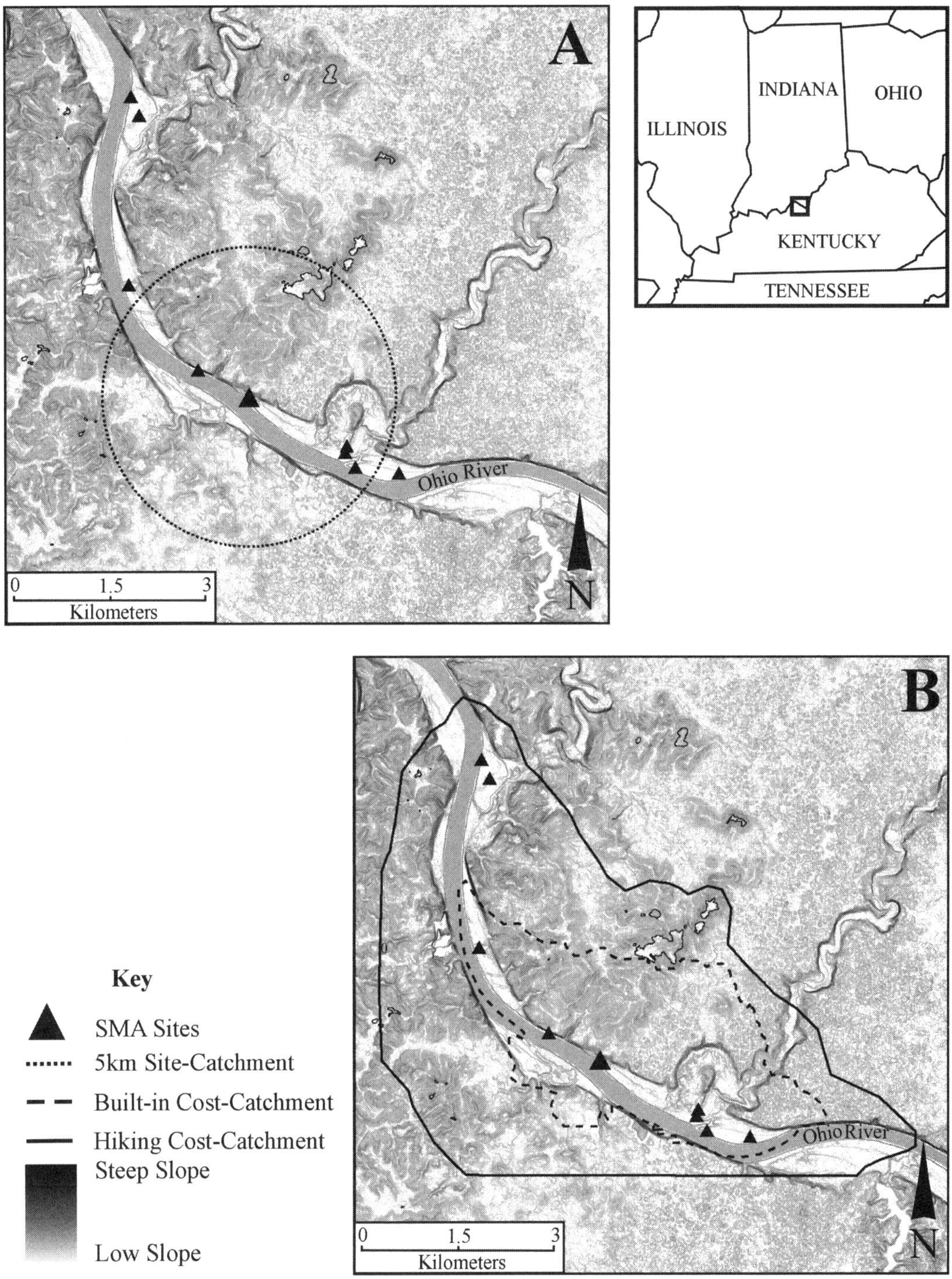

FIGURE 8.8. Comparison of (A) 5-km site catchment and (B) built-in and hiking cost catchments for site 12-Hr-96, with the locations of other SMA sites shown.

TABLE 8.2. Counts of Resources within Catchments.

	5-km Site Catchments				Built-in Cost Catchments				Hiking Cost Catchments			
Site No.	Physiography	Bedrock Geology	Karst Springs	SMA Sites	Physiography	Bedrock Geology	Karst Springs	SMA Sites	Physiography	Bedrock Geology	Karst Springs	SMA Sites
12-Cl-1	3	4	0	2	3	4	0	2	2	4	0	2
12-Cl-10	2	3	0	0	2	3	0	1	2	3	1	1
12-Cl-3	2	4	0	2	3	4	0	1	2	4	0	2
12-Cl-78	2	3	0	0	2	3	0	1	2	3	1	1
12-Fl-1	4	2	0	3	4	2	0	4	4	2	0	4
12-Fl-13	3	2	0	2	3	3	0	5	4	3	0	6
12-Fl-46	3	2	0	2	4	3	0	4	4	3	0	6
12-Fl-73	4	2	0	2	4	2	0	4	4	2	1	7
12-Hr-103	1	1	8	2	3	1	1	1	2	1	19	5
12-Hr-11	3	1	0	2	1	1	7	3	4	1	1	5
12-Hr-113	3	1	0	1	3	1	0	2	4	2	1	5
12-Hr-115	1	1	8	2	3	1	0	3	2	1	21	5
12-Hr-12	1	1	8	0	1	1	8	3	1	1	14	3
12-Hr-13	2	1	7	3	1	1	5	0	2	1	13	6
12-Hr-3	2	1	8	6	2	1	6	5	2	1	14	6
12-Hr-36	2	1	7	6	2	1	4	6	2	1	13	6
12-Hr-4	2	1	8	6	2	1	5	6	2	1	14	6
12-Hr-5	2	1	7	3	2	1	3	6	2	1	16	9
12-Hr-6	2	1	4	4	2	1	9	8	2	1	18	9
12-Hr-9	2	1	4	4	2	1	8	7	2	1	17	8
12-Hr-96	2	1	6	0	2	1	6	6	2	1	7	0
12-Ws-30	2	3	1	2	2	1	9	0	2	3	1	2
15-Jf-10	3	2	0	2	2	3	1	2	4	2	1	4
15-Jf-217	4	2	1	1	3	2	0	4	4	1	2	7
15-Jf-237	2	3	1	2	4	3	0	6	2	3	1	2
15-Jf-267	2	3	0	1	2	3	2	2	2	3	1	2
15-Jf-36	3	1	0	0	2	3	2	2	3	1	0	3
15-Jf-60	2	3	0	0	3	1	0	3	2	4	0	0
15-Jf-92	2	1	3	0	1	2	1	0	3	1	1	2
Total	68	53	81	60	70	55	77	97	75	56	178	124
Mean	2.3	1.8	2.8	2.1	2.4	1.9	2.7	3.3	2.6	1.9	6.1	4.3

TABLE 8.3. T-tests of Resource Accessibility for 5-km Site Catchments, Built-in Cost Catchments, and Hiking Cost Catchments.

		t	df	Sig (2-tailed)
Pair 1	Physiography: 5-km catchment & built-in cost corridor	-1.684	28	0.103
Pair 2	Geology: 5-km catchment & built-in cost catchment	-1	28	0.326
Pair 3	Springs: 5-km catchment & built-in cost catchment	0.394	28	0.696
Pair 4	SMA sites: 5-km catchment & built-in cost catchment	-5.192	28	**0**
Pair 5	Physiography: Tobler vs. 5-km catchment	-2.544	28	**0.017**
Pair 6	Geology: Tobler vs. 5-km catchment	-1.361	28	0.184
Pair 7	Springs: Tobler vs. 5-km catchment	-3.876	28	**0.001**
Pair 8	SMA sites: Tobler vs. 5-km catchment	-5.975	28	**0**

Note: Results in boldface indicate statistically significant differences.

not one specific type of locale in which shell mounding occurred. Rather, the largest shell mound sites are found in three settings: (1) on Ohio River lowlands near the intersection of multiple physiographic zones, (2) inland near karst springs and wetlands, and (3) on Ohio River lowlands near backwater lakes or tributary confluences with marshes. Each of these landscape contexts is distinct, yet all of them fit the expectation that SMA peoples are locating their base camps to access multiple, abundant resources.

Surprisingly, the dissimilar shape of standard 5-km site catchments and the built-in cost catchments produced no statistically significant difference in the environmental resources (physiography, bedrock types, or karst springs) accessible from SMA sites (see Table 8.3). However, according to the T-test results, there is a significant difference in how cost catchments derived from the hiking function characterize accessibility to diverse physiographic zones and karst springs compared with the 5-km site catchments. This significant difference is due, in part, to the fact that the hiking cost catchments cover a greater average area. While hiking cost catchments are larger, if we consider that the area selected in the hiking cost catchment is only the lowest 10 percent of the costs to walk in this terrain, then it seems reasonable that this cost catchment could accurately represent a minimum daily foraging area. This becomes more meaningful if we consider what the lowest 10 percent of cost is in terms of travel time. Because the total costs differ for each site due to local factors of terrain, the lowest 10 percent is not exactly the same for each site, but the average travel cost is 4.8 km/hr. This value seems quite realistic to expect given that Archaic peoples were accessing resources within this distance and time-frame on a daily basis.

The hiking cost catchments are less than 30 km^2 larger than the 5-km site catchments, so I find it surprising that the slightly larger area is the sole factor that makes springs and physiographic provinces more abundant in the hiking cost catchment. Alternatively, the greater access to springs and different physiographic zones in the hiking cost catchments may refer to the spatial patterning of these resources. Cost catchments preferentially select lowland settings and certain environmental features that generally cluster within the lowlands. If Archaic hunter-gatherers were in fact choosing site locations with the goal of increasing access to these environmental features, as proposed by Janzen (1977), then hiking cost catchments may be a good method for capturing the expected spatial relationship between sites and the resources offered by these environmental proxies.

In other words, the location of springs and characterization of physiography may be good proxy indicators for environmental diversity and resource abundance in this region. In addition, the method of hiking cost catchments seems to capture a culturally significant relationship between SMA sites and these resources. The analysis also shows that there is little correlation between bedrock exposure and site location. This result might indicate that the variable of bedrock geology is not a good proxy for access to lithic sources. Alternatively, it may be that lithic raw material was not a significant contributing factor affecting site placement for Archaic peoples in this area.

8.4.2. Built-in Catchments vs. Hiking Catchments

In addition to comparing the cost catchments with traditional site catchments, it was important to consider how the two cost-catchment models differed from each other and to consider whether one model provided any advantages over the other. Another T-test was conducted to compare how the built-in and hiking methods characterized accessibility and to evaluate the following hypotheses:

H_1 = There is a difference in the mean number of physiographic provinces (geological units, springs, or other SMA sites) within the built-in cost catchments and hiking cost catchments.

H_0 = There is no difference in the mean number of physiographic provinces (geological units, springs, or other SMA sites) within the built-in cost catchments and hiking cost catchments.

The results of the T-tests are shown in Table 8.4. There was no significant difference in the number of bedrock outcrops and physiographic zones that were accessible in each cost-catchment

TABLE 8.4. T-tests of Resource Accessibility for Built-in Cost Catchments and Hiking Cost Catchments.

		t	df	Sig (2-tailed)
Pair 1	Physiography: Built-in cost catchments vs. hiking cost catchments	-1.722	28	0.096
Pair 2	Geology: Built-in cost catchments vs. hiking cost catchments	-0.328	28	0.745
Pair 3	Springs: Built-in cost catchments vs. hiking cost catchments	-3.854	28	**0.001**
Pair 4	SMA sites: Built-in cost catchments vs. hiking cost catchments	-4.7	28	**0**

Note: Results in boldface indicate statistically significant differences.

model. However, there was a significant difference in the number of other SMA sites (see Figure 8.8) and karst springs (Figure 8.9) accessible in the hiking cost catchments and built-in cost catchments. This difference is due, at least in part, to the fact that the hiking cost catchments have a larger average catchment area.

However, the differences in the shape of built-in and hiking cost catchments may also be responsible for the significant difference in the number of springs and SMA sites within these two different cost catchments. The shape of the cost-catchment boundaries is determined by the constraints of the terrain and how the algorithm responsible for determining cost weighs these constraints. The boundaries of the hiking cost catchments are larger and appear to be less affected by terrain than the built-in method (see Figures 8.8 and 8.9). While Rademaker et al. (this volume) note that the built-in algorithm produced a model that overexaggerated the effects of terrain, I found that the "exaggeration" may be desirable for this study. In some ways, Tobler's hiking function excessively minimizes the impact of landscape travel barriers. The defined boundaries of the hiking cost catchments do not appear to adequately weigh the cost of traveling across particular landscape features. For example, Figure 8.8 shows that the predicted corridor crosses the Ohio River and traverses steep bluffs that separate the uplands and lowlands, a boundary that seems unrealistic for modeling walking transit costs. In comparison, the built-in cost catchment (also in Figure 8.8) is much more constrained by the topography of the landscape and seems more intuitively reasonable.

These observations suggest that the methods of calculating hiking cost catchments may not accurately weight the costs of traveling in rugged terrain. The pathways, corridors, and cost catchments derived from the hiking algorithm trivialize topographic constraints more than the built-in models. I have hiked much of the Falls region myself, and I cannot stress enough the difficulty of the terrain and the fact that Ohio River flood basins are separated from the uplands by bluffs more than 200 feet high in many places. Even today, there are few roads into the Ohio River lowlands from the uplands. While these bluffs are not impassable, they are significant impediments to foot traffic. It seems highly unlikely that Archaic foragers would trek up the bluffs on a daily basis. Therefore, the built-in algorithm results in smaller corridors and models reduced accessibility; however, the built-in models appear to *fit* the landscape constraints of the Falls region better than the cost-catchment models developed from Tobler's hiking function.

These results underscore the fact that there is not a "one size fits all" algorithm for LCA. Researchers should consider conducting test analysis with several different algorithms, much like exploratory data analysis in statistics, before settling on a method of analysis. Factors that may influence which algorithm is most productive include the type of LCA (whether pathways or cost catchments), the scale of analysis (local or regional), the nature of the regional topography, the research question, and the type of culture being studied.

8.5. Future Directions

The research described here was an initial step in the development of a cost-based method for examining resource accessibility. The cost-based catchment models were successful at delineating catchment areas that considered the geographic restrictions of the study area. However, there are several ways in which the study could be improved. First, as noted above, the weighted

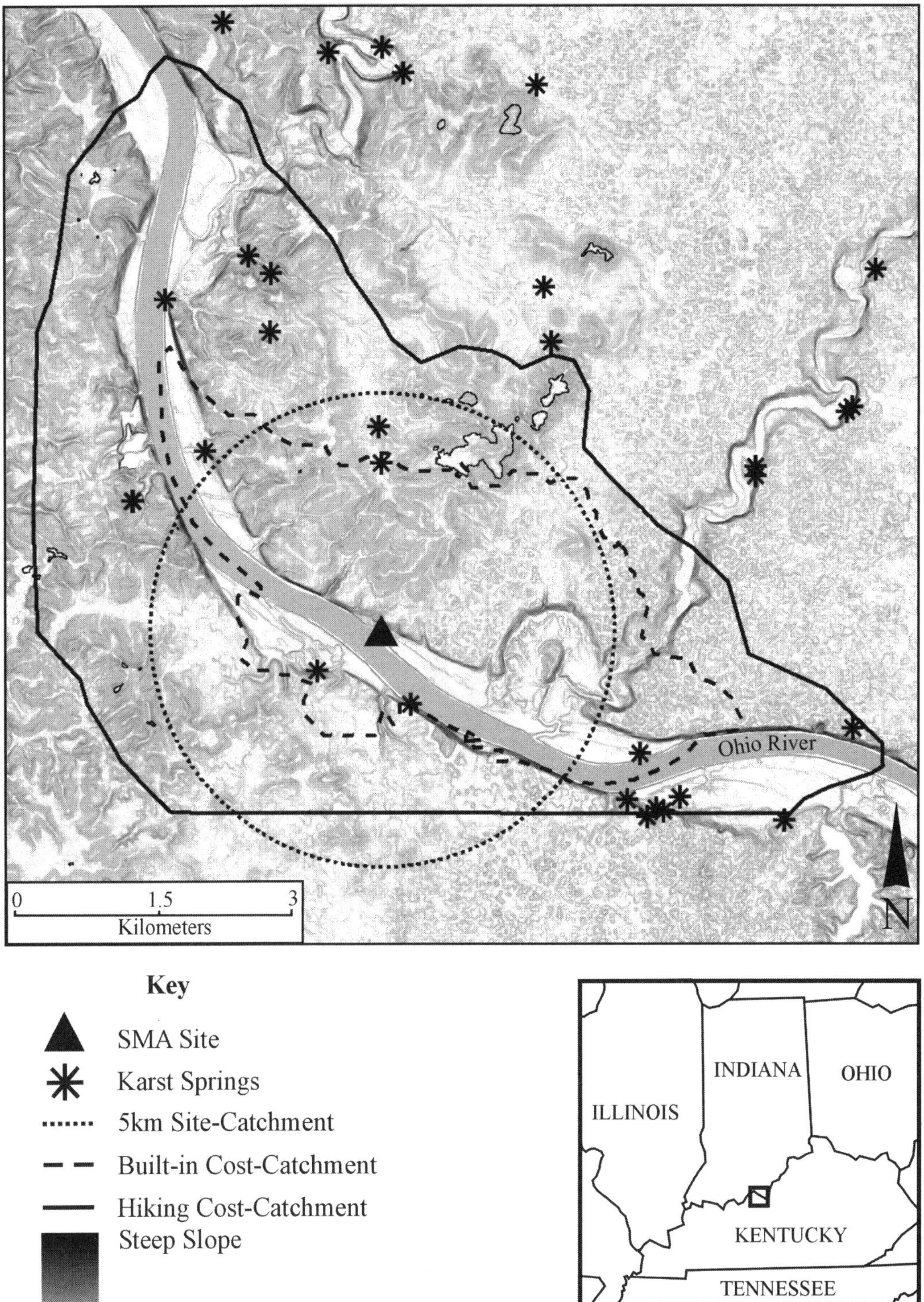

FIGURE 8.9. Comparison of all three catchment models for site 12-Hr-96 with the locations of karst springs shown.

value for the Ohio River should be adjusted to reflect its probable role as a barrier to foot traffic for Archaic travelers. The current hiking cost catchments used here were not sensitive enough to factor in the costs of traversing the river and the nearby bluffs. Revised hiking cost catchments could provide better insights into the relationship between SMA sites and landscape resources.

Second, as Rademaker and colleagues (this volume) and also White (this volume) demonstrate, using an energetic algorithm might provide a more insightful way to examine site accessibility than slope or travel time. Cost catchments developed from caloric or other energetic measures could provide meaningful and quantifiable results that have some basis in ethnographic literature. Energetic algorithms could be particularly useful if coupled with more detailed paleoenvironmental data (Surface-Evans 2006b) that could allow an estimate of return rates for particular catchments (e.g., Kelly 1995).

Third, owing to the geographic and environmental data available at the time of this study, only four resource-related variables were modeled. In addition, several of these variables were rather simplistic proxies. For example, the variable of bedrock geology may not have been a reasonable proxy for access to lithic resources. As additional environmental datasets are published and made available for the study region, this problem should be easily rectified. In addition, through a program of pollen sampling at or near SMA sites throughout the region, it may be possible to refine computer models with information concerning the mosaic of vegetation that once existed in the Falls region (Surface-Evans 2006b). Currently, there are no detailed paleoenvironmental reconstructions for this region that could be used for this type of modeling applications.

Fourth, it would also be interesting to generate cost catchments for different scales of resource procurement, such as maximum foraging areas. While cost catchments appear to be useful for daily foraging areas, the question remains whether cost-based catchments provide any measurable benefit over traditional Euclidean catchments when dealing with larger-scale catchments.

Lastly, for the results to be entirely meaningful, the patterns of accessibility observed for the SMA sites in the study area should be tested against a randomly generated set of hypothetical sites. In other words, are resources actually any more accessible or diverse surrounding SMA sites than in other locations on the landscape? Based on my firsthand knowledge of the study region, I suspect that SMA locales are significantly different from randomly chosen locations, but this is an intuitive conclusion, not statistically demonstrated. The resources accessible within cost catchments for a set of random sites should be compared with the SMA sites in future studies.

8.6. Conclusion

In this chapter, I presented an alternative cost-based method for developing site catchments. This method of LCA proceeded from my previous research (Surface-Evans 2009b), which examined the spatial configuration of Shell Mound Archaic sites in the Ohio Falls region to assess whether these sites fit the expectations for residential base camps in a collector mode of subsistence and settlement. Two different cost algorithms were compared: one derived from a built-in algorithm in ArcGIS and one derived from Tobler's hiking function. How resource accessibility was characterized by both cost-catchment methods was tested against traditional circular site catchments with a radius of 5 km. Each of these catchment methods provided models of resource accessibility that were designed to approximate a daily foraging area.

Comparison of the catchments developed from these three methods determined that traditional site catchments may be adequate for reconstructing accessibility to static, and perhaps random, environmental features of a landscape such as bedrock outcrops. Cost catchments appear better suited for characterizing access to variables that occur in discrete areas (such as karst springs and SMA sites) or that are clustered (such as SMA sites). Cost catchments were particularly sensitive to identifying relationships to resources within lowland settings, which were of particular interest to SMA peoples. The resulting catchments clearly indicate that the SMA peoples were choosing locations on the landscape that have high accessibility to riverine and other aquatic resources. This in itself is not unexpected or surprising but lends support to the view that SMA sites represent a shift from highly mobile foragers to more sedentary collectors who "map onto" particularly rich zones (Binford 1980; Crothers 1999; Janzen 1977). The accessibility characterized in all three catchment methods indicates a relationship between SMA sites and access to highly diverse resources. Indeed, sites with the longest and most intensive occupations are found in areas with the greatest degree of resource diversity (Surface-Evans 2009b).

The methodology developed here has wide-ranging applications in archaeological research. Cost catchments can be modeled in any regional-scale study where the archaeological record can be assumed to be fairly representative. By creating catchments that are based on the "costs" of the landscape, rather than arbitrary distances, archaeologists can begin to approximate past conditions more accurately. The more closely a model resembles the ancient landscape, the more insight it can offer about past human behavior.

Acknowledgments

I am grateful to George Crothers of the Kentucky Office of State Archaeologist and Rick Jones of the Indiana Division of Historic Preservation and Archaeology for allowing me extensive access to the state site files from which I compiled the archaeological data used in this study. I thank Ashton Shortridge for his countless hours of technical advice and for being a steadfast sounding board for my ideas. I also express gratitude to Lynne Goldstein and Alison Rautman for all their valuable editorial comments.

References

Bader, A. T., and Joseph E. Granger
1989 *Recent Archaeological Investigations on the Kentucky Air National Guard Site (15Jf267), Jefferson County, Kentucky*. Granger Consultants, Louisville, Kentucky.

Bailey, T. C., and Anthony C. Gatrell
1995 *Interactive Spatial Data Analysis*. Prentice Hall, Essex, U.K.

Bell, T., and G. Lock
2000 Topographic and Cultural Influences on Walking the Ridgeway in Later Prehistoric Times. In *Beyond the Map: Archaeology and Spatial Technologies*, edited by G. Lock, pp. 85–100. NATO Science Series A: Life Sciences, vol. 321. IOS Press, Amsterdam.

Bellis, J. O.
1981 Test Excavations at the Breeden Site 12-Hr-11, in Harrison County, Indiana. *Proceedings of the Indiana Academy of Science* 91:78–92.

Bettinger, R. L.
1991 *Hunter-Gatherers: Archaeological and Evolutionary Theory*. Interdisciplinary Contributions to Archaeology. Plenum Press, New York.

Binford, L. R.
1980 Willow Smoke and Dogs' Tails: Hunter-Gatherer Settlement Systems and Archaeological Site Formation. *American Antiquity* 45(1):4–20.
2001 *Constructing Frames of Reference: An Analytical Method for Archaeological Theory Building Using Hunter-Gatherer and Environmental Data*. University of California Press, Berkeley.

Burdin, S. R.
1998 The Archaeology of the Falls Area of the Ohio River Valley: Paleoindian and Archaic Periods (12,000–2,750 BP). Manuscript on file, Department of Anthropology, Indiana University Southeast, New Albany.
2002 *An Archaeological Assessment of Site 12Fl-73 in Floyd County, Indiana*. University of Kentucky, Department of Anthropology, Lexington.
2008 *Archaeological Investigations of the Overflow Pond Area in Harrison County, Indiana: The 2007 Survey and Excavations at the Breeden (12Hr11) and Overflow Pond (12Hr12) Sites*. University of Kentucky, Department of Anthropology, Lexington.

Campbell, R. K.
1997 Limestone Ledges and "Crawfish Flats": The Bluegrass Natural Region. In *The Natural Heritage of Indiana*, edited by M. T. Jackson, pp. 161–166. Indiana University Press, Bloomington.

Cantin, M.
1994 *Provenience, Description, and Archaeological Use of Selected Cherts in Indiana*. Technical Report. Indiana State University, Terre Haute.

Crothers, G. M.
1999 *Prehistoric Hunters and Gatherers, and the Archaic Period Green River Shell Middens of Western Kentucky*. Ph.D. dissertation, Washington University. University Microfilms International, Ann Arbor, Michigan.

Dye, D. H.
1996 Riverine Adaptation in the Midsouth. In

Of Caves and Shell Mounds, edited by K. C. Carstens and P. J. Watson, pp. 140–158. University of Alabama Press, Tuscaloosa.

Foley, R.
1985 Optimality Theory in Anthropology. *Man* 20: 222–242.

Gorenflo, L. J., and Thomas L. Bell
1991 Network Analysis and the Study of Past Regional Organization. In *Ancient Road Networks and Settlement Hierarchies in the New World*, edited by C. D. Trombold. Cambridge University Press, Cambridge.

Granger, J.
1976 *A Preliminary Regional Research Design for Jefferson County, Kentucky*. Archaeological Survey, University of Louisville, Louisville, Kentucky.
1988 Late/Terminal Archaic Settlement in the Falls of the Ohio River Region of Kentucky: An Examination of Components, Phases, and Clusters. In *Paleoindian and Archaic Research in Kentucky*, edited by C. D. Hockensmith, D. Pollack, and T. N. Sanders, pp. 153–204. Kentucky Heritage Council, Frankfort.

Gray, H. H.
1997 The View from the Window: Physiography. In *The Natural Heritage of Indiana*, edited by M. T. Jackson, pp. 28–44. Indiana University Press, Bloomington.
2000 *Physiographic Divisions of Indiana*. Special Report 61. Indiana Geological Survey, Bloomington.

Homoya, M. A., and H. Huffman
1997 Sinks, Slopes, and Stony Disposition: The Highland Rim Natural Region. In *The Natural Heritage of Indiana*, edited by M. T. Jackson, pp. 167–171. Indiana University Press, Bloomington.

Hunt, E.
1992 Upgrading Site-Catchment Analysis with the Use of GIS: Investigating the Settlement Patterns of Horticulturalists. *World Archaeology* 24(2):283–309.

Janzen, D. E.
1971 Excavations at the Falls of the Ohio River Region. Paper presented at the Kentucky Archaeological Association, Eastern Kentucky University, Richmond.
1977 An Examination of Late Archaic Development in the Falls of the Ohio River Area. In *Anthropological Papers, Museum of Anthropology No. 61*, edited by C. Cleland, pp. 123–143. University of Michigan, Ann Arbor.
2008 *Unearthing the Past: The Archaeology of the Falls of the Ohio River Region*. Butler Books, Louisville, Kentucky.

Kantner, J.
1996 An Evaluation of Chaco Anasazi Roadways. Paper presented at the 61st Annual Meeting of the Society for American Archaeology, New Orleans.

Kelly, R. L.
1995 *The Foraging Spectrum: Diversity in Hunter-Gatherer Lifeways*. Smithsonian Institution Press, Washington, D.C.

Marquardt, W. H., and P. J. Watson (editors)
2005 *Archaeology of the Middle Green River Region, Kentucky*. Institute of Archaeology and Paleoenvironmental Studies, Monograph No. 5. Florida Museum of Natural History, University of Florida, Gainesville.

Mocas, S. T., and Edward E. Smith
1995 *Archaeological Subsurface Investigations at the Swan's Lansing Site (12Hr304), Harrison County, Indiana*. Reports of Investigations 95-36. Glenn A. Black Laboratory of Archaeology, Bloomington, Indiana.

Parslow, C.
2006 Social Interaction in the Prehistoric Natufian: Generating an Interactive Agency Model Using GIS. Unpublished Ph.D. dissertation, University of Toronto, Toronto.

Phillips, J. L., and James A. Brown (editors)
1983 *Archaic Hunters and Gatherers in the American Midwest*. Academic Press, New York.

Powell, R. L.
1999 *Geology of the Falls of the Ohio River*. Circular 10. Indiana Geological Survey, Bloomington.

Price, T. D., and James A. Brown (editors)
1985 *Prehistoric Hunter-Gatherers: The Emergence of Cultural Complexity*. Academic Press, Orlando, Florida.

Sanger, D.
1996 Testing the Models: Hunter-Gatherer Use of Space in the Gulf of Maine, U.S.A. *World Archaeology* 27(3):512–526.

Sassaman, K. E.
2008 The New Archaic, It Ain't What It Used to Be. In *The New Archaic*, edited by K. E. Sassaman. Special issue, *SAA Archaeological Record* 8(5):6–8.

Stafford, C. R., and M. Cantin
1996 *Archaeological Phase I Surface and Subsurface Investigations in the Caesars World Development, Harrison County, Indiana*. Anthropology Laboratory Technical Report 32. Indiana State University, Terre Haute.
2005 Archaic Period Chronology in the Hill Coun-

try of Southern Indiana. Paper presented at the Urbana Conference on the Archaic Societies of the Midcontinent, University of Illinois, Champaign-Urbana.

Styles, B. W.

1986 Aquatic Exploitation in the Lower Illinois River Valley: The Role of Paleoecological Change. In *Foraging, Collecting, and Harvesting: Archaic Period Subsistence and Settlement in the Eastern Woodlands*, edited by S. W. Neusius, pp. 145–174. Center for Archaeological Investigations, Occasional Paper 6. Southern Illinois University, Carbondale.

Surface-Evans, Sarah

2005 Contextualizing the Shell Mound Archaic Adaptation: The Utility of Existing Archaeological Data for Regional Model Building. Paper presented at the 51st Midwest Archaeological Conference, Dayton, Ohio.

2006a Visualizing Social Complexity: The Role of Landscape on Shell Mound Archaic Settlement Patterns in the Central Ohio River Valley. Paper presented at the 71st Annual Meeting of the Society for American Archaeology, San Juan, Puerto Rico.

2006b *Report of Pollen Collection from the Breeden Shell Midden (12-Hr-11) and Overflow Pond, Harrison County, Indiana*. Report of Investigations for Indiana Department of Natural Resources, Division of Historic Preservation and Archaeology, under Permit No. 200561. Division of Historic Preservation and Archaeology, Indianapolis.

2007 "Where Many Paths and Errands Meet": Modeling Archaic Hunter-Gatherer Movement in the Ohio Falls Landscape. Poster presentation for the 72nd Annual Meeting of the Society for American Archaeology, Austin, Texas.

2009a This Way or That Way: Hypothetical Least-Cost Models of Regional Interaction among Shell Mound Archaic Settlements in the Central Ohio Valley. Paper presented at the 74th Annual Meeting of the Society for American Archaeology, Atlanta, Georgia.

2009b Hunter-Gatherer Cultural Landscapes: A Case Study for a GIS-Based Reconstruction of the Shell Mound Archaic in the Falls of the Ohio Region of Indiana and Kentucky. Unpublished Ph.D. dissertation, Michigan State University, East Lansing.

Tacon, P.

1999 Identifying Ancient Sacred Landscapes in Australia: From Physical to Sacred. In *Archaeologies of Landscape: Contemporary Perspectives*, edited by W. Ashmore and A. B. Knapp. Blackwell, Malden, Massachusetts.

Tobler, W.

1979 Smooth Pycnophylactic Interpolation for Geographic Regions. *Journal of American Statistical Association* 74(367):519–536.

1993 Three Presentations on Geographical Analysis and Modeling. In *Technical Report 93-1*. National Center for Geographic Information and Analysis, University of California, Santa Barbara.

Ullah, Isaac I. T.

2011 A GIS Method for Assessing the Zone of Human-Environmental Impact around Archaeological Sites: A Test Case from the Late Neolithic of Wadi Ziqlâb, Jordan. *Journal of Archaeological Science* 38(3):623–632.

Vita-Finzi, C., and E. S. Higgs

1970 Prehistoric Economy in the Mount Carmel Area of Palestine: Site Catchment Analysis. *Proceedings of the Prehistoric Society* 36:1–37.

Wheatley, D., and M. Gillings

2002 *Spatial Technology and Archaeology: The Archaeological Applications of GIS*. Taylor and Francis, New York.

White, D. A.

2007 Transportation, Integration, and Facilitation: Prehistoric Trail Networks of the Western Papaguería. Unpublished Ph.D. dissertation, University of Colorado, Boulder.

PART 3

Custom Applications and Emerging Methods

CHAPTER 9

Modeling the Consequences of Village Site Location

Least Cost Path Modeling in a Coupled GIS and Agent-Based Model of Village Agropastoralism in Eastern Spain

Isaac I. Ullah and Sean M. Bergin

9.1. Introduction

Agent-based modeling (ABM) of social actors or social systems is a new and exciting tool kit in the arsenal of computational archaeology. However, traditional ABM techniques alone are not adequate for those computational archaeologists interested in modeling human-environment interaction. This goal can be accomplished by coupling agent-based modeling of human land use decision making with a dynamic and spatially explicit landscape process model (LPM). Agent movement across the landscape is often only implicitly considered during the construction of hybrid models of social and environmental systems (coupled ABM-LPMs). In fact, movement modeling is a vitally important component of this coupling, and determines the scope of possible locations of specific human interactions with the environment.

Least cost analysis (LCA) and least cost path (LCP) modeling offer a convenient empiric method for achieving movement modeling in a coupled ABM-LPM, and allow for agents to consider the cost of movement across a landscape without requiring them to actually traverse that landscape. This chapter reports on one method of incorporating LCA and LCP modeling into a coupled ABM-LPM of village-based subsistence agropastoralism by the Mediterranean Landscape Dynamics (MedLanD) project at Arizona State University (Barton, Ullah, and Bergin 2010; Barton, Ullah, and Mitasova 2010). While there are a wide variety of interesting experiments that can be undertaken with such a model, here we examine the effects of LCP modeling on agent-environment interaction in our coupled ABM-LPM when villages are located in different physiographic settings.

In the past, village sites were chosen for a variety of reasons. Social or cultural reasons (such as defensibility, viewshed, or seclusion) may outweigh pragmatic or economic ones (such as access to fields or centrality). Thus it is difficult to predict the actual location of a village in an area, even though it may be possible to determine a range of "optimal" village locations for a specific suite of economic and ecologic locational factors. Furthermore, traditional archaeological methods are unable to assess the trade-offs villagers make when they decide to place a village in a location that is less than optimal for the type of subsistence activities they routinely undertake. The incorporation of LCA and LCP modeling in a coupled ABM-LPM, however, provides a method for making this assessment.

This chapter first examines the various ways in which movement modeling has been used in ABM and coupled ABM-LPM, then describes the model developed by the MedLanD project, and finally lays out the experimental case study. The case study models the long-term effects of four contrasting village locations in the Penaguila Valley in eastern Spain (Figure 9.1), using a 5-m spatial resolution DEM for the region and

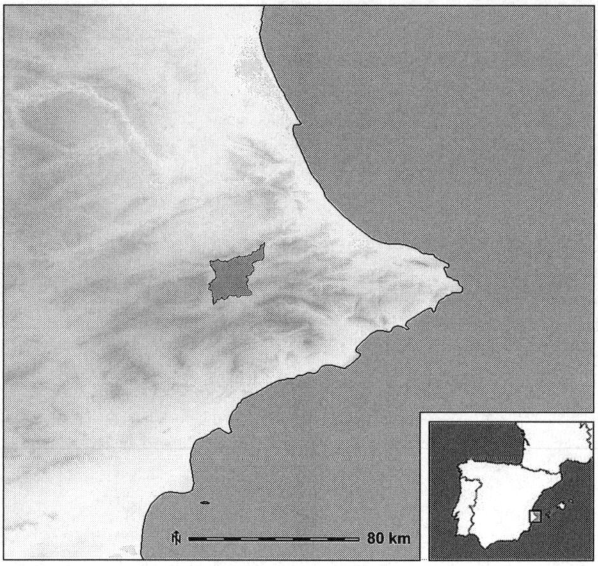

FIGURE 9.1. Location of Penaguila Valley study area in eastern Spain.

modeling the economic, social, and environmental trajectories of the four different village locations over a hundred years of village-based agropastoralism.

9.2. Background

There are several ways to model the movement of agents on a landscape. The implementation and efficacy of different methods depends largely on the way the landscape is digitally represented (i.e., vector or raster data models) and the level of intricacy present in that representation (i.e., the spatial resolution and extent). For example, it is possible to create a schematic representation of a landscape or social network using a digital map of vector lines and nodes (Calafell and Bertranpetit 1993; Lansing et al. 1998). This is an abstract representation of the real landscape, but because travel along vector lines can take place in only two directions, it is relatively simple to route agent movement along the various paths between nodes. Depending on the modeling goals, agents can be programmed to take the most direct path, the path that passes through a specific node or nodes, or the path that is the most convoluted. Regardless of the modeling goals, a vector representation of the landscape limits the possibilities of agent movement to travel only along those vector lines that have been included in the model.

If a more realistic digital representation of a landscape is needed to model agent movement, a raster data model can be used. Because a raster is spatially contiguous, calculating movement paths is more complex than in our vector example. Spatially explicit ABMs often use non-intricate (low spatial resolution) raster landscape representations that are usually unidimensional—most often representing the spatial spread of only a single resource or hazard (Dean et al. 1999; Lake 1999; Axtell et al. 2002). Agent movement in these types of models is usually calculated stepwise on an opportunistic basis. In any given time-step, agents can "see" only the neighboring cells and choose which of these cells to move to using a benefit-maximizing and/or a cost-minimizing decision algorithm. The resultant "path" of agent movement is built one cell at a time and represents the ad hoc history of agent movement choices based on a very limited set of information. This type of spatially explicit ABM is more correctly thought of as a "cellular automata" (CA) movement model, with the agent being represented as a cell on the landscape (i.e., the center cell of a nine-cell-square neighborhood), and it works very well for simple landscapes. However, when the cell resolution and/or map extents are increased, so too is the computational intensity of the CA routine. The complexity of the routine, as well as the calculation time, continues to increase as more layers of landscape data are added. Hence, the incorporation of CA movement modeling into a coupled agent-landscape model becomes less and less practical as the landscape model becomes more and more realistic.

Also, while the ad hoc nature of the CA movement modeling method may accurately reflect agent movement choices under certain conditions, in most real-world cases movement choices will be made using information from beyond the immediately neighboring cells. In reality, agents will have some a priori knowledge of their environment, and thus movement choices may also be shaped by a prior understanding of the "best" way to move from point A to point B. LCP modeling provides an excellent source of such a priori knowledge, especially for raster maps of high spatial resolution. The least cost surfaces produced by LCA represent the travel time along the path of minimum accrued cost from an agent's "home base" (e.g., a village) to all other locations in the surrounding landscape, and thus can serve as a digital representation of an agent's "mental template" of how best to move through the landscape. Framing least cost surfaces in this perspective also addresses Branting's (this volume) second problem with LCP models (i.e., that they represent a "perfect knowledge" of the environment). If the least cost calculations are based on a relatively constant variable (e.g., topography), and the costs are empirically calculated (e.g., the time it takes an average adult human to walk over a given distance of given slope), then these maps can serve as long-term inputs into a raster-based coupled ABM-LPM.

An additional consideration to bear in mind when one is implementing an ABM-LPM is that certain aspects of landscapes are more dynamic than others and thus may change rapidly throughout the time-span of a single model run. For example, landscape attributes such as vegetation cover, local soil depths, and soil fertility, as well as spatially explicit social attributes such as parcel ownership and territorial boundaries, all change on a much faster time-scale than does the general topography of a region. If a coupled ABM-LPM is to incorporate such variables, it must somehow reconcile agents' a priori knowledge of minimum cost routes (i.e., as determined by LCA) with an ability for the agents to find, comprehend, and incorporate the more dynamic landscape variables into their movement decisions. This chapter describes one method of producing a coupled ABM-LPM that calculates agent movement and subsistence decisions by incorporating LCA with these types of dynamic variables.

Researchers are now beginning to integrate GIS capabilities into their agent-based models in order to better incorporate landscape dynamics. However, the integration of two distinct modeling paradigms into a single software tool is not easily implemented. Our modeling approach therefore uses two separate, publicly available modeling platforms—the DEVS-Suite agent-based modeling platform (Kim et al. 2009) and the GRASS GIS platform (GRASS Development Team 2010)—and bridges them via a third, custom-developed "interaction module" (IM) that spans the divide between the two softwares. The DEVS-Suite agent-based Agropastoral Simulation (AP-Sim) model emulates non-irrigated subsistence cereal farming and site-tethered pastoralism, while the GRASS GIS Landscape Dynamics (LandDyn) model simulates surface process dynamics (e.g., erosion and deposition) at a regional spatial extent and at an annual temporal scale using process-based erosion-deposition equations (Warren et al. 2005; Barton, Ullah, and Bergin 2010; Barton, Ullah, and Mitasova 2010), and it simulates simple vegetation succession dynamics based on field data from southern Spain (Bonet and Pausas 2004). The IM acts as both an interpreter and a messenger between the two models and also serves as a manager for the coupled agent/landscape simulations by explicitly modeling the interactions between AP-Sim and LandDyn (Mayer and Sarjoughian 2007; Barton, Ullah, and Bergin 2010; Barton, Ullah, and Mitasova 2010). Using the IM to model interactions within a hybrid model also provides the researcher with the ability to manage a number of disparities between the agent and environment subsystem models that are inherent to this kind of hybrid model, including disparities of timing, structure, scale, and resolution.

9.3. Building a Coupled Agent-Landscape Model

9.3.1. Integrating Agents and Landscapes

The IM is responsible for the interchange between the DEVS-Suite-based AP-Sim and GRASS-based LandDyn. When a household agent requires information about a landscape patch, it sends a coordinate to the IM, which then caches all such information requests from agents to request information about each specific coordinate. The output from those queries is processed and sent to the requesting household agent. After agents have formalized their subsistence plans, the IM then relays these decisions as impacts at specific locations, and executes LandDyn scripts that appropriately modify land cover and soil fertility values, calculate the amount of erosion and deposition, and create new maps to be used as input into the next model cycle.

9.3.2. Modeling Agropastoral Consumption Patterns and Population Dynamics

The agent-based component of the hybrid model (i.e., AP-Sim) is made up of two agent types, villages and households. Villages represent a collection of households and are responsible for sending information to the appropriate households from the interaction model.

Household agents represent a family of agropastoralists acting as a cohesive unit. Household agents farm wheat and barley and raise sheep and goats to acquire food, which they require for survival and growth. The number of people represented by a household strongly influences what it can do. The maximum amount of land that can be planted and subsequently harvested is based on the percentage of each household that is available to do work. The desired amount of farmed land is also based on population, since the number of plots to be farmed by a household in a given year is based on the kilocalorie needs of that household and yield expectations based on the average yield of the previous year. The amount of grazed land is dependent on the number of sheep and goats possessed by a household. Households consume most of the farmed wheat and barley directly and use the rest as supplemental fodder for the goats and sheep. Sheep and goats also need to graze wild vegetation to round out their diet, and they provide households with additional kilocalories derived from meat and milk products. The yearly consumed kilocalorie need of each household is thus translated to a yearly gross farming kilocalorie need as the sum of wheat and barley kilocalories that will be directly consumed with those that will be used as supplemental fodder. The ratio of supplemental fodder kilocalories to consumed meat/milk calories is calculated explicitly based on an ethnoarchaeologically derived model of supplemental foddering in which village-based herd animals also graze natural vegetation (Thomson et al. 1986). The amount of supplemental fodder needed for herd animals in a village-based pastoral system is calculated as the deficit between the amount of kilocalories typically obtained from grazing wild vegetation and the total kilocalorie need of those herd animals.

The consequences of farming and herding decisions for households are incorporated in the model as changes to birth and death rates. Household birth and death probabilities are set by the modeler, and change based on household need/kilocalorie yield ratios. If a household fails to meet its energy requirements, the probability that a household member will die increases and the probability of a new household member being born decreases. Likewise, if a household exceeds its energy needs, death rates decrease and birth rates increase.

9.3.3. Modeling Agent Movement with Least Cost Paths

In the AP-Sim model, the information available to households for their subsistence/movement decisions includes knowledge of the walking costs across the landscape, knowledge of the soil fertility and depth, and knowledge of the land cover. AP-Sim uses GRASS's r.walk LCA module to calculate a map of walking-time costs, using the center of each household's village as the starting point. This map represents each household's a priori knowledge of the time it takes to walk from their village to each cell of the landscape following the path of least effort. Least cost maps produced by r.walk are *anisotropic*, which means that the algorithm takes direction into account when calculating walking time (for a more complete discussion of isotropy and anisotropy in LCA, see Kantner this volume). It takes more effort to walk up a slope than it does to walk down that same slope, so the calculated cost of walking along a path in one direction may be lower or higher than it would be to walk along that same path in the opposite direction. These time costs are calculated using a variation of Naismith's rules (Langmuir 1984), and the least cost routes are determined using Dijkstra's algorithm (Dijkstra 1959; Rees 2004; Surface-Evans and White this volume). The cost approximation algorithms are implemented in our model using values derived for human walking in moderately steep terrain. For a more complete description of how these walking costs are implemented within our modeling environment, see Ullah (2011). Kantner (this volume) describes Naismith's rules in more depth, provides the formula, and details the input variables. More general information about the r.walk module and about cost-distance modeling in GRASS can be found in Neteler and Mitasova (2007). The resultant cost-surface map is in units of seconds of walking time (i.e., the default output units of the Naismith algorithm [Kantner this volume]) for the average human traveling at an average stride to reach a given spot on the landscape from the point of origin.

Although we could have modeled other costs (see Surface-Evans and White, Rademaker et al., and Kantner in this volume for more information about alternative cost measures), we feel that walking time is appropriate here, as minimization of daily travel time is likely to be an important aspect of subsistence movement considerations for small-scale agropastoralists. Also, because the LCA routine is computationally intensive, and because the difference in walking costs from year to year resulting from annual erosion and deposition are extremely minimal, the LCP model is calculated only once, and households continue to use the same cost-distance information throughout the entire model run. This, in effect, adds a degree of "social memory" to the land valuation procedure, in that households' a priori knowledge of walking costs is not always perfectly current, and enables the (very human) potential for agents to make somewhat less than optimal decisions over time due to landscape knowledge that is inherited and integrated without yearly reconfirmation.

9.3.4 Modeling Soil and Vegetation Dynamics

Soil fertility, soil depth, and land cover are dynamic variables and are updated from year to year by the LandDyn model. LandDyn uses a simple soil fertility routine, r.soil.fertility, to keep track of the fertility of farmed plots. The routine accepts a map of agent impacts to soil fertility. Fertility is tracked on a linear scale where 0 is complete infertility and 100 is full fertility. Farming a plot will reduce the fertility in that plot by a value of 2 per model year. If agents relinquish a plot or let it lie fallow (unfarmed), then the soil fertility begins to regenerate by a value of 1 per model year. The depth of the soil is also tracked by LandDyn, but this feature is integrated into the surface processes module r.landscape.evol, which is described in detail below.

The LandDyn vegetation succession model, r.landcover.update, maintains an annual land cover map in which vegetation types are coded as numerical values in a vegetation succession scheme. In this model, the vegetation on a plot with soil at maximum fertility and sufficient depth would grow from bare ground through grass to scrub and eventually to fully matured forest in 50 years if no human impact has occurred (Pardo and Gil 2005). The r.landcover.update module accepts agent impacts from the IM as a map of land cover reduction values and will modify the land cover map by the amounts specified in this impacts map. Initially, farming reduces the land cover value of a plot to that of a vegetation cover of grassland (wheat agriculture) with a large reduction value, and then will subsequently maintain a farmed plot with a reduction value of 0. Animal grazing reduces the land cover value of a grazed plot of land by a value of 1, which represents the effect of grazing a plot at a sustainable density (in the future, unsustainable grazing densities will be allowed). The amount to which vegetation is regenerated each year in any cells that are unmodified by agents in that year is calculated via the average of the quadratic regression functions for soil depth and fertility:

$$R = \frac{\begin{pmatrix} ((-0.000118528 \cdot SF^2) + (0.0215056 \cdot SF) \\ +0.0237987) + ((-0.000118528 \cdot SD^2) \\ +(0.0215056 \cdot SD) + 0.0237987) \end{pmatrix}}{2} \quad (1)$$

In this function, R is the calculated regrowth amount, SD is the current soil depth, and SF is the current soil fertility. This function allows plots of full fertility and sufficiently deep soils to regrow at the optimum 50-year succession rate (regrowth rate of 1), but scales the regrowth rates of plots with low fertility and/or soil depths so that as soil fertility and/or soil depth approaches 0, the vegetation regrowth rate also approaches 0, but does so along a quadratic fall-off curve.

9.3.5. Incorporating Movement Costs and Landscape Variables in Household Subsistence Choices

At the beginning of each simulation year, households must decide on a subsistence plan consisting of the total amount of farming and grazing plots they will need. Households seek to minimize walking costs while maximizing subsistence returns when deciding on the locations of these plots, and do so using a land valuation equation specific to a particular subsistence task (in the current model, this is only either wheat/barley agriculture or pastoralism). Thus in any given year, agent movement is governed by the balance of a unique series of cost minimization and benefit maximization decisions, which provides a dynamic and constantly changing pattern of movement over time.

For farming plots, households want relatively level land that has deep fertile soil but is not too far away from the village. Thus the potential plots are evaluated using the following equation:

$$LandValue = P_{max} \cdot SV \cdot \left(\frac{F \cdot SD}{F_{max} \cdot SD_{max}} \right) - \left(\frac{D \cdot D_w}{D_{max}} \right) \quad (2)$$

Here P_{max} is a constant (500) that represents the maximum amount of energy that can be produced from any one plot, SV is a slope modification value (0°–10°, $SV = 1$; 11°—20°, $SV = 0.75$; 21°—60°, $SV = 0.25$; 60°—90°, $SV = 0$) that makes lower slopes more valuable than higher ones, F is the current soil fertility value, F_{max} is a constant that represents the maximum possible fertility (100), SD is the current soil depth, and SD_{max} is a constant that represents the maximum soil depth needed for full yields (1 m). D is the least cost distance of the current cell from the village, D_w is the least cost distance weight (currently set to 1), and D_{max} is the maximum calculated least cost distance value on the least cost map. This equation selects for deep fertile plots but scales the attractiveness of such plots based on their slope and reduces their value based on their distance from the village along the least cost route.

Once all the households have chosen agricultural plots, they must decide on locations for grazing. For pastoralism, slope and soil attributes are unimportant when it comes to deciding on grazing locations. Households prefer grazing patches that have abundant fodder but are not too far from the village. Thus potential grazing patches are valuated according to the following equation:

$$LandValue = \left(\frac{LC}{LC_{max}} \right) \cdot \left(\frac{1}{D} \right) \quad (3)$$

LC is the land cover of a particular patch, LC_{max} is the land cover value that produces the maximum amount of fodder, and D is the least cost distance of a particular cell from the village. In Mediterranean environments, slightly immature, grassy open oak and pine woodland produces the largest amount of edible vegetation, and fodder production actually decreases as these woodlands mature into denser oak forests (Al-Jaloudy 2006). To account for this, LC_{max} is set to 40, which corresponds with open grassy woodland, and land cover values greater than 40 are rescaled by a Boolean function in descending order (e.g., 41 is changed to 39, 42 to 38, etc.) before being entered into the equation. This equation favors patches of land with high fodder production but scales their attractiveness according to their distance from the village along the route of least accumulated costs. Thus grazing patches that have large amounts of edible vegetation but are very far from the village are less attractive to agents than are grazing patches that have less edible vegetation but are very close to the village.

Once the households valuate their land, they send requests to farm and graze individual plots through the village. AP-Sim does not calculate specific least cost routes from the village to each chosen plot (i.e., the physical "paths" that a farmer would walk along) but rather uses the walking-time cost calculated by r.walk in the raster cell of a particular plot as the cost that the agent must account for if the agent is to use that plot. This concept is similar to Surface-Evans's "cost corridors" (this volume), as it uses the calculated cost surface as a proxy for the mental list of the general travel time from the village to every plot owned by an agent, thus addressing Branting's (this volume) assertion that cost-surface models are best understood as "idealized patterns" of movement in a particular locale. A list of the coordinates of the utilized plots, along with the type of subsistence activity, are passed to the IM, which translates them into maps of agent impacts on land cover and/or soil fertility and then initiates the LandDyn modules.

9.3.6. Modeling Surface Processes

LandDyn, our landscape evolution model, consists of a suite of GRASS scripts that simulate natural landscape processes. Core among these is r.landscape.evol, which is a process-based cellular automata simulation of surface runoff flow, sediment detachment (i.e., erosion), sediment transport, and sediment deposition. It is based on well-known and broadly applied algorithms that calculate sediment flux at points on the landscape given the supply of sediment at that point, the topographic characteristics in the vicinity of that point, and the location of that point in the hydrological system (Braun et al. 2001; Mitasova et al. 2002; Dietrich et al. 2003; Hancock 2004; Willgoose 2005; Tucker and Hancock 2010). R.landscape.evol works by first calculating the sediment transport capacity at each point (raster cell) on the landscape, using the proper process equations for the position of that cell in the drainage. Cutoff breakpoints in the amount of accumulated flow are used to determine where on the digital landscape each equation is to be used. We use two variants of the (inaptly named) Unit Stream Powered Erosion Deposition equation—a three-dimensional modification of the Revised Universal Soil Loss equation—to model sediment transport processes on hillslopes (Warren et al. 2005). We also use a transport-limited process equation based on the reach-average shear stress to model sediment transport processes in streams (Howard 1980; Tucker and Hancock 2010). In order to avoid confusion due to nomenclature, we refer collectively to the hillslope process equations as the hillslope erosion deposition (HED) model, and we refer to the stream process equation as the stream erosion deposition (SED) model.

For areas of low flow accumulation, slope is the determining factor for sediment transport and LandDyn simulates hillslope diffusion (soil creep). In this case, HED takes the form of:

$$T = \kappa \cdot \sin(\beta) \qquad (4)$$

T is the calculated sediment transport capacity at a cell, κ is the diffusion constant for a particular climatic regime (in m/millennium), and β is the topographic slope of the cell in degrees.

For flow farther down on the hillslopes, HED replaces the diffusivity constant with a function of upslope accumulated area and local environmental attributes, and takes the form:

$$T = R \cdot K \cdot C \cdot A^m \cdot \sin(\beta)^n \qquad (5)$$

Once again, T is the calculated sediment transport capacity at a cell. R is the rainfall intensity factor; it is computed by an equation that combines monthly precipitation amounts and is expressed as a map of rainfall intensity for each cell (in tons/ha/mm/hr) (Renard and Freimund 1994; Renard et al. 1997). K is a unit-less soil erosion resistance factor based on the percentage of sand, silt, clay, and organic matter in the soil. This can be represented as a constant throughout the landscape or expressed as a dynamic soil map. K is scaled from 0 (not erodible) to 1 (highly erodible). C is a unit-less vegetation erosion protection factor based on the overall ability of different kinds of vegetation to hinder raindrops and surface flow and to bind soil in place. C, like K, is also scaled from 0 to 1. The ability for r.landscape.evol to accept new C maps at every year is very important, as land cover is what is affected by agent activity. The annual land cover values derived by r.landcover.update from the yearly agent land use impacts are translated into C values based on a linear regression of the known relationship between "classic" land cover types and C values in Mediterranean environments. In our case, these known points are for bare land (land cover value 0, C 0.8), climax grassland (land cover value 5, C 0.4), climax shrub/maquis (land cover value 19, C 0.08), and climax Mediterranean woodland

(land cover value 50, C 0.005). The C value for the transitional vegetation communities between these points is interpolated via linear regression. A is the upslope contributing area (flow accumulation per unit cell resolution), and β is again the topographic slope of the cell. The exponents m and n are empirically derived and vary depending on the process being modeled. For sheetwash (flow accumulation large enough for diffuse laminar flow of water), $m = n = 1$, whereas for rill/gully flow (flow accumulation large enough for concentrated turbulent flow of water), $m = 1.6$ and $n = 1.3$.

On parts of the landscape where water is concentrated into proper streams (flow accumulation values higher than those for hillslopes), LandDyn implements the SED equations. This is done by first calculating the reach-average shear stress (τ), here estimated for a cellular landscape simply as:

$$\tau = g \cdot \sin(\beta) \cdot D \tag{6}$$

In this equation, g is a constant related to the gravitational acceleration of water (9806.65), β is the slope of the cell in degrees, and D is the instantaneous depth of flowing water in the cell. D is here assumed to be roughly equivalent to the depth of flow during the average minute of rainfall, calculated by:

$$D = \frac{(R_e \cdot A)}{R_t} \tag{7}$$

R_e is the total annual rainfall excess (precipitation minus infiltration) in meters, A is the upslope accumulated area in square meters, and R_t is the total time (in minutes) that it rained in a one-year period. The sediment transport capacity T of a stream cell can then be calculated by:

$$T = Kt \cdot (\tau)^n \tag{8}$$

Kt is the transport efficiency factor related to the character of the stream bed (0.001 for normal sediment to 0.000001 for bedrock), and n is an empirically determined exponent related to the dominant type of transport in the stream system (1.5 for bed load transport or 2.5 for suspended load transport).

After the local sediment transport capacity is calculated for each cell of the landscape, LandDyn calculates the amount of erosion or deposition that has occurred in each cell. This is done via the following differential equation:

$$ED = \frac{\delta T \cdot \cos(\alpha)}{\delta x} + \frac{\delta T \cdot \sin(\alpha)}{\delta y} \tag{9}$$

ED is the calculated erosion or deposition rate for sediment, α is the topographic aspect (the direction of slope), and T is the sediment transport capacity as calculated by one of the HED or SED equations. Whether flowing water will erode or deposit sediment in a particular cell is determined by the *change* in sediment flux (transport capacity) from one cell to the next. If the transport capacity increases (e.g., due to an increase in the steepness of the slope or amount of flowing water), more sediment will be entrained and erosion will occur; if the transport capacity decreases (e.g., due to a decrease in slope or water flow), sediment will be deposited.

The soil creep equation (Eq. 4) calculates T (and thus ED) in units of vertical meters of sediment and thus can be used directly to calculate the elevation change per year. The USPED (Eq. 5) and stream flow (Eq. 9) equations, however, calculate T (and thus ED) in units of mass/area (tons/ha for USPED, and kg/m^2 for streamflow), and therefore it must be converted in order to calculate the change in elevation at each cell (m_{vert}). The conversion is calculated via a simple algorithm that uses the density of the soil and the cell resolution:

$$m_{vert} = \frac{(ED \cdot A_c)}{S_d} \tag{10}$$

Here Ac is a conversion constant to translate ED to units of tons/m^2 ($Ac = 0.00001$ for USPED, and $Ac = 1000$ for streams), and Sd is the density of the soil (tons/m^3). These values are used to create a new digital elevation model, and the depth of soil is calculated by subtracting these new elevation data from a map of bedrock elevations. When a cell has been eroded to bedrock, no more erosion can take place and no vegetation can grow.

9.4. Case Study

9.4.1. Study Location, Goals, and Expectations

The Penaguila Valley, located in eastern Spain, is one of the study areas for the MedLanD project (Figure 9.1). It is a primary watershed of the

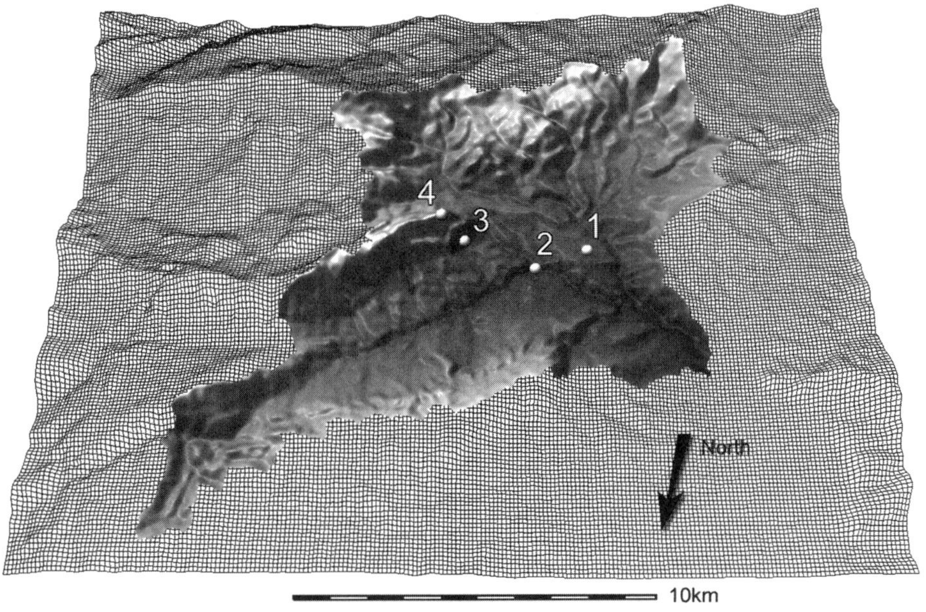

FIGURE 9.2. Three-dimensional perspective view of the Penaguila watershed showing the locations of the villages used in the four simulations. The viewpoint for this visualization is oriented from the north.

Río Serpis, whose headwaters originate in the Penaguila and the adjacent valleys. The MedLanD project has accrued substantial data on the ecology, geology, climate, geomorphology, and archaeology of the Penaguila (Barton, Ullah, and Bergin 2010; Barton, Ullah, and Mitasova 2010), making it an ideal location for the experiments presented in this chapter. The geometry of the study region was delimited by hydrological watershed modeling, and thus the DEM used in the experiments is constrained by the "natural" boundary of the valley's edges.

Using the AP-Sim model, we conducted an experiment to better understand the long-term effects of village location on population, economy, vegetation cover, and erosion/deposition rates. Four separate 100-year-long simulations were conducted in which each village was located in a different physiographic setting. Each of the chosen village locations might result from different social, economic, or practical considerations. In each simulation, the village contained three households, and each household began with a population of six people. Over the course of each 100-year simulation run, we tracked the population of each village, the amount of food collected by each household within the village, the spatial extent of land cover types, and the amount and location of erosion and deposition. To account for any between-run stochastic variability, we replicated each simulation five times. In the results section below, we use the five-run average of the tracked values to compare the outcome of the four simulations.

In the first model (Figure 9.2, location 1), the village is located for optimal access to farm fields and pasture lands. In this simulation, the village is centrally located in the Penaguila watershed, in the middle of a broad flat plain between the two forks of the Río Serpis. This means that farmers and herders can move to their fields and pastures with equal ease in any direction and that all the nearest land is flat enough to farm. We expect this village to be the most successful with respect to population (and harvest returns) and to have the smallest impact on the landscape due to its wide area and the flat surrounding area.

In the second model (Figure 9.2, location 2), the village is located for optimum seclusion. In this simulation, the village is located deep in the *barranco* (incised gorge) of the eastern fork of the Río Serpis. Locating the village deep in the *barranco* reduces the number of places from which the village can be seen. However, farmers and

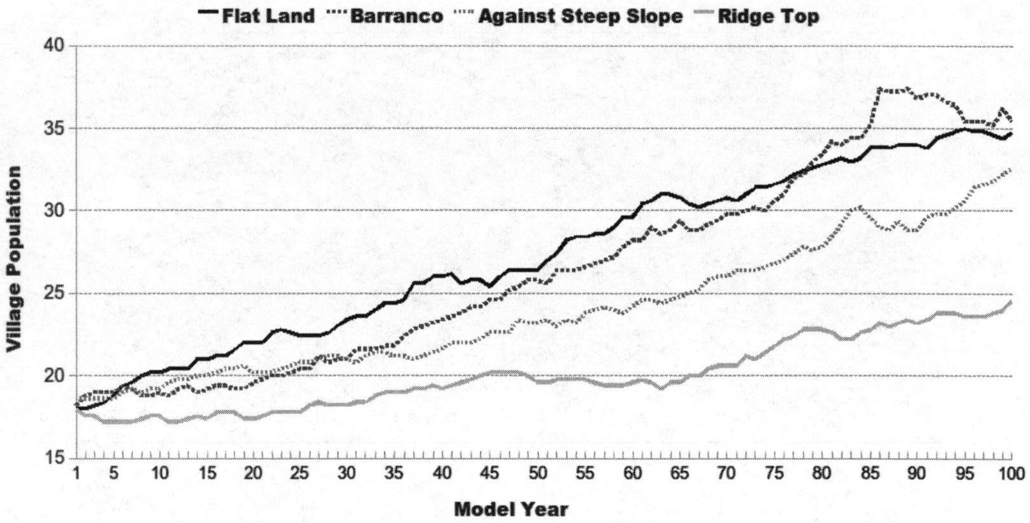

FIGURE 9.3. Village population totals over 100 years of simulation.

herders face a steep climb to their fields and pastures regardless of which direction they wish to go, and much of the nearest land is too steep to farm. The proximity of the second village to the areas of flat land at the edges of the *barranco* should provide adequate farmable area for subsistence needs (although less than the first village), but its proximity to so many steep slopes in an active channel suggests that this village would have higher rates of erosion and deposition than the first village.

The village in the third model (Figure 9.2, location 3) is located for optimum defensibility. This village is located against a steep slope, south of the eastern fork of the Río Serpis. Positioning the village against a steep slope limits the direction of access to the village by potential enemies, making it easier to defend. This also means that it is easier for farmers and herders to farm and graze their animals on one side of the village than on the other, and that only half the nearest land is flat enough to farm. We expect this third village location to provide less farmland for subsistence, and thus it should sustain a significantly smaller population than the preceding models. Moreover, due to the adjacent steep slope, we also expect high amounts of erosion and deposition in this scenario.

In the fourth model (Figure 9.2, location 4), the village is located for an optimum viewshed. The village in the fourth simulation is located high on a ridge top, to the south of the confluence of the two forks of the Río Serpis. This location maximizes the amount of the landscape that will be visible from the village when compared with the other village locations. In this scenario, farmers and herders will have to walk down a relatively steep slope regardless of which direction they need to move, and most of the nearby land is too steep to farm. We expect that this village location will likely result in a much smaller population than the other models. In addition, we expect the abundance of nearby steep slopes to amplify the deleterious effects of subsistence practices on the landscape, and erosion and deposition rates should be higher than in all the other models.

9.4.2. Results: Demography and Economy

The differing physiographic locations were chosen so we could better understand the long-term effects of village placement on landscape, demographic, and economic variability. In each case, the results reported in this section derive from the average of five model runs with the same parameters.

The average village harvest returns are closely related to average village populations, so therefore we will examine only the patterns of population change for each village. The population in each village begins at 18 (i.e., three households of six people each), and the population of each household varies each year based on its economic success. In Figure 9.3, we can see that a fair amount of variability occurs between the models in only a short time. In the first 35 years, the village located on flat land was the most successful, while the village located on the ridge top was the least successful. The villages located in the *barranco* and against the steep slope were both moderately successful.

Of all the villages, the one positioned on the ridge was the only one to experience an initial population decline. The lack of nearby flat land suitable for farming resulted in very low wheat yields during the first few years. The population remained steadily low for approximately 65 years, at which point the village experienced a small increase in growth. Because there was less farmable land within this village's LCP catchment, the total population never grew to the size of any of the other villages.

The village located on the flat land was initially very successful at farming, which led to elevated rates of population growth (Figure 9.3). The increasing population meant that each year there were more mouths to feed, inducing the households to cultivate more and more new fields. By year 80 of the simulation, it appears that the village reached the limit of its LCP catchment and therefore could farm only on plots that had been under almost continuous cultivation and now had greatly reduced yields. The previous rate of population growth became unsupportable and began to decline to a steady state by the end of the simulation.

The village in the *barranco*, on the other hand, experienced initially low farming returns and therefore had an initially low population growth rate (Figure 9.3). This is likely because the nearest flat farmable land was located above the steep slopes of the *barranco* and therefore was more costly to reach than were the first farmed plots of the village located on the flat land. The smaller population required less land, and since all land was harder to get to, the households in the village in the *barranco* farmed plots for longer periods (i.e., they used shorter fallow periods) than did the households in the village on the flat land. Nevertheless, the population increased over time, and by year 75 of the simulation there was enough demand (and excess labor) that households finally found it profitable to farm the virgin plots at the extreme edge of their LCP catchment. These virgin plots produced extremely high yields, leading to the population boom that reached its zenith at year 85.

The village located against the steep slope experienced 80 years of steady population growth, the longest period of population growth rate stability of any of the villages. After year 80, the village experienced a decade of fluctuating population before continuing to increase at a higher rate. This population readjustment is likely due to the same factors as the readjustment that occurred for the village located in the *barranco*. However, it occurred much later in the simulation for this village because its LCP catchment was even more constrained than that of the village in the *barranco*.

9.4.3. Results: Land Cover and Erosion/Deposition

We can examine the impact of villages on land cover by comparing the vegetation composition of the watershed for each village location. Figure 9.4 depicts the spread of land cover types for each village location after 100 years of the simulation: in each of the simulations, over 70 percent of the landscape within the range of agent activities had been reduced to a land cover of less than 20. Most of this land cover reduction is due to the clearing and subsequent planting of plots for agricultural use, which produces a signature "curve" in the middle range of the graph of the frequency of land cover values. The main characteristic of the "agricultural" portion of the land cover curve is that it results in higher amounts of grassland and shrubs than would naturally occur on the landscape (Figure 9.4). We refer to this type of land cover disturbance as "bottom up," as plots are immediately reduced to a low land cover value and are left to slowly regrow only when agents are finished using them (Barton, Ullah, and Mitasova 2010). A larger agricultural catchment means that more new fields are available, and thus there will be more fallow fields regrowing at any given time, resulting in a less steep fall-off curve. A smaller agricultural catchment means that the demand for farmable plots will be higher and will result in

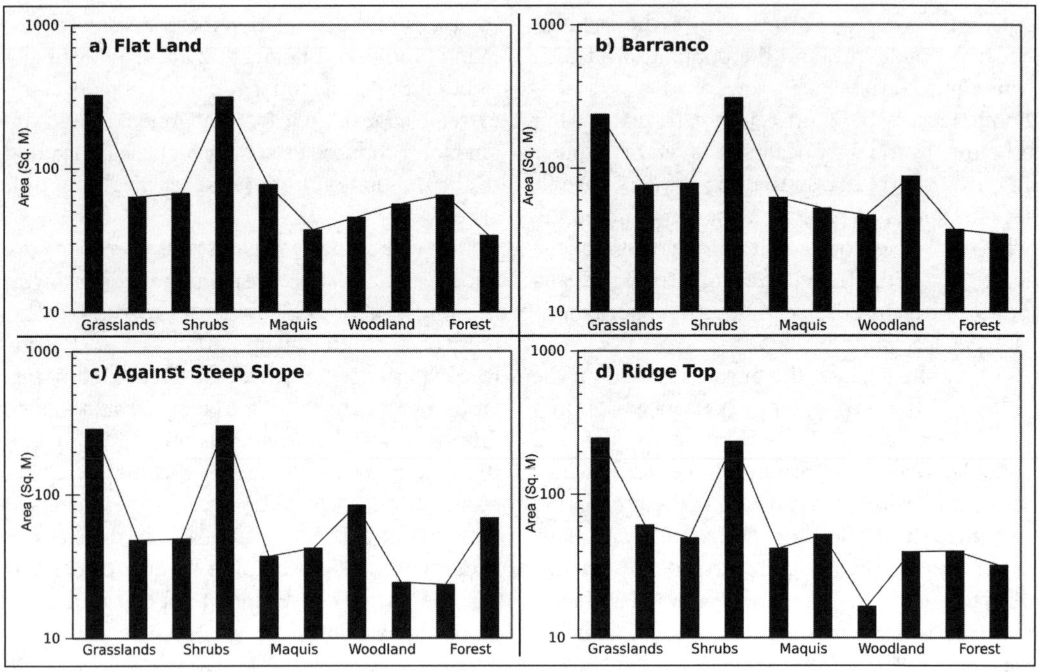

FIGURE 9.4. The area covered by the different land cover types for each of the four simulations after the 100th year of simulation. Note that the fifty raw land cover types have been condensed into ten bins and that the *y*-axis has been log transformed to ease intermodel comparison.

the agricultural portion of the land cover curve possessing a steeper fall-off because fields will not be allowed to fallow for very long before being reused.

Pastoralism, on the other hand, is a "top down" disturbance. As goats and sheep graze the existing vegetation, they slowly reduce the land cover value of a patch over time (Barton, Ullah, and Mitasova 2010). The signature land cover curve for grazing of natural vegetation occurs in the top range of the graph of land cover values and is very different from that of agriculture. Because open grassy woodland (the second bar from the right in Figure 9.4a) produces the highest grazing returns, there will be a large number of patches with land cover in that range, with the "pastoral" portion of the land cover curve falling off to the left due to grazing pressure on favored patches and to the right due to regrowth on less favorable patches (Figure 9.4). A larger grazing catchment allows for a more balanced use of grazing patches, and the curve will be smoother and more centered at the optimum grazing range (i.e., in the grassy woodland bar). A smaller grazing catchment will force more frequent reuse of grazing patches, and thus the pastoral portion of the curve will be skewed to the right.

In the AP-Sim model, most grazing takes place at the outer areas beyond the farthest extent of the farming zone. However, AP-Sim also allows grazing on fallowed fields, and therefore many of the closest fields will be used for grazing after agents release them from cultivation. When this occurs, the "bottom up" effect of agriculture is combined with the "top down" effect of grazing to create a signature in the bottom range of the land cover frequency graph (Figure 9.4). Over time, many of the fallow fields, which have already had their land cover values reduced by farm clearing, will be grazed to bare land. Therefore, the major characteristic of the land cover curve for grazing on fallowed fields is a large amount of bare land (i.e., the first bar in each part of Figure 9.4) and with a convex curve to the right, which first falls low before rising to the value for farmed fields (i.e., the fourth bar in each part of Figure 9.4). As mentioned above, a larger agricultural catchment means that there will be more fallow fields, meaning that there will more of these fields that have been grazed to bare land.

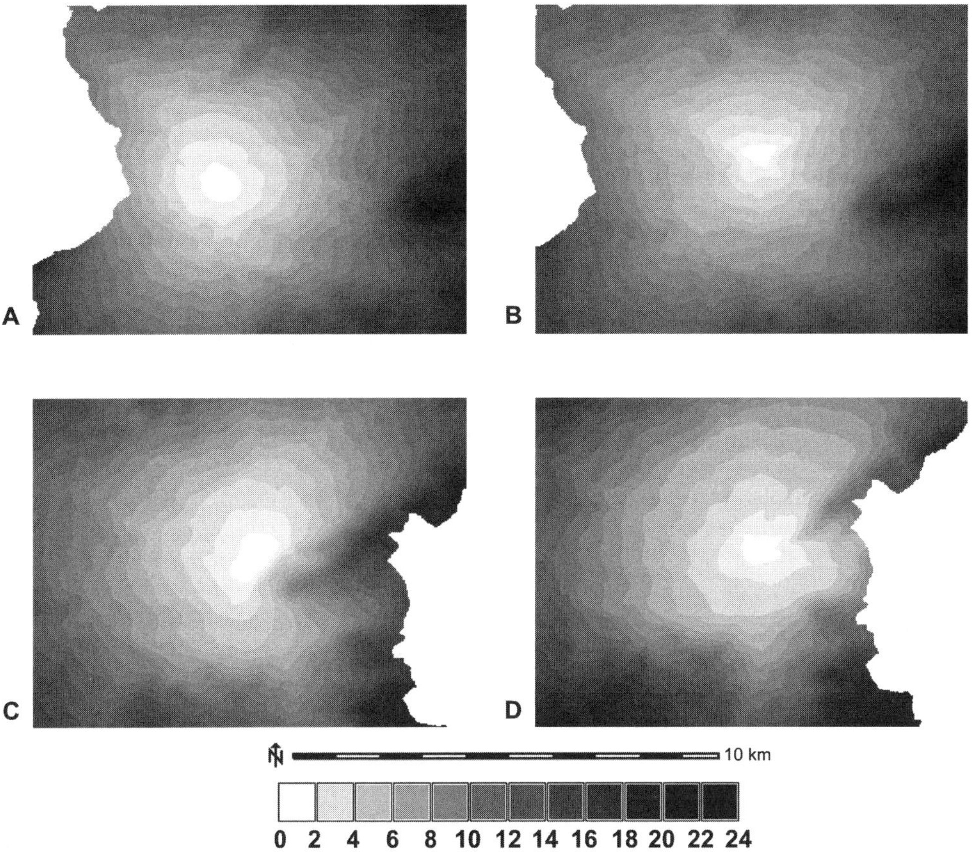

FIGURE 9.5. Maps of walking efforts (least cost maps) for (a) the flat-land village, (b) the village located in a *barranco*, (c) the village located against a steep slope, and (d) the village located on a ridge. Maps in this figure are shown in units of hours of walking time to ease intermodel comparison. The least cost maps used in the simulations were in units of seconds of walking time.

Comparing the land cover curves produced by each of the four village locations yields some interesting insights into how village location affects the resulting vegetation profile. Subsistence practices around the village located on the flat plain resulted in a vegetation profile that closely resembles an "idealized" vegetation-response curve for small-scale mixed agropastoralism (Figure 9.4a). This makes sense, because most of the farm plots and grazing patches are almost equally easy to get to from the village, as can be seen in the map of walking costs from this village (Figure 9.5a). This results in a very even land use pattern, and thus a very even pattern of land cover (Plate 8a). We can use the vegetation profile for model 1 as a sort of "control model" against which to compare the deviation in the curves of the other village locations.

While the agricultural portion of the land cover curve for the village located in the *barranco* closely matches that of model 1, the natural and fallow grazing portions of the land cover curve do not (Figure 9.4b). The natural grazing portion of the curve is skewed slightly to the left (i.e., third bar from the right in Figure 9.4b), and the fallow grazing portion of the curve is less deeply convex. This is likely because, although it was sufficiently easy to access the farmland nearby, *all* the nearest land was used for farming, leaving only the areas that were farther away and more difficult to get to for grazing. Indeed, the pattern of walking costs becomes more and more convoluted farther from the village (Figure 9.5b), and the pastoral portions of the catchment have a more complicated geometry (Plate 8b). The skew in the natural grazing portion of the curve indicates that most grazing patches were being used heavily and were not always allowed to fully regrow back to the optimal value for fodder production before being regrazed. The shallowness of the convexity in the fallow grazing area of the curve indicates that there were fewer fallow fields and that they were reclaimed for farming on a regular basis, thus preventing them from being grazed all the way down to bare land.

The pattern is different yet again for the village located against the steep slope. The agricultural portion of the land cover curve has a much steeper fall-off than in model 1, the natural grazing portion of the curve is severely skewed to the left (i.e., the fourth bar from the right in Figure 9.4c), and the fallow grazing portion of the curve is more deeply convex (Figure 9.4c). The cost surface for this village shows that it is easier for the agents to access both potential farm plots and grazing patches that are far away on the plain than it is for them to access plots that are quite close by up on the slope (Figure 9.5c), and indeed, the 100-year land cover map shows that almost all subsistence activity was located out on the plain (Plate 8c). The steep fall-off in the agricultural curve is due to the constriction of the agricultural catchment to a smaller area than in the "control model." Because most of the nearby land is kept under almost constant cultivation, the natural grazing catchment is located very far from the village. Thus grazing patches were reused very frequently, preventing regrowth to optimal forage production levels before they were regrazed. However, it seems that fallowed fields were grazed more frequently than in the other models. This is probably because the natural grazing catchment was so constricted that even the relatively poor fodder available on fallow fields that were fairly far from the village provided better grazing returns than those available on the farthest portions of the natural grazing catchment.

The vegetation curve for the village located on the ridge top is again very different from the others (Figure 9.4d). The apex of the agricultural portion of the curve is less high than in the other models, and the fall-off of the curve is also relatively steep. The natural grazing portion of the curve is bimodal (one peak in the fifth bar from the right and the other in the second bar from the right in Figure 9.4d) and has lower amounts than in the other models. The fallow grazing portion of the curve is relatively close to the "control model," although with a slightly lower amount of shrub vegetation. The cost surface shows that it is easy for the agents to travel in any direction (Figure 9.5d), but the fact that agents prefer not to farm on steep slopes means that a large portion of the farmed plots have to be located quite far from the village (Plate 8d). Consequently, there is also an abundance of unfarmed virgin forest in very close proximity to the village, which is an excellent situation for grazing. Having such a large and easily accessible grazing catchment results in a very low amount of grazing pressure, which accounts for the low values in the natural grazing portion of the vegetation profile curve (the second bar from the right in Figure 9.4d). The second mode in the upper portion of the vegetation curve (the fifth bar from the left in Figure 9.4d) is likely due to regrowth on relatively steep plots that were farmed in the early years of the simulation but lost soil due to erosion and thus became unattractive for farming. These plots can be seen as rings of brown and orange in the map of land cover for year 100 of this model (Plate 8d).

We can also examine village impacts on the landscape by comparing the amounts of erosion and deposition (Plate 9 and Figure 9.6). Comparing the effects of human land use on erosion and deposition rates between each model is complicated by the fact that each village is situated in a different location in the watershed. Each village affects a maximum area of 60–120 hectares, while the watershed itself is over 15,000 hectares. This means that much of the watershed remains unaffected by agent subsistence practices. The geometry and geographic location of agent impacts (Plate 9) is different from model to model, and thus the unaffected areas cannot easily be "masked out" when we calculate statistics. The large area of very low amounts of erosion and deposition in the greater watershed skews comparative statistics that rely on central tendencies, rendering them ineffective. Thus the mean amounts of erosion and deposition produced are quite similar for all the models. However, differences in the pattern of erosion and deposition exist between the models and become apparent when the

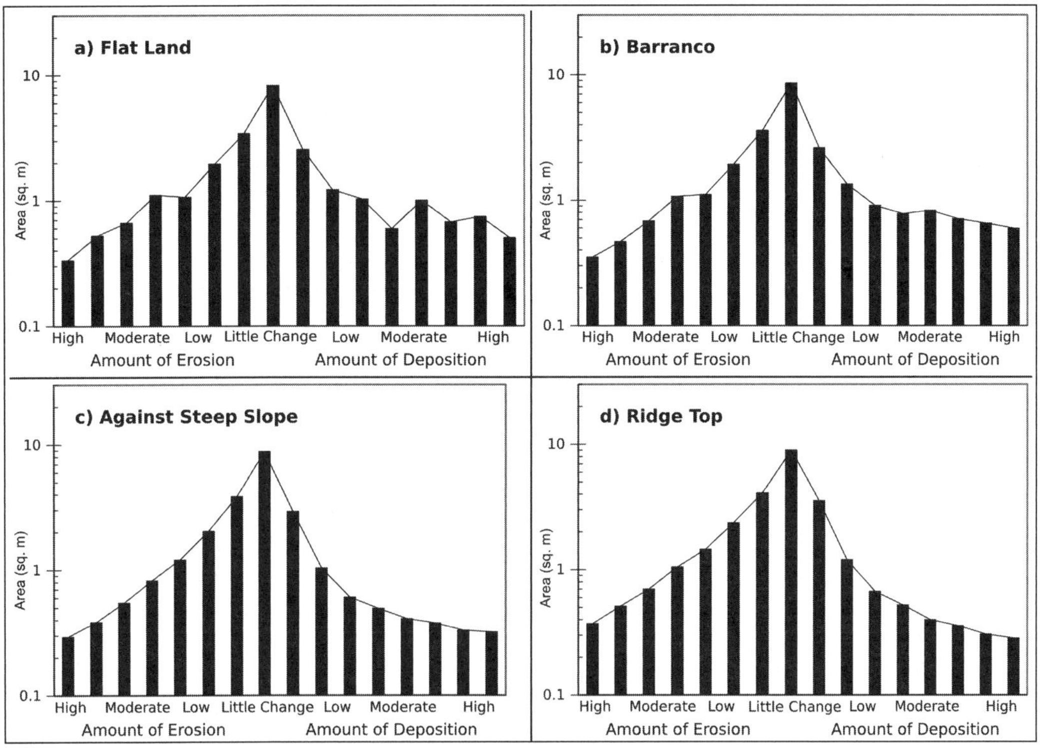

FIGURE 9.6. The total amount of area covered by sixteen categories of cumulative vertical change (cumulative erosion or deposition) for each of the four simulations after 100 years of simulation. The *y*-axis has been log transformed in order to better show small variations in area totals for those categories of vertical change with small total areas. Values of vertical change between -0.1 and 0.5 have been left out of this graph because of the large amount of area this category covers (i.e., most of the map).

log of the area totals for binned categories of erosion depths and deposition heights are plotted as a histogram (Figure 9.6).

Model 1, the village located on the flat plain, affected the largest area of all the models (Plate 9a) and also resulted in the largest amount of areas of high erosion and deposition (Plate 8a). Model 2, the village located in the *barranco*, affected less overall area than model 1 (Plate 9b) but also had large amounts of areas with high values of erosion and deposition (Plate 8b). Model 3, the village located against the steep slope, affected the smallest overall area of all the models (Plate 9c) and also resulted in the lowest amount of areas with high erosion and deposition values (Plate 8c). However, it had more areas of low erosion and deposition values than the first two models. Model 4, the village located on the ridge top, affected a fairly large area (Plate 9d) and had the most interesting pattern of erosion and deposition values (Plate 8d). It resulted in relatively small amounts of areas with high deposition values but had high amounts of areas with high erosion values. The results of model 4 are more similar to those of model 3 for deposition but are more similar to those of models 1 and 2 for erosion.

9.4.4. Discussion

Using a coupled ABM-LPM, we simulated the effects of multiple agropastoral village locations within the Penaguila Valley. Each village location was chosen with different physiographic variables so we could better understand the effect of village location on socioeconomic and environmental variability. LCP modeling is an integral component of this type of modeling and adds a degree of realism that cannot otherwise be easily achieved. The use of LCP allows for the integration of realistic agent movement calculated empirically and logically—without the need to physically route agents across the landscape. This not only decreases the run time of this type of model (hours as opposed to days or weeks), it also greatly simplifies the programming of this type of model without sacrificing reliability or accuracy.

The village location chosen for model 1 maximized the amount of surrounding flat land. The resulting LCP model (Figure 9.5a) made a large amount of land available for farming and grazing. As we expected, this village location resulted in one of the highest average populations (and thus the least variable average annual harvest). Its effect on the landscape, however, produced the highest recorded amount of erosion and deposition for any village. We believe that the large impact on the landscape resulted from two unpredicted factors. First, the wide area of available farmland resulted in impacts to a larger area of the landscape than in any of the other models. Second, the high population led to intensive land cover modification within this region, which resulted in greater and more dynamic changes in an area of the landscape that would have remained static without human impacts. In economic and social terms, this village location could be viewed as the most successful location for its occupants; however, it had the most deleterious impact on its surrounding environment.

The village position in model 2 was located deep in a *barranco*, effectively surrounding the village with steep slopes that provided a fair amount of seclusion from the surrounding area. As we expected, the village located in the *barranco* was economically successful, although not as successful as the village located on flat land. The differing success between these two models resulted from the high cost of movement from the bottom of the *barranco* up to the flat plain, as calculated by the LCP model (Figure 9.5b), which limited access from the village to the best farmland. The village's effect on the landscape was large, resulting in the second-highest rates of erosion and deposition and greatly reducing the surrounding land cover.

The village in the third model was located against a steep slope—a location that might be chosen to optimize defensibility. As we predicted, the amount of farmland available to the village was reduced due to the fact that one side of the village was bounded by the steep slope. However, we believed that the sloped areas near the village would be heavily used for grazing, which was not the case (Plate 8c). As shown in the LCP model (Figure 9.5c), households chose grazing lands located far away to the west because these areas were effectively "closer" in terms of walking costs than were the physically close areas nearby in the sloped area. Essentially, even though the eastern sloped area contained perfectly edible fodder, the costs of obtaining that fodder were higher than the costs of obtaining fodder from the grazing areas out past the edge of the farming catchment. The village in model 3 grew to the third-highest average population and had only a limited effect on erosion and deposition in the valley.

Our fourth village was located on a high ridge top for an optimum viewshed. With respect to population (and harvest returns), the village was the least successful of the four. In part, this was due to the lack of low-slope areas suitable for farming in the nearest LCP proximity to the village. Interestingly, and as opposed to the results of the village located against a steep slope, in this model, grazing *did* occur on the slopes near the village. This is because the village was located *above* these slopes, and according to the results of the anisotropic LCP model (Figure 9.5d), these areas were evenly and easily accessible to agents, despite the steepness of their slope. Because these areas were not used for farming, herds of goats and sheep had their pick of grazing patches in a large and nearby grazing catchment. Consequently, grazing had less of an effect on land cover than it did in all the other models, and although the village was located in an area at high risk for erosion, the dispersed nature of the agropastoral activity surrounding the village resulted in low overall values of erosion. Interestingly, the model also resulted in a large amount of high deposition, likely due to highly localized redeposition of sediment deriving from small-scale but widespread erosion on the slopes. This effect would likely be increased by longer occupation or more intensive subsistence practices.

The various locations of village sites in these models had marked and sometimes surprising effects on the subsistence economy and population growth rates of each village and on the local landscape. Larger areas of low-cost access led to easier attainment of farmable land. On the other hand, the larger agropastoral catchments that resulted from this ease of access also created a more marked impact on the landscape, which resulted in greater changes to the natural vegetation and larger overall erosion-deposition values. The employment of an anisotropic LCP model added a realistic element in the behavior of agents toward the use of sloped areas. Rather than avoid all sloped areas, households avoided using high-slope areas only if they had to walk *up* the slope, especially if there was a lower-cost alternative (as in model 3). If there was no lower-cost alternative, then agents were forced to use these slopes, regardless of the costs associated with doing so (as in model 2). However, agents did not avoid the use of steep slopes if they only had to walk *down* them (as in model 4). The use of an isotropic LCP routine would change this behavior, as the cost of movement over a landscape cell of a given slope would be the same regardless of the direction of movement across that cell. Isotropy introduces a degree of simplification, however, because directionality *does* matter. The best solution would therefore be to continue to calculate costs using an anisotropic algorithm, but do so for "there-and-back" trips rather than just in the outbound direction. Although it would be much more computationally expensive, the integration of a round-trip anisotropic LCP routine into our coupled ABM-LPM would mean that agents would likely treat downslope areas similarly to upslope areas, so this is a promising avenue of investigation for future experiments.

9.5. Conclusion

Coupled ABM-LPM models such as the MedLanD project's AP-Sim/LanDyn model simulate socioecological systems that are dynamic and complex. While such modeling has the potential to be highly intricate and complicated, we have tried to maintain the highest possible ratio of simplicity to realism in our model. A complicated model produces complicated results with no guarantee of increased accuracy. We have therefore consciously avoided the tendency to "model reality" with a model that "does everything" and instead attempted to make a model that simulates those few social and environmental processes that we found most interesting. The model output serves as a hypothesis generator against which to test data from the archaeological record.

The scenarios presented in the case study, however, are all potentially "real" scenarios. The land use decision-making logic in the model is based on ethnoarchaeological data on how self-subsisting small-scale agropastoralists actually use their landscapes, and the landscape model is based on empirically derived equations that model the way real landscapes behave. Thus the results of the case study provide realistic and empiric hypotheses related to the effects of village location on agropastoral economy, village population, and landscape change. These hypotheses would be extremely difficult or impossible to generate from archaeological data — even extraordinarily well preserved archaeological data — alone.

The increasing use of computer modeling to test archaeological hypotheses represents a shift in the way researchers approach archaeological questions. Computer modeling provides an alternative framework for addressing anthropological questions that may be difficult or impossible to answer with available archaeological or ethnographic evidence. Models are not themselves evidence and should not be treated as such, but they are a new and useful way to develop and test hypotheses. Coupled ABM-LPM has enormous potential for archaeological research, especially for the investigation of socioeconomic questions and questions of human-environmental interaction. The potential applications for anthropologically based computer modeling greatly increase with the inclusion of LCP modeling. As agent-based and landscape models become individually more complex, however, it becomes more important to link them in a meaningful way. LCP modeling provides a very good method for achieving this linkage. On its own, LCP modeling provides a sophisticated method for identifying likely routes through a landscape. The inclusion of LCP modeling into a coupled ABM-LPM model uses this characteristic as an effective and authentic platform for agent decision making that shapes agent interaction with the landscape in an efficient and realistic manner.

Acknowledgments

We recognize the support of the National Science Foundation (grant BCS-0410269), which has made this research possible. We also thank C. Michael Barton, Gary Mayer, Hessam Sarjoughian, and Alexandra Miller for their contributions and counsel. Additionally, we thank the editors for organizing this book and inviting us to contribute.

References

Al-Jaloudy, Mahmud Abusetta
2006 Country Pasture/Forage Resource Profiles: Jordan. Country Profile. *Food and Agriculture Organization of the United Nations Country Pasture Profiles*. Electronic document, http://www.fao.org/ag/AGP/agpc/doc/Counprof/Jordan/Jordan.htm#4.%20RUMINANT%20LIVESTOCK%20PRODUCTION%20SYSTEMS.

Axtell, R. L., J. M. Epstein, and J. S. Dean et al.
2002 Population Growth and Collapse in a Multiagent Model of the Kayenta Anasazi in Long House Valley. *Proceedings of the National Academy of Sciences* 99(Suppl. 3), May 14.

Barton, C. Michael, Isaac I. Ullah, and Sean Bergin
2010 Land Use, Water and Mediterranean Landscapes: Modelling Long-Term Dynamics of Complex Socio-Ecological Systems. *Philosophical Transactions of the Royal Society A: Mathematical, Physical and Engineering Sciences* 368(1931):5275–5297.

Barton, C. Michael, Isaac I. Ullah, and Helena Mitasova
2010 Computational Modeling and Neolithic Socioecological Dynamics: A Case Study from Southwest Asia. *American Antiquity* 75(2):364–386.

Bonet, Andreu, and Juli G. Pausas
2004 Species Richness and Cover along a 60-Year Chronosequence in Old-Fields of Southeastern Spain. *Plant Ecology* 174:257–270.

Braun, Jean, Arjun M. Heimsath, and John Chappell
2001 Sediment Transport Mechanisms on Soil-Mantled Hillslopes. *Geology* 29(8):683–686.

Calafell, Francesc, and Jaume Bertranpetit
1993 The Genetic History of the Iberian Peninsula: A Simulation. *Current Anthropology* 34(5): 735–745.

Dean, J. S., G. J. Gumerman, and J. M. Epstein et al.
1999 Understanding Anasazi Culture Change through Agent-Based Modeling. In *Dynamics in Human and Primate Societies*, edited by Timothy Kohler and G. J. Gumerman, pp. 179–205. Oxford University Press, Oxford.

Dietrich, William E., Dino G. Bellugi, and Leonard S. Sklar et al.
2003 Geomorphic Transport Laws for Predicting Landscape Form and Dynamics. *Geophysical Monograph* 135:103–132.

Dijkstra, E. W.
1959 A Note on Two Problems in Connexion with Graphs. *Numerische Mathematik* 1(1):269–271.

GRASS Development Team
2010 Geographic Resources Analysis Support System. Electronic document, http://grass.osgeo.org/.

Hancock, G. R
2004 Modelling Soil Erosion on the Catchment and Landscape Scale Using Landscape Evolution Models — A Probabilistic Approach Using Digital Elevation Model Error. In *Super Soil 2004: 3rd Australian New Zealand Soils Conference*. University of Sydney, Australia.

Howard, A. D.
1980 Thresholds in River Regimes. In *Thresholds in Geomorphology*, edited by D. R. Coats and J. D. Vitek, pp. 227–258. George Allen and Unwin, London.

Kim, S., Hessam S. Sarjoughian, and V. Elamvazhuthi
2009 DEVS-Suite: A Simulator for Visual Experimentation and Behavior Monitoring. In *High Performance Computing and Simulation Symposium: Proceedings of the Spring Simulation Conference*. ACM Press, New York.

Lake, Mark Winter
1999 MAGICAL Computer Simulation of Mesolithic Foraging. In *Dynamics in Human and Primate Societies*, edited by Timothy Kohler and G. J. Gumerman, pp. 107–143. Oxford University Press, Oxford.

Langmuir, E.
1984 Mountaincraft and Leadership. *The Scottish Sports Council/MLTB*. Cordee, Leicester, U.K.

Lansing, J. Stephen, James N. Kremer, and Barbara B. Smuts
1998 System-Dependent Selection, Ecological Feedback and the Emergence of Functional Structure in Ecosystems. *Journal of Theoretical Biology* 192:377–391.

Mayer, Gary R., and Hessam Sarjoughian
2007 Complexities of Simulating a Hybrid Agent-Landscape Model Using Multi-formalism Composability. In *Spring Simulation Conference: Agent-Directed Simulation*, pp. 161–168. IEEE Press, Norfolk, Virginia.

Mitasova, Helena, W. M. Brown, and D. Johnston
2002 *Terrain Modeling and Soil Erosion Simulation Final Report*. Geographic Modeling Sys-

tems Lab, University of Illinois, Urbana-Champaign.

Neteler, M., and H. Mitasova
2007 Open Source GIS: A GRASS GIS Approach. *International Series in Engineering and Computer Science* 773. New York.

Pardo, Fernando, and Luis Gil
2005 The Impact of Traditional Land Use on Woodlands: A Case Study in the Spanish Central System. *Journal of Historical Geography* 31(3): 390–408.

Rees, W. G.
2004 Least-Cost Paths in Mountainous Terrain. *Computers and Geosciences* 30(3):203–209.

Renard, K. G., G. R. Foster, G. A. Weesies, D. K. McCool, and D. C. Yoder
1997 *Predicting Soil Erosion by Water: A Guide to Conservation Planning with the Revised Universal Soil Loss Equation (RUSLE).* USDA Agriculture Handbook 703. U.S. Department of Agriculture, Washington, D.C.

Renard, Kenneth G., and Jeremy R. Freimund
1994 Using Monthly Precipitation Data to Estimate the R-factor in the Revised USLE. *Journal of Hydrology* 157(1–4):287–306.

Thomson, E. F., F. Bahhady, A. Termanini, and M. Mokrebel
1986 Availability of Home-Produced Wheat, Milk Products and Meat to Sheep-Owning Families at the Cultivated Margin of the NW Syrian Steppe. *Ecology of Food and Nutrition* 19:113–121.

Tucker, G. E., and G. R. Hancock
2010 Modelling Landscape Evolution. *Earth Surface Processes and Landforms* 35(1):28–50.

Ullah, Isaac I. T.
2011 A GIS Method for Assessing the Zone of Human-Environmental Impact around Archaeological Sites: A Test Case from the Late Neolithic of Wadi Ziqlâb, Jordan. *Journal of Archaeological Science* 38(3):623–632.

Warren, Steven D., Helena Mitasova, and Matthew G. Hohmann et al.
2005 Validation of a 3-D Enhancement of the Universal Soil Loss Equation for Prediction of Soil Erosion and Sediment Deposition. *Catena* 64:281–296.

Willgoose, Garry
2005 Mathematical Modeling of Whole Landscape Evolution. *Annual Review of Earth and Planetary Sciences* 33(1):443–459.

CHAPTER 10

No Crows Made Mounds

Do Cost-Distance Calculations of Travel Time Improve Our Understanding of Southern Appalachian Polity Size?

PATRICK LIVINGOOD

Digital technology and the widespread availability of geographic datasets are enabling researchers to compute cost distances using ever-increasing resolution and model complexity. However, no matter how fast the hardware or elegant the software, cost distance will never be as simple to calculate as straight-line or geodesic distance. Least cost analysis (LCA) will become a regular part of the archaeologist's tool kit only if it provides better insights: better predictions, better understanding of the data, or a better fit to our theoretical models. In this chapter, I calculate cost distances between Mississippian mounds in the southern Appalachians to show that LCA can contribute to an improved understanding of the data by providing a better match to theoretical models.

This case study is based on a famous discovery by David Hally (1993, 1999, 2006). During the Mississippian period in the southern Appalachians (AD 1000–1600), residents built earthen flat-top pyramidal mounds at their civic-ceremonial centers (Figure 10.1). The societies that constructed these mounds are typically referred to as chiefdoms and were kin-based, hierarchical societies with hereditary inequality. The mounds are highly visible features on the landscape, and Georgia archaeologists have engaged in an excellent long-term program of survey, which means we have fairly high confidence that the locations of all the mound sites are known. Hally (1999) compiled chronological data for the 45 known mound sites and computed the pairwise straight-line distance between every set of contemporaneous mounds (Figure 10.2). He identified a major break in the distribution: most mounds are located less than 22 km from each other or more than 32 km. He interpreted this pattern as evidence of the bounds of Mississippian polities, in which secondary centers are located no more than 22 km from their primary center, and primary centers of competing polities are located at least 33 km from each other. The only mound pairs that fell in this intermediate distance ranges were 9Ck1/9Ck2, located 27 km apart, and 9Ge5/46Mg46, located 28 km apart. These are interpreted as being secondary centers of one polity that are located an intermediate distance from the primary center of a different polity.

The goal of this chapter is to convert the units of Hally's analysis from straight-line distance to cost distance in order to see if it improves our understanding of the southern Appalachian Mississippian. One way the cost-distance approach could be judged a success is if it identifies a clearer modal break in the data, implying that cost distance was the true variable underlying the patterning of sites and that straight-line distance was only an approximation.

Another goal is to compare cost distance with the theoretical expectation that chiefdom-type societies worldwide are usually limited in their capacity to control territory greater than a half-day's journey from the center (Bauer and Covey 2002:847–848; Cohen and Schlegel 1968:136;

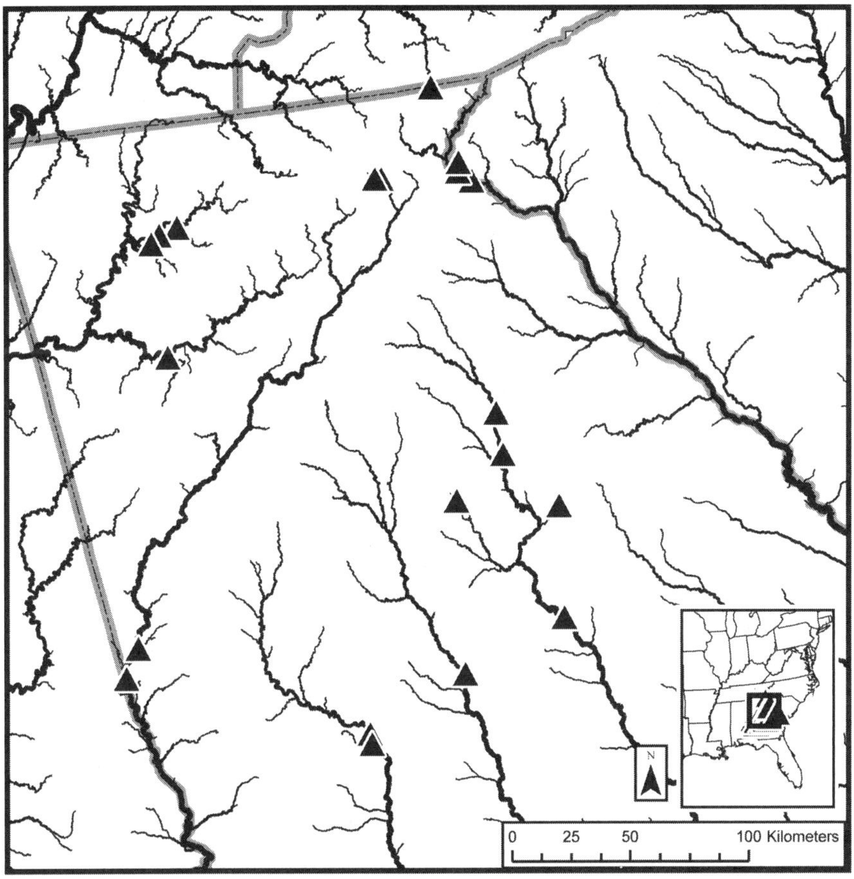

FIGURE 10.1. Map of the 45 southern Appalachian mounds in the case study. Note that they are not contemporaneous.

Helms 1979:51–53; Johnson 1987; Little 1967:240; Spencer 1990:6–8). This limit exists because of the lack of internally specialized administrative units (Wright 1977, 1984), which inhibits chiefs from extensive delegation of authority and imposes on them the requirement to manage their domain from the center (Spencer 1987, 1990, 1993). When polities are within this limit, chiefs can visit members of their communities without having to impose on their hospitality because they can return home at the end of the day (also important for state-level administrators in Mexico, as discussed in Bell et al. 1988:178). It also permitted the chief to respond with coercive force quickly if such action was required. From a bottom-up perspective, communities wanting to be integrated within a polity would choose to live closer to decrease the costs of participation. A community desiring autonomy would opt to live more distantly from a potentially meddlesome or threatening chief or the other apparatus of the polity.

10.1. Selecting the Unit of Cost-Distance Analysis

The work of geographers, urban planners, psychologists, and others makes the case that people use their evaluations of the cost of travel all the time in order to make decisions. There is a substantial body of research on this topic because it is of special interest to urban planners: people's evaluations of the cost of travel influence which route they take, which store they choose to go to, which mode of travel they choose, where they choose to live, where they choose to work, and many other aspects of life. Researchers have examined the way people evaluate the costs of

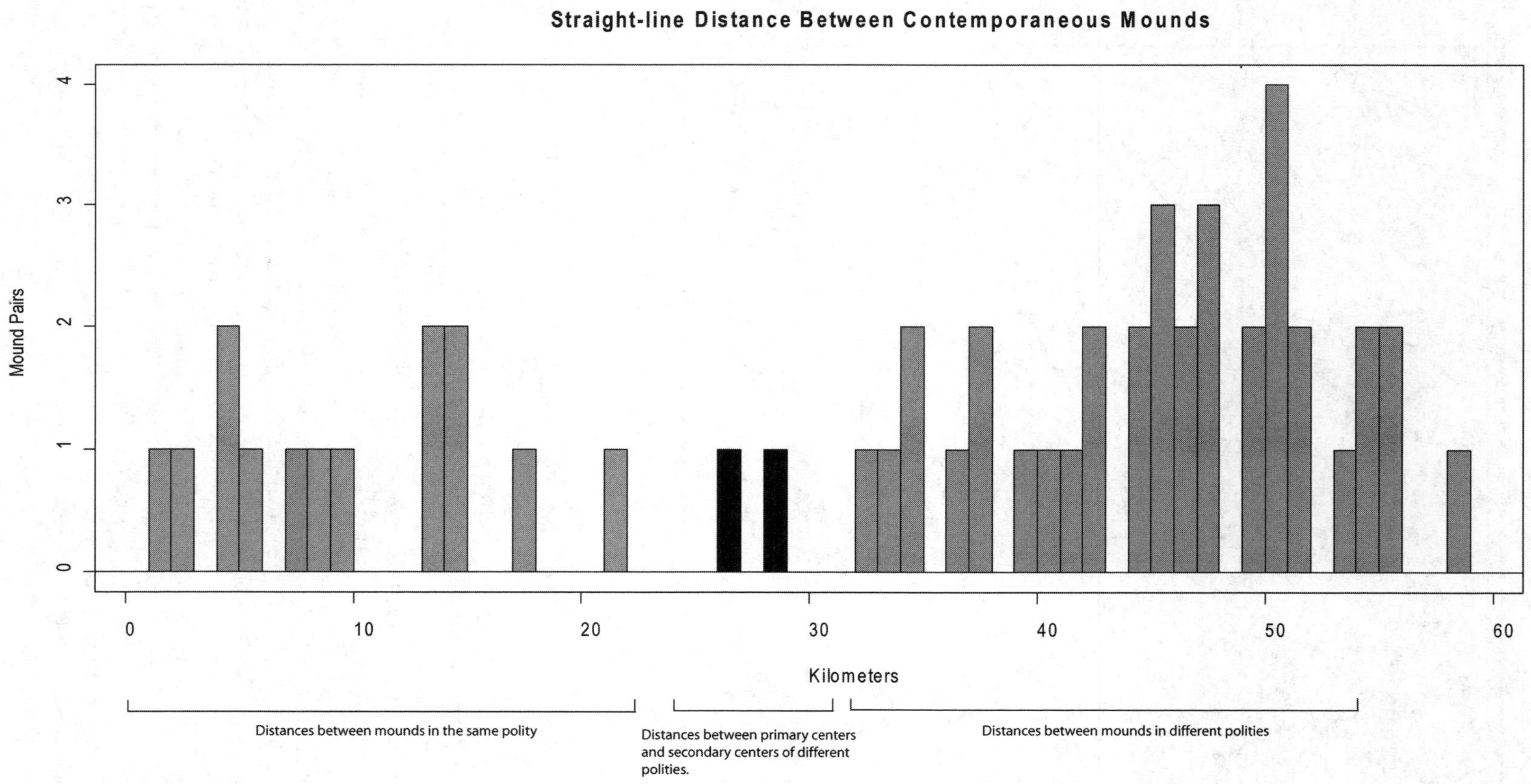

FIGURE 10.2. The straight-line distances between contemporaneous mounds in the study area for which the travel cost is less than 13 hours. A total of 52 mound pairs are represented here.

travel, and it is influenced by many factors, not just the objective distance. This opinion about the costs of a journey is referred to as the subjective distance (Montello 1997).

In Montello's (1997) framework, the subjective distance is influenced by three aspects: environmental features, travel time, and travel effort. Experimental research shows that environmental features are particularly important, but unfortunately these are nearly impossible to model in a prehistoric archaeological context. For example, modern urban travelers will judge a route to be longer despite its objective distance or duration if there are more turns (Sadalla and Staplin 1980), if there are no visible landmarks they are navigating toward (Nasar et al. 1985), or if there are simply more vistas (Montello 1997; Nasar et al. 1985). It is also clear that in many studies, cognitive, experiential, and cultural biases distort subjective distance in sometimes surprising ways. Studies have found that people overestimate distances of nearby destinations and underestimate distances to faraway destinations (McCormack et al. 2008), that people overestimate the distances of routes with which they are more familiar (Crompton 2006), and that people overestimate costs of travel into a city and underestimate costs of travel out of a city (Lee 1970).

Archaeologists are rarely able to incorporate environmental features into our attempts to approximate subjective distance, but we can calculate the other two aspects: travel time and cost effort. While a few archaeological studies use a flow model in which relative costs are computed that are not tied to any real-world units (e.g., Anderson and Gilliam 2000; Limp 1990; Sijia et al. 2007), most recent studies calculate costs in the units of time or calories. Caloric expenditure (e.g., Hare 2004; Hollenbach 2005; Jones and Madsen 1989; Wood and Wood 2006) has the advantage of decades of physiological research that permits one to anticipate the calories burned while factoring in variables such as sex, weight, speed, burden, and slope. Such simulations are especially useful when one is employing an optimal foraging model as the underlying theoretical paradigm, because one can compare the cost of a trip with the anticipated caloric return.

For this case study, time has been selected as the unit for cost distance. As experimental studies have found, travel time is a major, if not the most important, factor in a traveler's evaluation of subjective distance (Burnett 1978; Golledge and Zannaras 1973; MacEachren 1980). One possible reason is that instruments to measure time (clocks, watches, the movement of the sun, meals, etc.) are more readily available than instruments to measure distance (Montello 1997:302). This is somewhat supported by anecdotes from ethnographic and linguistic studies that show that time is used more often than geographic distance as a basis for subjective distance. For example, the basic unit of distance among the twentieth-century Tofa of Siberia is *kösh*, which is the distance one can travel on reindeer-back in one day (Rassadin 1995:23 as cited in Harrison 2007:105). It is approximately 25 km but is affected by terrain, snowfall, and other factors (Harrison 2007). Among the Sherpa and Bantawa Rai people of mountainous Nepal, maps and language emphasize the vertical dimension of place, which is the best predictor of travel time and effort (Harrison 2007:113–114). Finally, a traveler among the Malays in the 1870s noted that the major references to distance included "as far as a gunshot can be heard," "the distance you can travel before your hair dries," "the number of times you chew betel between locations," "the distance covered in a day's walk," and, for boatmen, the number of turns in the river (Bird 1883; Mitra 1910).

I would argue that travel time, more than caloric cost or straight-line distance, was the cost most likely to be perceived by the Mississippian traveler and most likely to be actively incorporated into decisions about routes, actions, and settlement patterns. Time is also useful in this study because the underlying theoretical expectation—that chiefly polities are typically no larger than a half-day's travel from the center—is framed in units of time.

10.2. Selecting the Parameters of the Cost-Distance Model

One of the challenges of using cost distance in any unit is that travel can be extremely variable and costs can be affected by many factors: weather, misfortune, man-made barriers, secondary activities (such as foraging for food, protecting against attack, or transporting goods), whether the traveler knows the route well enough to select the optimal path, and the speed and physical ability of the travelers. As others have done in similar research, but not always explicitly, the cost-distance model must usually simplify matters by specifying a traveler of median speed and ability who is making an optimal trip without any major unforeseen obstacles or delays.

Indigenous peoples of eastern North America had two primary modes of travel: they could walk or they could use a canoe. This case study permits both. For calculating travel times on foot, I use the commonly employed hiker's formula developed by geographer Waldo Tobler (Tobler 1993). The formula has been used in numerous archaeological calculations of cost distance (e.g., Aldenderfer 1998; Gorenflo and Bell 1991; Hare 2004; Jennings and Craig 2001; Kantner 1997; and Phillips and Leckman, Surface-Evans, White, and Kantner in this volume); for further discussion of the application of this formula to archaeological datasets, see Aldenderfer (1998:12), Gorenflo and Gale (1990), and Leusen (2002). The formula calculates walking speed as a function of slope and is expressed

$$Speed\ (km/hr) = 6e^{-3.5|x+0.05|} \qquad (10.1)$$

It predicts a speed of 5 km/hr on level terrain and a maximum speed of 6 km/hr on a 5 percent downslope. This corresponds well to recorded instances of modern and ethnographic rates of travel (Aldenderfer 1998:11–15; Lee 1979), assuming relatively clear terrain and light burdens.

Tobler further suggested that under less than ideal conditions we could reduce the predicted speed by a certain factor. For example, we would apply a 40 percent reduction in speed if travel is off-trail and a 20–40 percent reduction for carrying moderate loads. I have chosen to assume all travel in this simulation is on-trail because the purpose is to predict settlement patterns over a long period of time. Presumably, if a route was important, a trail would have been established. This is supported by accounts of the historic Creeks, who mostly stayed on established, if sometimes obscure, trails (Ethridge 2003:122).

The only additional impediment to walking I have chosen to implement is the barrier presented by crossing waterways. In accounts of historic travelers' overland journeys across the Southeast, often the only landmarks mentioned consistently were rivers and streams (Bartram 1996; Charlevoix [1761] 1966; Clayton et al. 1993; Lawson [1709] 1967; Tanner 1989:16; Waselkov and Braund 1995). They were both notable and all too often notably challenging to cross when there were no bridges (Ethridge 2003:124) or shoals available. In this model, these barriers are modeled by a simple cost (Table 10.1) that is dependent on the size of the waterway as measured in cubic feet of water per second (cfs). In historic accounts, these crossings are extremely variable, from a few minutes for well-prepared travelers with canoes stashed or those crossing at shoals or bridges to multiple days or weeks if unprepared travelers had to wait for flood-swollen rivers to subside. The goals of this simulation are to model a typical optimal trip for a well-prepared traveler who knows the terrain, and thus the costs were arbitrarily and conservatively chosen. Currently, this simulation does not model any other barriers to overland travel, such as wetlands, which were almost certainly a major impediment to overland travel, especially below the fall line (Hudson 1976:314), or shoals, which would have facilitated water crossings. These may be important, but there are no widely available datasets on their Precolumbian locations.

TABLE 10.1. Cost Penalty for Crossing a Waterway.

Water flow (cubic feet per second)	Delay
<10	0
10–100	3 seconds
100–1,000	5 minutes
1,000–10,000	10 minutes
>10,000	30 minutes

Note: This is implemented in the simulation by assessing half the cost in this table in addition to the cost calculated by Tobler's formula for any movement from a land cell to a water cell or vice versa. A complete crossing is assessed the penalty listed here.

Unfortunately, there are no well-established analogues to the Tobler formula for estimating the speed of canoe travel. The most comprehensive study of canoe speeds along eastern North American waterways was done by Little (1987). The single most important factor in determining the speed of canoe travel is the speed and direction of the current: upstream trips take twice as long as downstream ones (Little 1987:59). For this simulation, I have modeled the speed of canoe travel

TABLE 10.2. Historic Canoe Speeds.

Group or Explorer	Citation	Speed (not Current-Adjusted)	Notes
Aztec	Hassig 1985:64	2.6–3.5 km/hr	Travel through canal system of central Mexico with large cargo canoes
French under Iberville	Bénard de La Harpe 1971:23	5.7 km/hr	Travel on March 27, 1700, over a 34-hour period in canoe
French explorer Roullet	Rowland and Sanders 1927	19.25 km/day downstream	Travel along Pearl River from near Leake and Neshoba County line to mouth over 24 days in pirogue
Champlain	Little 1987:59	40–45 km/day upstream, 90–110 km/day downstream	Travel along St. Lawrence and Great Lakes in canoe
Average of Marquette, Joliet, LaSalle, Colden, Celoron	Little 1987:59	16–32 km/day upstream, 45 km/day downstream	Travel on Mississippi, Mohawk, and Ohio Rivers in canoes

as a base speed, which a canoeist could achieve on calm water, plus or minus the speed of the current, depending on the direction of travel. In order to determine this base speed, I collected several accounts of canoe travel in both historic and modern contexts. Unfortunately, most accounts of canoe travel in the historic Southeast are only moderately helpful because they fail to specify the number of hours spent each day in the water.

Table 10.2 lists several historic canoe journeys. In most cases we have only estimates of trip speed in days, but in some cases we have enough information to confidently calculate speeds in kilometers per hour. These records show that trips varied in speed from 16–45 km/day upstream and 19–110 km/day downstream. If one were to assume 8 hours a day on the water on average, it would provide a calculation of gross speeds of 2–5.6 km/hr upstream and 2.4–13.75 km/hr downstream.

Another, more refined estimate of possible canoe speeds comes from the group of people who have canoed down the entire Mississippi River in relatively recent times for adventure and occasionally profit (Table 10.3). At the extreme end, three pairs of canoeists have set progressively faster world records for travel down the Mississippi River since 1984. These attempts are useful for our purposes because they provide an obvious upper bound for our speed estimates, they are well documented, and the journeys are long enough that small variations in weather and conditions do not overly affect the averages. These record holders traversed the approximately 3700 km of the river in 18–23 days, which represents a gross speed of 6.5–8.5 km/hr. Since they had to spend some time making portages and navigating numerous locks and dams, their actual speed in the water was even higher. However, when we simulate a run taking into account the current speeds of the various segments of the Mississippi River, we can calculate the base speed needed to generate their results. Their current-adjusted speeds were 4.7–6.4 km/hr.

Obviously, many aspects of these attempts do not make good analogues for Mississippian canoe travel: the canoes were modern and ultralightweight; the canoeists were assisted by crews that provided them food they did not have to carry themselves; and they had the advantages of well-marked channels and no natural obstructions. On the other hand, these modern canoeists had to contend with locks and dams and barge traffic, which were not a problem for Precolumbian canoeists. Mike Schnitzka, one of the record holders, characterized these trips as significant demonstrations of endurance, not necessarily speed, because a sprinting speed would be impossible to maintain for three weeks. During world record attempts of this type, the canoeists never leave the canoe except for portages, and they sleep in shifts in the boat for only about three hours a night while their partner keeps paddling (Mike Schnitzka, personal communication 2007). If we assume that the modern barriers of locks and dams are equivalent in magnitude to the ancient ones of beaver dams and log accumulations, the

TABLE 10.3. Modern Canoe Travel with Current-Adjusted Speeds.

Explorer	Reference	Distance (km)	Avg. Speed of Current (km/hr)	Days of Travel	Current-Adjusted Speed at 8hrs/day (km/hr)	Notes
Captain Willard Glazier	Glazier 1891	3766.9	3.1	86	3.6	Traveled down the Mississippi river from Elk Lake to the Gulf of Mexico between July 22 and November 15, 1881. Used multiple birch-bark canoes and a small crew. Assume 8 hours/day of travel
Matthew Mohlke	Mohlke 2001	3583.2	3.3	79	3.9	Traveled solo down Mississippi from Lake Itasca to New Orleans between May 15 and August 1, 1999. Assume 8 hours/day of travel
John Pugh and Jessica Robinson	Pugh and Robinson 2005; Simmons 2005	3484.4	3.1	71	4.4	Traveled as a pair down the Mississippi from Lake Itasca to Sweetwater Bay, La., via the Atchafalaya between May 14 and July 27, 2005. Assume 8 hours/day of travel
Michael Schnitzka and William Perdzock (Wisconsin River)	Schnitzka and Perdzock 2007	725	2	4 days, 2 hours, 22 minutes	12.3	World record holders for canoeing down the Wisconsin River; canoed as a pair in 1989 in a racing canoe
Michael Schnitzka and William Perdzock (Mississippi River)	Schnitzka and Perdzock 2007	3766.9	3.1	23 days, 9 hours, 51 minutes	4.7	1989 world record journey for canoeing the Mississippi River
Verlen Kruger and Valerie Fons (Mississippi River)	Peterson 2006:297	3766.9	3.1	23 days, 10 hours, 20 minutes	4.7	1984 world record journey for canoeing the Mississippi River
Bob Bradford and Clark Eid (Mississippi River)	Peterson 2006:297	3766.9	3.1	18 days, 4 hours, 51 minutes	6.4	2003 world record journey for canoeing the Mississippi River
Average					**5.8**	

Note: Distances were calculated from NHDPlus data (Horizon Systems Corporation 2006). Current-adjusted speeds were calculated from NHDPlus: for the actual speed down the river, take the current-adjusted speed and add or subtract the flow of the channel. In the case of the Mississippi River attempts since 2000, the NHDFlow table data were altered slightly. Attempts were made using the pool and dam location data to identify modern reservoirs and to set the current of those channels to zero for calculating the current-adjusted speed.

most significant difference between these attempts and what the Mississippians would have been capable of is dictated by the technology of the watercraft.

There were also three well-published modern attempts to canoe the Mississippi in which the canoeists were not trying to set any records. From their books it was possible to determine how many days were spent on the water and to calculate their current-adjusted speeds, assuming an average of 8 hours a day on the water. The result is speeds of 3.6–4.4 km/hr. These compare favorably with Aztec cargo canoes that traversed Lake Texcoco and the surrounding canals at 2.5–3.5 km/hr and a French trip down the Mississippi in 1700. These results show a broad range of base speeds, and probably any value between 3.5 and 5 km/hr could be defended, with a range between 4 and 4.5 km/hr as the most likely. For this case study, I decided to use a value of 4 km/hr for the base speed of canoe travel plus or minus the speed of the current and to permit canoe travel only on waterways with flows of 100 cfs or more. It would be ideal to also simulate waterfalls, shoals, and other barriers that would have required portages. Unfortunately, there are no widely available datasets on their locations. Also, like land travel, canoe travel would have been variable from season to season and easier in the high waters of spring and early summer than in fall or winter (Little 1987:57).

To summarize, this case study permits its simulated travelers to use two modes of travel. Land travel speed is calculated using the hiker's formula, which takes into account slope plus an additional penalty for crossing waterways, depending on the size of the water body as measured by volume of flow. Water travel speed is calculated by adding or subtracting the velocity of the current from the base speed of 4 km/hr, and no water travel is permitted on waterways with flows less than 100 cfs.

10.3. Implementation of the Cost-Distance Model

The data used to calculate the cost distances come from two sources. Elevations are based on digital elevation model (DEM) data produced by the U.S. Geological Survey. The data's original resolution is such that each raster square corresponds to a spot of land 30 m by 30 m. I coarsened the data to 180-m^2 blocks in order to compute the results in adequate time. The hydrography data come from the NHDPlus dataset (Horizon Systems Corporation 2006) and include average annual current volume and velocity for each segment of waterway, which were themselves calculated from catchment sizes and modern rainfall data (Research Triangle Institute 2001). These data are accurate for our purposes insofar as we are willing to assume that rainfall amounts in the Mississippian period were similar to the average rainfalls recorded between 1960 and 1999 and we are willing to consider only mean annual flows for waterways.

To calculate least cost distances, I wrote a custom software application because available software was inadequate to handle the complex factors considered by my model (a problem also recorded by White this volume). The algorithm required for this simulation needs to be anisotropic, meaning the costs vary by direction, and generalizable, meaning the costs in each direction are independent of each other. With ArcGIS or IDRISI it is possible to implement a simulation of Tobler's anisotropic walking function alone, but it is currently impossible to also factor in costs associated with waterways.

In order to analyze the case study, I wrote custom code in the Visual Basic module within ArcGIS and in Visual Basic.NET. The code I wrote is an implementation of Dijkstra's algorithm (Dijkstra 1959) and the A* algorithm (Hart et al. 1972). These algorithms are simple, and I have chosen not to optimize the algorithm, which is referred to as the "brute-force" solution. As implemented, these algorithms are slow, but they are guaranteed to find the optimal solution rather than settle on a local optimum, which is a concern with the implementation in the widely available GIS packages (see Kantner this volume). Both algorithms are described in Chapter 1 of this volume.

10.4. Results

Figure 10.3 shows the least cost times between mounds in the sample. As in the distance measurement, there are two primary modes and a pair of outliers. The least cost distance calculations inform us that most secondary centers were less than 4 hours' travel from the administrative center of their polity and all are located less than 5 hours away. A single secondary center, Wilbanks (9Ck5), was located 22 km from its administrative center, Etowah (9Br1). That is a trip that would have taken 4.7 hours downstream from 9Ck1 to Br1 and 4.9 hours upstream. On average, contemporaneous mounds belonging to the same polity were located 2.2 hours or 9.9 km from each other. If we exclude the outlier, 9Ck5, these averages are 2.0 hours and 9.0 km.

Mounds from different polities are located at least 26 km or 5.6 hours from each other, and no competing administrative centers are closer than 33 km or 7.5 hours distant. Most mounds from competing centers are located a minimum of 8–10 hours from each other. A comparison of the straight-line distances and average speeds show that there is a high degree of correlation, which is to be expected: the greatest predictor of travel time is travel distance. On average, these journeys were made with a net speed of 4.6 km/hr. However, there is a significant range of variation with a low of 3.3 km/hr for the 4-km trip from 9St3 to the nearby 38Oc47. The fastest journey is the 46 km between 9Hk1 and 9Bl1, which averaged 4.9 km/hr.

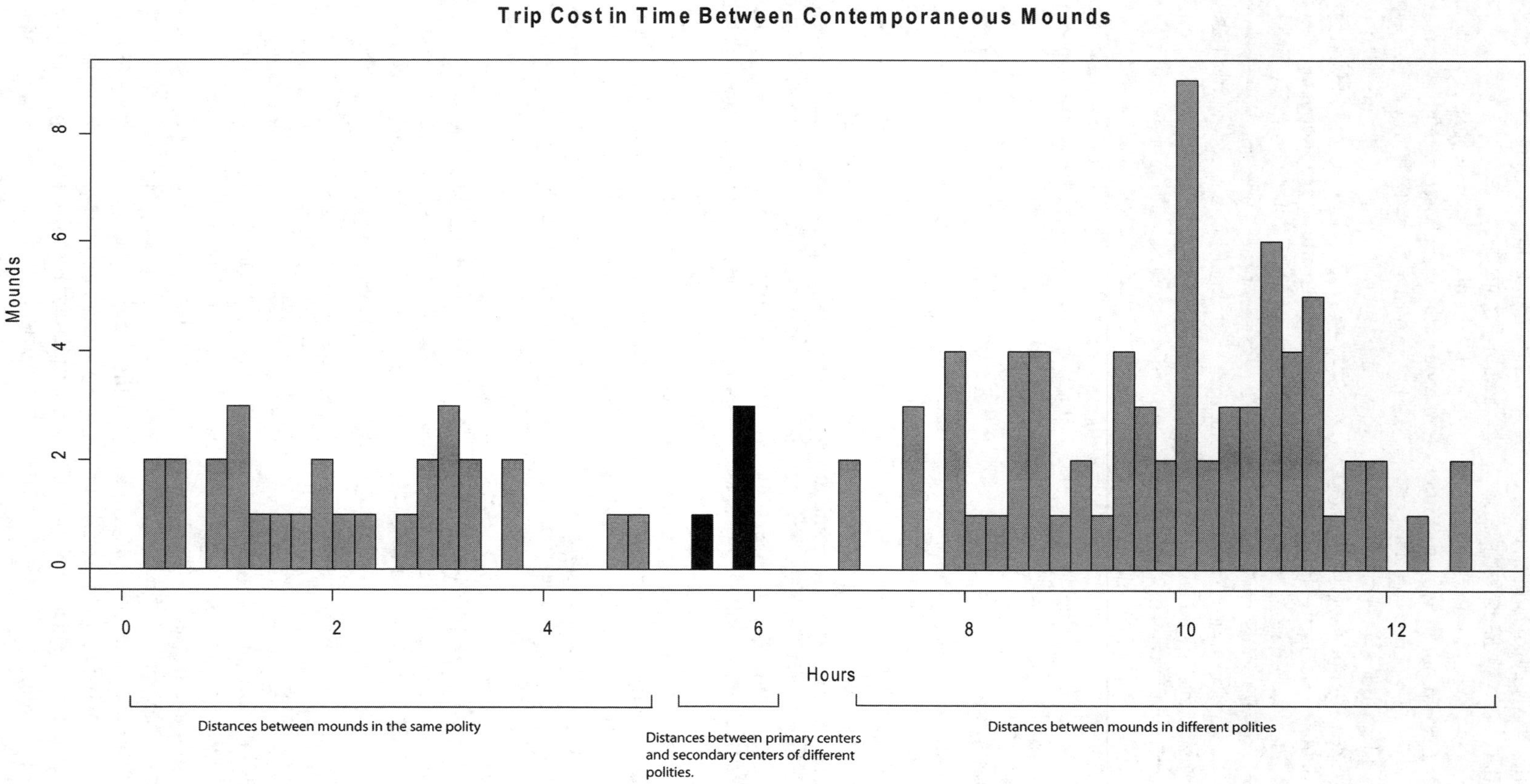

FIGURE 10.3. The travel time between contemporaneous mounds in the study area for which the travel cost is less than 13 hours. A total of 104 journeys between 52 mound pairs are represented here.

One of the reasons for this result is that under the parameters in this simulation, water travel is utilized very rarely. If one calculates the optimal path between pairwise combinations of all 45 mound pairs (ignoring, for now, contemporaneity), water is on average used to cover only 2 percent of the distance or 1.8 percent of the travel time. In most trips, water travel would not have been used at all for an optimal journey. In only 94 of the trips between mounds (out of 1,980 possible mound pairs, or 4.7 percent) is water used to travel more than 10 percent of the needed distance, and it is used to cover more than half the distance for only 6 of the possible trips (.3 percent). If we focus on the mounds that have been the subject of most of the discussion here, that is, pairs of mounds that were contemporaneous and located less than 13 hours' travel from each other, water travel was important between only one pair of sites: 9St1 and 9St3. Water travel was negligible (<6 percent of the travel distance) in all other cases.

Clearly, pedestrian travel is most significant, and given the relatively low relief of much of the study area, the most significant influence on travel times is the number and size of the river crossings. This corresponds fairly well with Hally's original estimation that canoe travel was a minor form of transportation in this region (David Hally, personal communication 2008). Of course, if one were to try simulating travel for the movement of heavy loads, it is likely that canoe travel would be more efficient in more cases.

Despite the high correspondence between travel time and distance, there are some subtle, but important, differences in the results. Namely, a visual inspection of the two histograms shows a stronger pattern of bimodality when travel time is used because of small shifts in some of the boundary cases. This can be statistically evaluated with Silverman's test (Baxter and Cool 2010; Silverman 1981, 1986), which helps quantify the probability of a certain number of modes being present.

For travel time, the test of the null hypothesis that there is just a single mode provides an associated probability of $p = .047$ (at bin width 2.253, or $p = .014$ with the calibrated version suggested by Hall and York 2001), which meets the traditional standard of statistical significant ($\alpha = .05$). Stated another way, the null hypothesis of a single mode is rejected, suggesting two modes. A subsequent test of more than two modes fails to hold ($p = .546$, calibrated $p = .423$, critical bin width = .746), which provides statistical evidence that there are two, but not more than two, modes. In contrast, the Silverman's test on pairwise distance has a $p = .059$ (calibrated $p = .051$, critical bin width = 10.244), which if we use a α of .05 says that we cannot reject the null hypothesis of a single mode. Phrased another way, we are fairly confident that there is more than one mode in the travel time data, with only a 1–5 percent chance that this is just statistical noise, but we have a little less certainty with the distance data, with a 5–6 percent chance that there is really just a single mode. The differences between these results are small ($p = .047/.014$ for time and $p = .059/.051$ for distance), but they provide empirical evidence that the histogram of travel time is more strongly bimodal that that of distance.

10.5. Evaluation

Straight-line distance certainly has many advantages as the unit of comparison in this study or in others like it. It is simple, it can be calculated rapidly, and it does not rely on any intervening models. The cost-distance simulation, in contrast, is built on multiple assumptions, such as the speed of canoes and the penalty to cross a water channel. The values used in this simulation are reasonable, but they are not the only reasonable values.

I argue that LCA, despite these disadvantages, can still be a useful tool for archaeologists if it helps demonstrate that our data meet theoretical expectations. In this case it does this in two ways. First, the underlying theoretical expectation is that secondary centers should be no more than a half-day's travel from the chiefly center. If we convert geodesic distance to travel time, we have a rather clear demonstration of this principle. Second, we expect that Mississippians had no direct way to measure straight-line distance and that instead their settlement patterns are shaped by a subjective distance evaluation of whether their mounds are either closer than a certain threshold to a polity center or more distant than another threshold from a neighboring polity; the best proxy we have for subjective distance in a prehistoric case study is travel time. It is subtle, but the Silverman's test confirms that travel time is more strongly bimodal than distance, which suggests it does a better job of explaining the patterning. This confirms our theoretical expectation that travel time was what was being perceived and manipulated by Mississippians when they located mound sites.

10.6. Future Research

There are several ways that the cost-distance simulation could be improved. As mentioned previously, we expect that swamps were a major barrier to overland travel below the fall line, that waterfalls were a major barrier to canoe travel, and that shoals were a barrier to canoe travel and an aid to river crossings. Unfortunately, we currently lack digitized datasets with the locations of these features, but they could be reconstructed and digitized from historic maps. Furthermore, the base canoe speed and penalties to water crossings used in this simulation are reasonable, but they are not the only reasonable values. In the future, one could sweep a range of possible parameter values to see what their effects are on the results. One approach would be to look for parameters that create stronger modes in the data as potentially useful.

In addition, it would be possible to extend this model to create a more nuanced understanding of the regional settlement system. For example, one could simulate waterways in different seasons instead of using waterway data based on average annual rainfall.

10.7. Conclusion

In summary, most Mississippian archaeologists are now familiar with the traditional Hally circles, showing 18-km as-the-crow-flies radii around presumed polity centers. I suggest that the measure of distance Mississippians were most aware of and the most likely to act on in determining settlement location was travel time. Therefore, the better representations of polity boundaries are isolines representing an estimate of 5 hours' travel. These lines are more complex to calculate and are based on more complex assumptions, but technology is catching up to the task and these arguably represent a better depiction of the dynamic underlying Mississippian regional systems. After all, one thing we know for sure is that crows didn't build the mounds.

References

Aldenderfer, Mark S.
1998 *Montane Foragers: Asana and the South-Central Andean Archaic.* University of Iowa Press, Iowa City.

Anderson, David G., and J. Christopher Gilliam
2000 Paleoindian Colonization of the Americas: Implications from an Examination of Physiography, Demography, and Artifact Distribution. *American Antiquity* 65:43–66.

Bartram, William
1996 *Travels, and Other Writings.* Library of America 84. Library of America, New York.

Baxter, M. J., and H. E. M. Cool
2010 Detecting Modes in Low-Dimensional Archaeological Data. *Journal of Archaeological Science* 37:2379–2385.

Bauer, Brian S., and R. Alan Covey
2002 Processes of State Formation in the Inca Heartland (Cuzco, Peru). *American Anthropologist* 104(3):846–864.

Bell, T. L., R. L. Church, and L. Gorenflo
1988 Late Horizon Regional Efficiency in the Northeastern Basin of Mexico—A Location-Allocation Perspective. *Journal of Anthropological Archaeology* 7(2):163–202.

Bénard de La Harpe, Jean Baptiste
1971 *Historical Journal of the Settlement of the French in Louisiana.* University of Southwestern Louisiana, Lafayette.

Bird, Isabella
1883 *The Golden Chersonese and the Way Thither.* J. Murray, London.

Burnett, P.
1978 Time Cognition and Urban Travel Behavior. *Geografiska Annaler Series B—Human Geography* 60(2):107–115.

Charlevoix, Pierre de
[1761] 1966 *Journal of a Voyage to North-America.* 2 vols. Readex Microprint, New York.

Clayton, Lawrence A., Vernon James Knight, and Edward C. Moore
1993 *The De Soto Chronicles: The Expedition of Hernando de Soto to North America in 1539–1543.* 2 vols. University of Alabama Press, Tuscaloosa.

Cohen, Ronald, and Alice Schlegel
1968 The Tribe as a Socio-Political Unit: A Cross-

Cultural Examination. In *Proceedings: Essays on the Problem of Tribe*, pp. 120–149. American Ethnological Society, Seattle.

Crompton, A.
2006 Perceived Distance in the City as a Function of Time. *Environment and Behavior* 38(2):173–182.

Dijkstra, Edsger Wybe
1959 A Note on Two Problems in Connexion with Graphs. *Numerische Mathematik* 1:269–271.

Ethridge, Robbie Franklyn
2003 *Creek Country: The Creek Indians and Their World*. University of North Carolina Press, Chapel Hill.

Glazier, Willard W.
1891 *Down the Great River: Embracing an account of the discovery of the true source of the Mississippi, together with views, descriptive and pictorial, of the cities, towns, villages and scenery on the banks of the river, as seen during a canoe voyage of over three thousand miles from its head waters to the gulf of Mexico*. Hubbard Bros., Philadelphia.

Golledge, R. G., and G. Zannaras
1973 Cognitive Approaches to the Analysis of Human Spatial Behavior. In *Environment and Cognition*, edited by W. H. Ittelson, pp. 59–94. Seminar, New York.

Gorenflo, L. J., and Thomas L. Bell
1991 Network Analysis and the Study of Past Regional Organization. In *Ancient Road Networks and Settlement Hierarchies in the New World*, edited by C. D. Trombold, pp. 80–98. Cambridge University Press, Cambridge.

Gorenflo, L. J., and Nathan Gale
1990 Mapping Regional Settlement in Information Space. *Journal of Anthropological Archaeology* 9(3):240–274.

Hall, Peter, and Matthew York
2001 On the Calibration of Silverman's Test for Multimodality. *Statistica Sinica* 11:515–536.

Hally, David J.
1993 The Territorial Size of Mississippian Chiefdoms. In *Archaeology of Eastern North America: Papers in Honor of Stephen Williams*, edited by J. B. Stoltman, pp. 143–168. Mississippi Department of Archives and History, Jackson.
1999 The Settlement Pattern of Mississippian Chiefdoms in Northern Georgia. In *Settlement Pattern Studies in the Americas: Fifty Years since Virú*, edited by B. R. Billman and G. M. Feinman, pp. 96–115. Smithsonian Institution Press, Washington, D.C.
2006 Nature of Mississippian Regional Systems. In *Light on the Path: The Anthropology and History of the Southeastern Indians*, edited by T. J. Pluckhahn and R. Ethridge, pp. 26–42. University of Alabama Press, Tuscaloosa.

Hare, Timothy S.
2004 Using Measures of Cost Distance in the Estimation of Polity Boundaries in the Postclassic Yautepec Valley, Mexico. *Journal of Archaeological Science* 31:799–814.

Harrison, K. David
2007 *When Languages Die: The Extinction of the World's Languages and the Erosion of Human Knowledge*. Oxford University Press, Oxford.

Hart, P. E., N. J. Nilsson, and B. Raphael
1972 Correction to "A Formal Basis for the Heuristic Determination of Minimum Cost Paths." *SIGART Newsletter* 37:28–29.

Hassig, Ross
1985 *Trade, Tribute, and Transportation: The Sixteenth-Century Political Economy of the Valley of Mexico*. Civilization of the American Indian Vol. 171. University of Oklahoma Press, Norman.

Helms, Mary W.
1979 *Ancient Panama*. University of Texas Press, Austin.

Hollenbach, Kandace Detwiler
2005 Gathering in the Late Paleoindian and Early Archaic Periods in the Middle Tennessee River Valley, Northwest Alabama. Unpublished Ph.D. dissertation, University of North Carolina, Chapel Hill.

Horizon Systems Corporation
2006 *NHDPlus User Guide, July 1, 2006*. Horizon Systems Corporation, Herndon, Virginia.

Hudson, Charles M.
1976 *The Southeastern Indians*. University of Tennessee Press, Knoxville.

Jennings, Justin, and Nathan Craig
2001 Politywide Analysis and Imperial Political Economy: The Relationship between Valley Political Complexity and Administrative Centers in the Wari Empire of the Central Andes. *Journal of Anthropological Archaeology* 20: 479–502.

Johnson, Gregory A.
1987 The Changing Organization of Uruk Administration on the Susiana Plain. In *The Archaeology of Western Iran: Settlement and Society from Prehistory to the Islamic Conquest*, edited by F. Hole. Smithsonian Institution Press, Washington, D.C.

Jones, Kevin T., and David B. Madsen
1989 Calculating the Cost of Resource Transportation: A Great Basin Example. *Current Anthropology* 30(4):529–534.

Kantner, John
1997 Ancient Roads, Modern Mapping: Evaluating Chaco Anasazi Roadways Using GIS Technology. *Expedition* 39(3):49–62.
Lawson, John
[1709] 1967 *A New Voyage to Carolina*. University of North Carolina Press, Chapel Hill.
Lee, Richard B.
1979 *The !Kung San: Men, Women, and Work in a Foraging Society*. Cambridge University Press, Cambridge.
Lee, Terence
1970 Perceived Distance as a Function of Direction in the City. *Environment and Behavior* 2(1): 40–51.
Leusen, Pieter Martijn van
2002 Pattern to Process Methodological Investigations into the Formation and Interpretation of Spatial Patterns in Archaeological Landscapes. Unpublished Ph.D. dissertation, Rijksuniversiteit, Groningen, the Netherlands.
Limp, Fred
1990 Continuous Cost Movement Models. In *Applications of Space-Age Technology in Anthropology*, edited by C. A. Behrens and T. L. Sever, pp. 231–250. NASA, Bay St. Louis, Mississippi.
Little, Elizabeth A.
1987 Inland Waterways in the Northeast. *Midcontinental Journal of Archaeology* 12(1):55–76.
Little, Kenneth
1967 The Mende Chiefdoms of Sierra Leone. In *West African Kingdoms in the Nineteenth Century*, edited by D. Forde and P. Kaberry, pp. 239–259. Oxford University Press, London.
MacEachren, Alan M.
1980 Travel Time as the Basis for Cognitive Distance. *Professional Geographer* 32(1):30–36.
McCormack, Gavin R., Ester Cerin, Eva Leslie, Lorinne Du Toit, and Neville Owen
2008 Objective versus Perceived Walking Distances to Destinations: Correspondence and Predictive Validity. *Environment and Behavior* 40(3): 401–425.
Mitra, Sarat Chandra
1910 Further Notes on the Primitive Method of Computing Time and Distance. *Journal of the Anthropological Society of Bombay* 9(2):83–92.
Mohlke, Matthew
2001 *Floating Down the Country*. Lone Oak Press, Red Wing, Minnesota.
Montello, Daniel
1997 The Perception and Cognition of Environmental Distance: Direct Sources of Information. In *Spatial Information Theory: A Theoretical Basis for GIS*, pp. 297–311. Springer, Berlin.
Nasar, J. L., H. Valencia, and Z. A. Omar
1985 Out of Sight, Further from Mind: Destination Visibility and Distance Perception. *Environment and Behavior* 17(5):627–639.
Peterson, Phil
2006 *All Things Are Possible: The Verlen Kruger Story; 100,000 Miles by Paddle*. Adventure Publications, Cambridge, Minnesota.
Pugh, John, and Jessica Robinson
2005 *Source to Sea: The Mississippi River Project*. Electronic document, http://www.sourcetosea.net, accessed July 24, 2007.
Rassadin, Valentin Ivanovich
1995 *Tofalarsko-russkii slovar' russko-tofalarskii*. Vostochno-Sibirskoe knizhnoe izd-vo, Irkutsk, Russia.
Research Triangle Institute
2001 *The National Water Pollution Control Assessment Model, (NWPCAM) Version 2 Draft Report, Prepared for the U.S. Environmental Protection Agency*. Research Triangle Institute, Research Triangle Park, North Carolina.
Rowland, Dunbar, and A. G. Sanders
1927 *Mississippi Provincial Archives: 1729–1740, French Dominion*. Press of the Mississippi Department of Archives and History, Jackson.
Sadalla, E. K., and L. J. Staplin
1980 The Perception of Traversed Distance: Intersections. *Environment and Behavior* 12(2): 167–182.
Schnitzka, Michael, and William Perdzock
2007 *Conquer the Rhine*. Electronic document, http://www.conquertherhine.com/team.html, accessed July 26, 2007.
Sijia, Wang, Cecil M. Lewis Jr., Mattias Jakobsson, Sohini Ramachandran, Nicolas Ray, Gabriel Bedoya, Winston Rojas, Maria V. Parra, Julio A. Molina, Carla Gallo, Giovanni Poletti, Kim Hill, Ana M. Hurtado, Damian Labuda, William Klitz, Ramiro Barrantes, Maria Cátira Bortolini, Francisco M. Salzano, Maria Luiza Petzl-Erler, and Luiza T. Tsuneto
2007 Genetic Variation and Population Structure in Native Americans. *PLoS Genetics* 3(11):e185. doi:10.1371/journal.pgen.0030185.
Silverman, B. W.
1981 Using Kernel Density Estimates to Investigate Multimodality. *Journal of the Royal Statistical Society, Series B (Methodological)* 43(1):97–99.
1986 *Density Estimation*. Chapman and Hall, London.
Simmons, Tim
2005 Via Detour, Paddling Students Reach the Gulf; Weather Took Pair off Mississippi. *News & Observer* (Raleigh, North Carolina), August 8.
Spencer, Charles S.
1987 Rethinking the Chiefdom. In *Chiefdoms in*

the Americas, edited by R. D. Drennan and C. Uribe, pp. 369–390. University Press of America, Lanham, Maryland.
1990 On the Tempo and Mode of State Formation: Neoevolutionism Reconsidered. *Journal of Anthropological Archaeology* (9):1–30.
1993 Human Agency, Biased Transmission, and the Cultural Evolution of Chiefly Authority. *Journal of Anthropological Archaeology* 12:41–74.

Tanner, Helen Hornbeck
1989 The Land and Water Communication Systems of the Southeastern Indians. In *Powhatan's Mantle: Indians in the Colonial Southeast*, edited by P. H. Wood, G. Waselkov, and M. T. Hatley, pp. 6–20. University of Nebraska Press, Lincoln.

Tobler, W.
1993 Non-isotrophic Geographic Modeling. In *Three Presentations on Geographic Analysis and Modeling*. Technical Report 93-1. National Center for Geographic Information and Analysis, University of California, Santa Barbara.

Waselkov, Gregory A., and Kathryn E. Holland Braund
1995 *William Bartram on the Southeastern Indians*. Indians of the Southeast. University of Nebraska Press, Lincoln.

Wood, Brian M., and Zoe J. Wood
2006 Energetically Optimal Travel across Terrain: Visualizations and a New Metric of Geographic Distance with Anthropological Applications. *SPIE Electronic Imaging*, January.

Wright, Henry T.
1977 Recent Research on the Origins of the State. *Annual Review of Anthropology* 6:379–397.
1984 Prestate Political Formations. In *On the Evolution of Complex Societies: Essays in Honor of Harry Hoijer*, edited by T. Earle, pp. 41–78. Undena Publications, Malibu, California.

CHAPTER 11

Prehistoric Trail Networks of the Western Papaguería

A Multifaceted Least Cost Graph Theory Analysis

DEVIN A. WHITE

11.1. Introduction

Trails are a significantly understudied part of the global archaeological record, especially in the Western Papaguería portion of the North American Southwest. Trails tell a great deal about how a particular society operated and what was important to its constituent members. The meanderings of trails, their end points, what is found along them, what is found near them, and how often they were used can provide invaluable information about the economic, political, social, and ideological relationships in a particular culture. Trails are also physical manifestations of a society's degree of interconnectedness with the landscape, with the environment, and with its people. Indigenous groups in the broader Southwest, such as the Hopi, Zuni, Acoma, Pima, Maricopa, and O'Odham, still use trails—and they serve a variety of purposes. Trails within these cultures connect villages to one another across the landscape, villages to sources of raw materials, and villages to places of social importance such as cemeteries and shrines; they also serve as routes for spiritual pilgrimages. In other words, trail systems embody economic, social, political, and ideological relations. They provide routes for the transport of goods and labor in economic systems of exchange and in political systems of tribute collection and regional integration. They provide routes along which people move and interact. Furthermore, they are important ceremonial networks that social and political institutions use to symbolize and make manifest their structural character. Trails channel movement and influence how the landscape is experienced.

Studies of trails within the Western Papaguería are not commonplace, but a small handful of scholars have documented several prehistoric and historic long-distance footpath networks in the broader Sonoran Desert region that revolve around salt procurement, shell trade, pilgrimage, and warfare (Becker and Altschul 2008; Darling 2006; Darling and Eiselt 2003; Hayden 1967; Lumholtz 1912; Riley and Manson 1991; Rogers 1945).

In particular, Julian Hayden's work in the Sierra Pinacate region of northern Mexico found that over 700 km of well-preserved trails were linked into a network that had reliable water sources as the hubs (Hayden 1967). These authors all point out that the networks they found were just a small sample of what is awaiting discovery in the southern Southwest (see Phillips and Leckman this volume for examples). A more detailed and systematic study of footpaths in this region could prove to be very useful for archaeologists seeking to learn more about the many and varied cultures that traveled through and lived in the region. In particular, a study that can take a holistic and comparative look at large sections of the Western Papaguería using methods and techniques borrowed from the remote sensing and geospatial analysis communities, specifically least cost analysis (LCA), holds great po-

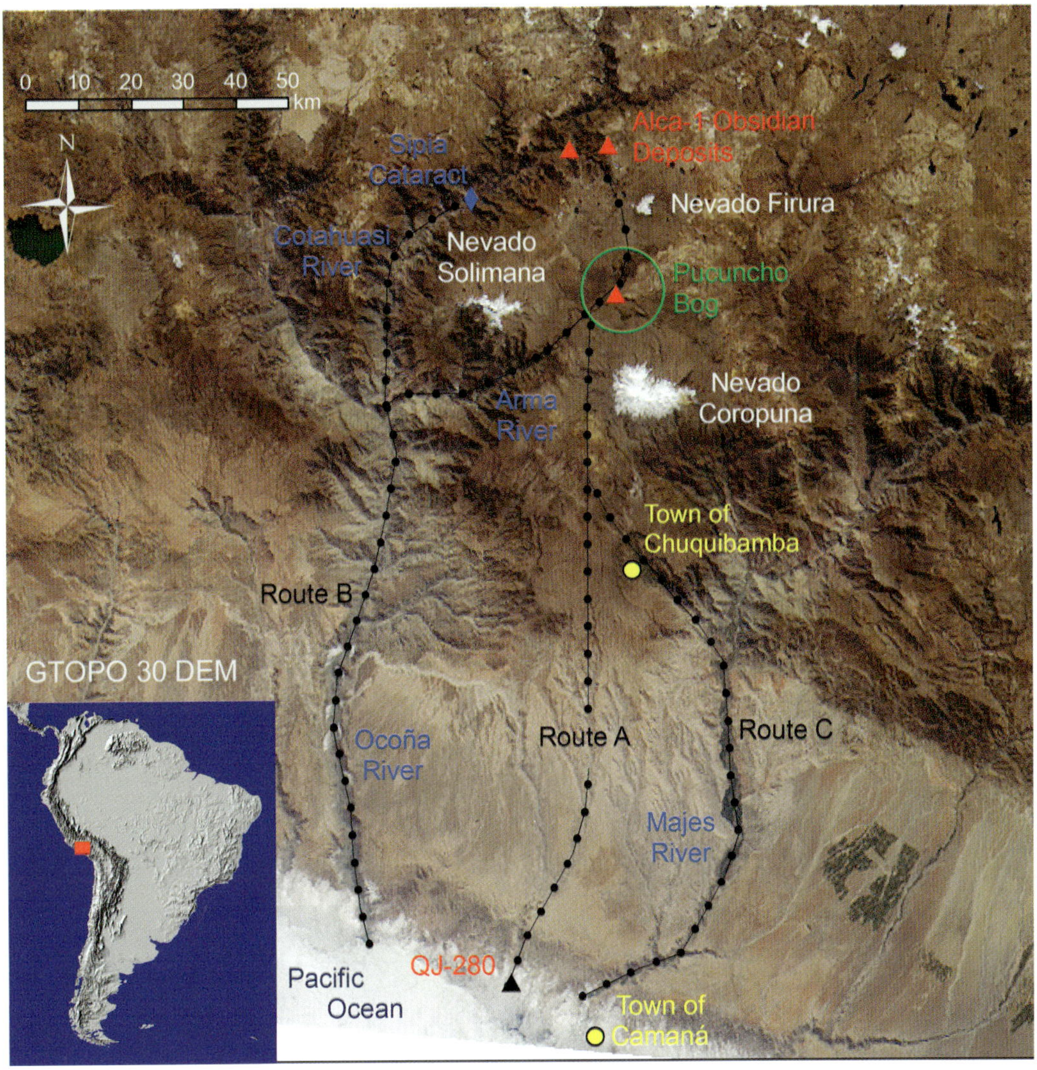

PLATE 1. Multi-angle Imaging SpectroRadiometer (MISR) satellite image taken July 17, 2000 (courtesy NASA Earth Observatory), with GTOPO 30 Digital Elevation Model (DEM) inset and places referred to in the text. Quebrada Jaguay (QJ-280), a coastal Paleoindian site (Sandweiss et al. 1998), and Alca-1 obsidian deposits (Rademaker 2006) are shown as triangles. The circle indicates the location of the Pucuncho Basin. Permanent ice caps, rivers, the Sipia Cataract, and modern towns are labeled. Dotted lines trace potential coast-highland corridor routes A, B, C, linking Quebrada Jaguay with Alca-1 obsidian deposits.

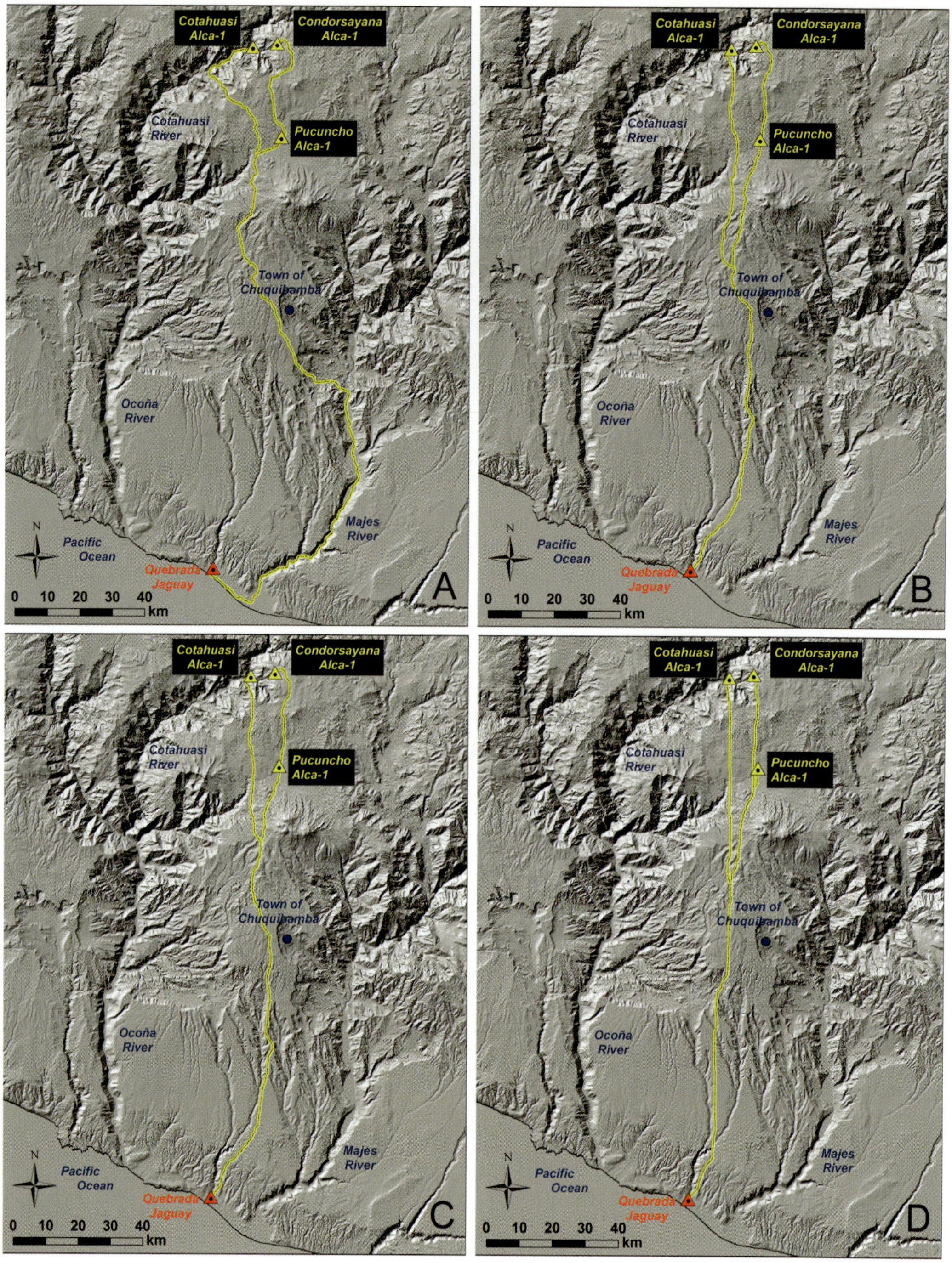

PLATE 2. Hillshaded SRTM DEM with LCPs between Quebrada Jaguay and Alca-1 obsidian deposits. (A) Slope-based Path Distance results; (B) Energy-based Path Distance Result 1; (C) Energy-based Path Distance Result 2; (D) Energy-based Path Distance Result 3.

PLATE 3. Hillshaded SRTM DEM of the plateau LCPs; locations of potentially early open-air and rockshelter archaeological sites and diagnostic LG and EH projectile points, lakes, wetlands; and paleogeographic reconstruction of LG ice and permafrost extents.

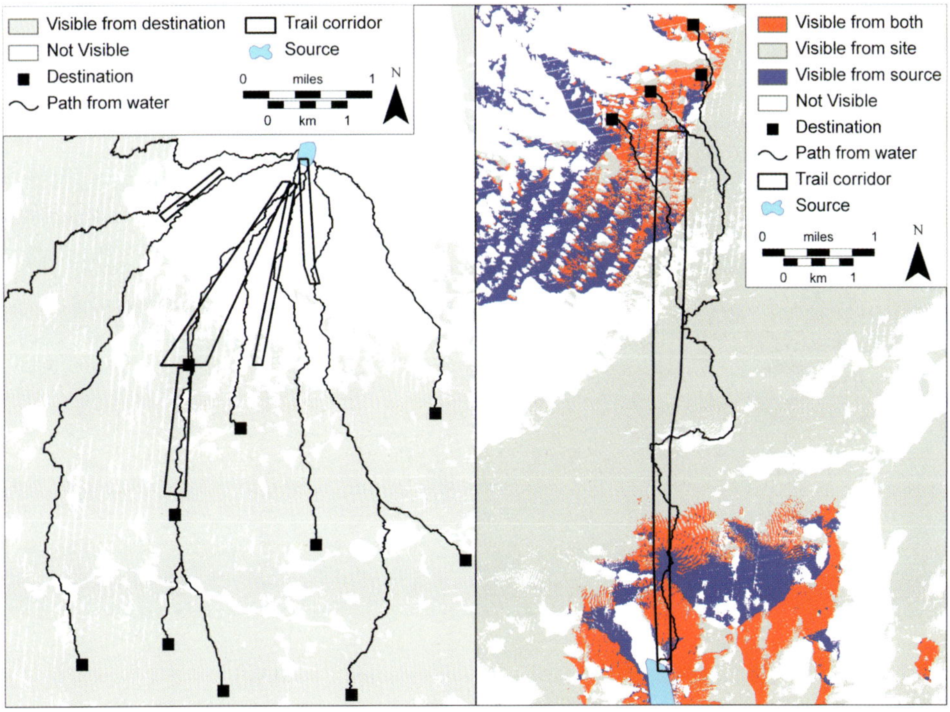

PLATE 4. Viewshed results for McGregor (*left*) and Doña Ana trails (*right*).

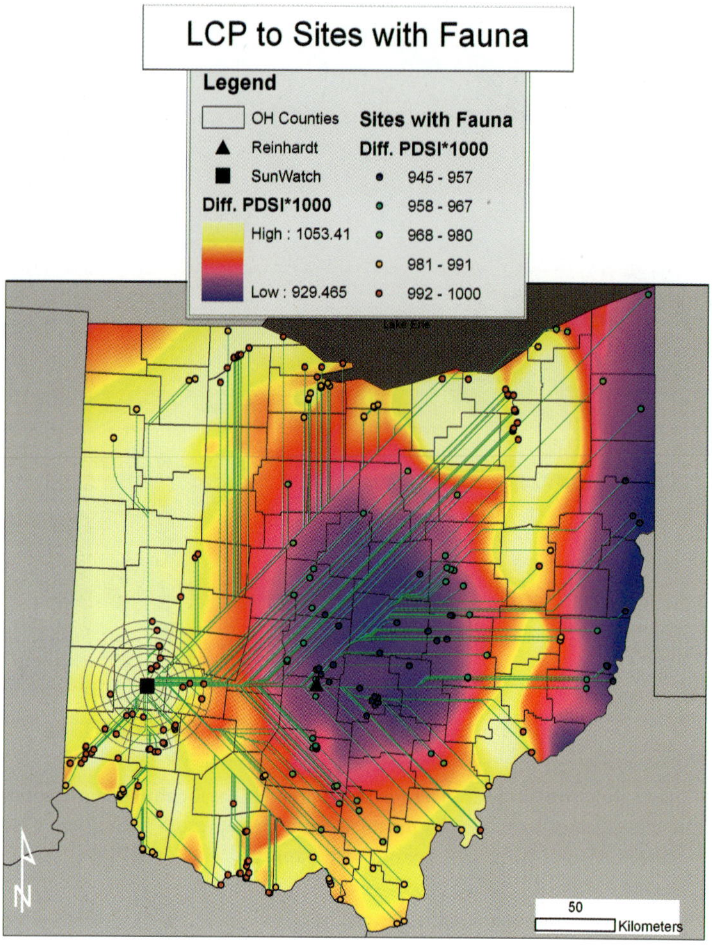

PLATE 5. Least cost paths from SunWatch for difference of PDSI for the period ad 1301–1350. Cost surface is the absolute value of the difference in PDSI from the value for SunWatch subtracted from 1 and multiplied by 1000. The higher the value, the greater the similarity in PDSI values for the period.

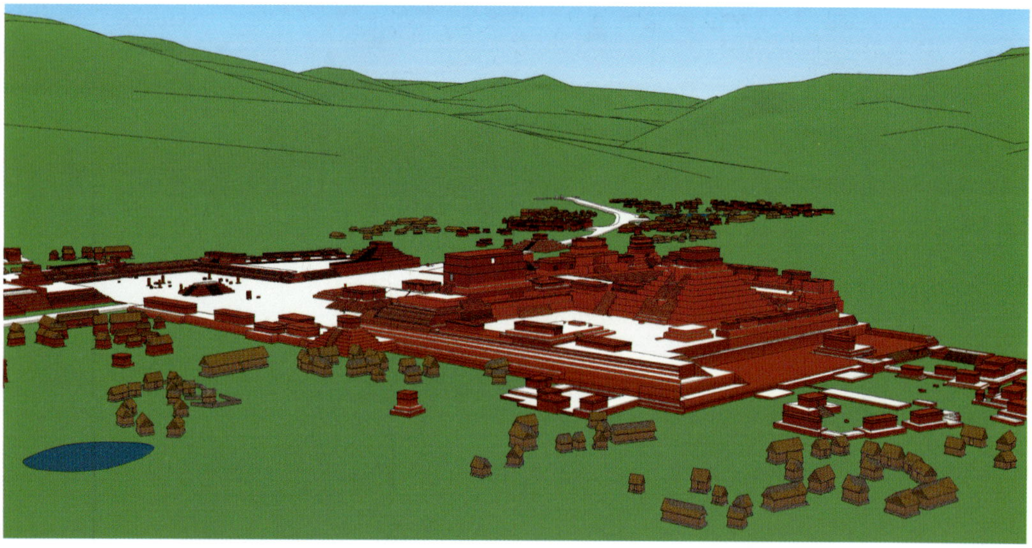

PLATE 6. Three-dimensional Google SketchUp reconstruction of Copán's urban core showing civic-ceremonial buildings in the Principal Group and surrounding residential sites. The reconstruction was derived from GIS data.

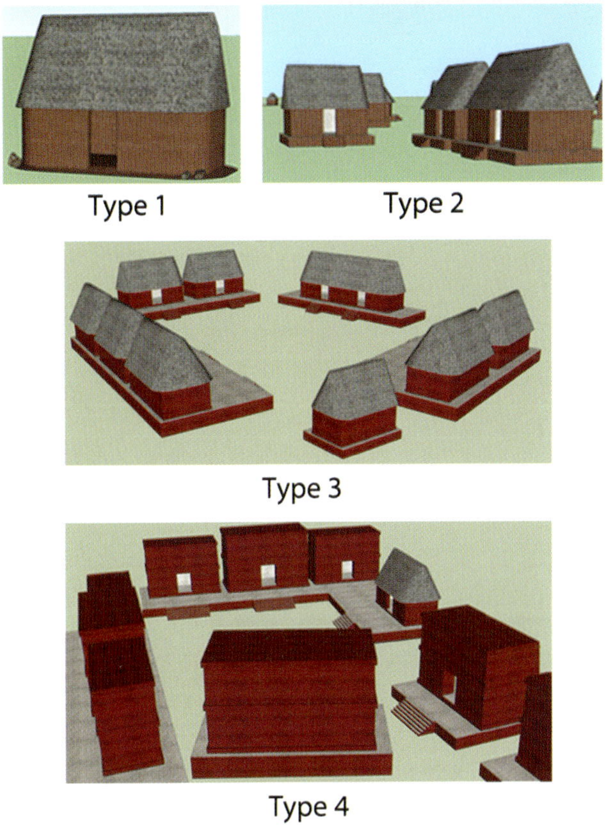

PLATE 7. Site types 1–4 of the Harvard Site Typology at Copán, Honduras, showing the variation in architecture, size, and complexity.

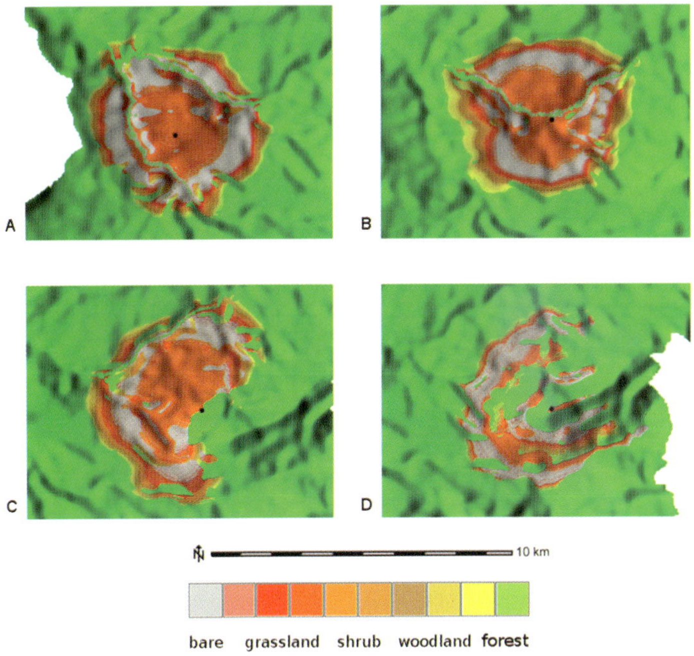

PLATE 8. Maps of land cover after the 100th year of simulation for (a) the flat-land village, (b) the village located in a *barranco*, (c) the village located against a steep slope, and (d) the village located on a ridge.

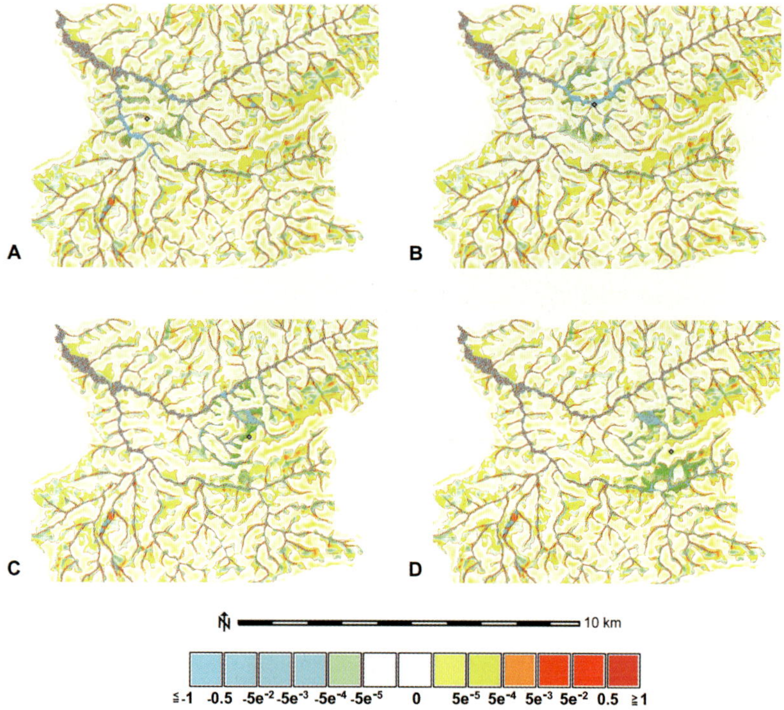

PLATE 9. Maps of cumulative vertical change (erosion and deposition) after 100 years of simulation for (a) the flat-land village, (b) the village located in a *barranco*, (c) the village located against a steep slope, and (d) the village located on a ridge.

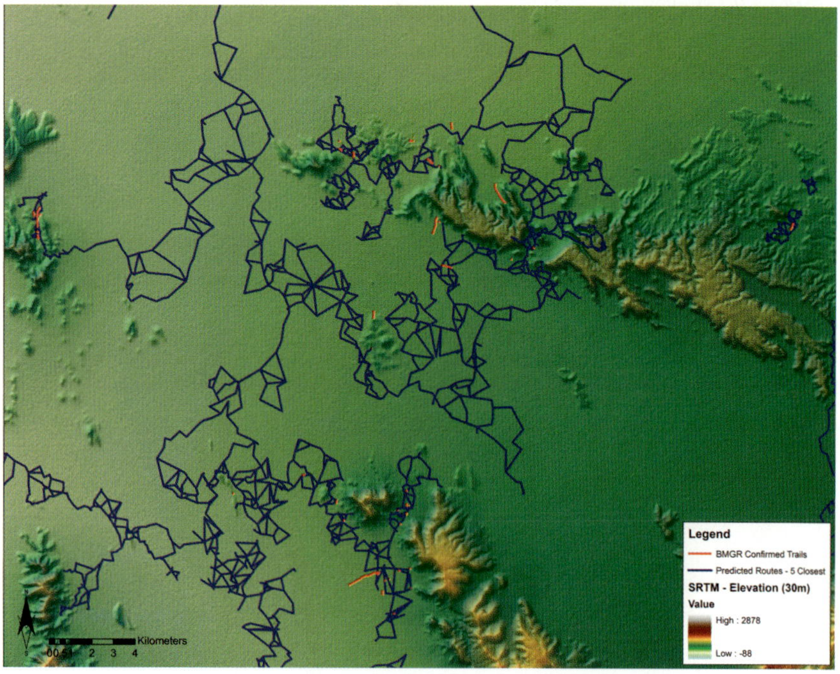

PLATE 10. Modeled prehistoric trails on NTAC and STAC. Results (*blue*) represent intraregional integration, in which all known archaeological sites are connected to their five closest neighbors via the most energy-efficient routes. Known, confirmed prehistoric trails (*red*) match up reasonably well on NTAC (*upper half of image*, 76 percent) and slightly better on STAC (*lower half of image*, 82 percent).

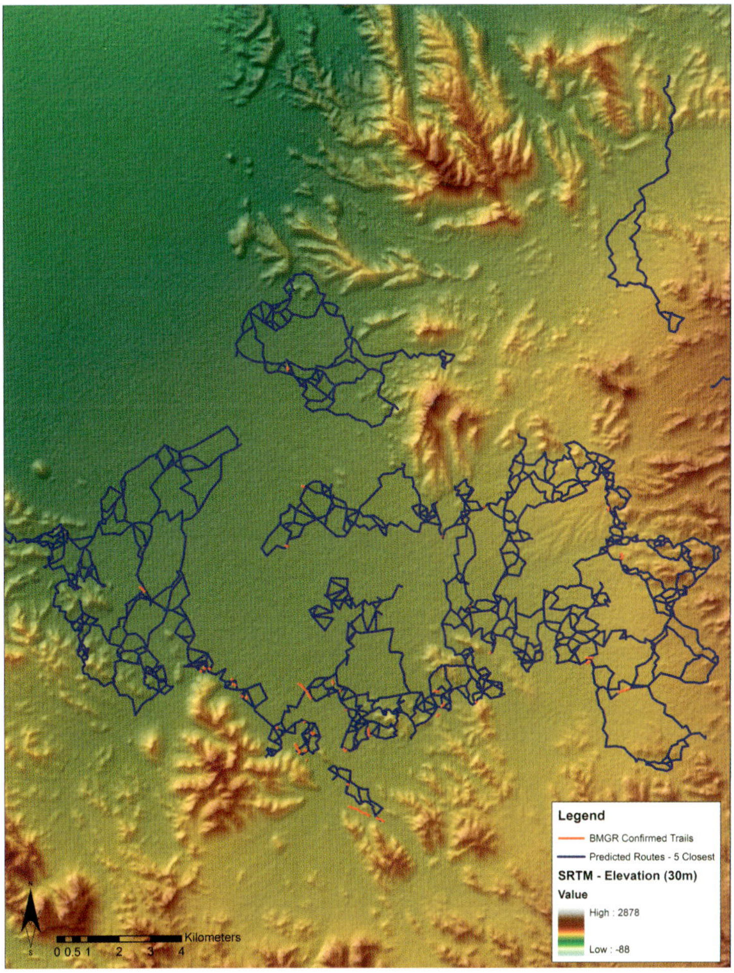

PLATE 11. Modeled prehistoric trails on ETAC. Results (*blue*) represent intraregional integration, in which all known archaeological sites are connected to their five closest neighbors via the most energy-efficient routes. Known, confirmed prehistoric trails (*red*) match up exceptionally well (94 percent).

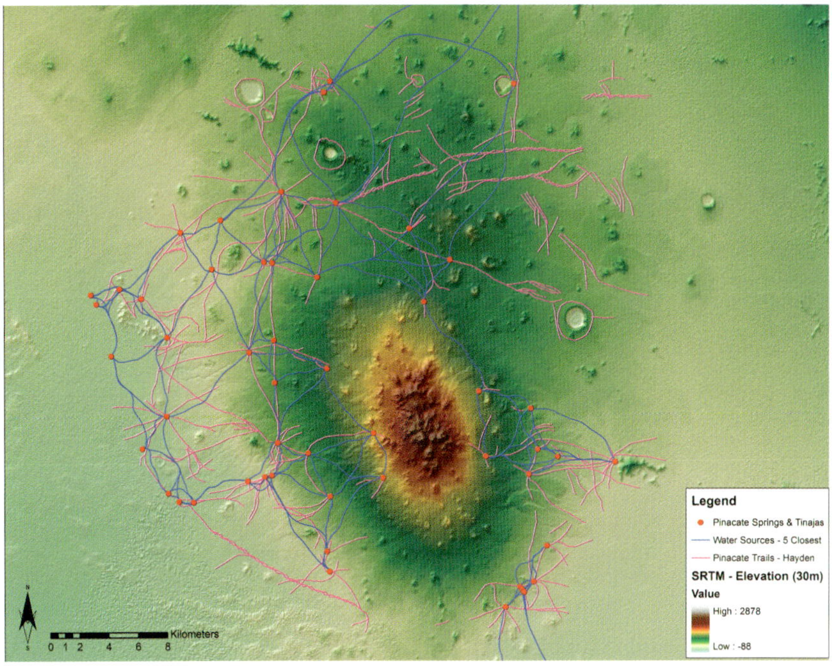

PLATE 12. Modeled prehistoric trails in the Sierra Pinacate. Results (*blue*) represent all known water sources connected to their five closest neighbors via the most energy-efficient routes. Known, confirmed prehistoric trails (*magenta*) do not match up well (46 percent). Note the large number of braided trails.

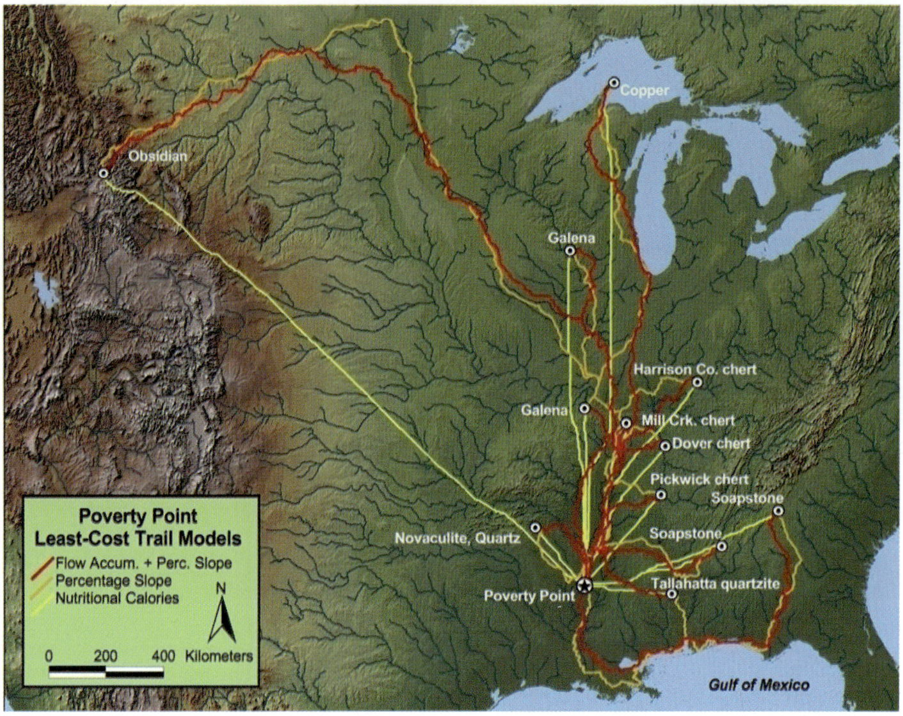

PLATE 13. Raw material acquisition routes into Poverty Point, Louisiana, ca. 3200 cal yr bp (image courtesy J. Christopher Gillam, from Anderson et al. 2007).

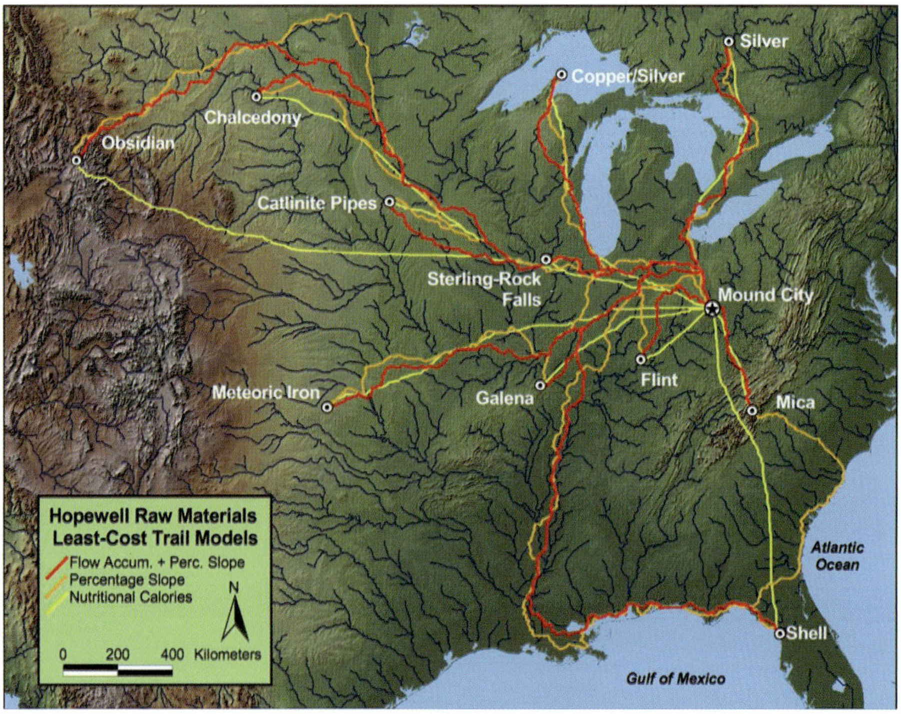

PLATE 14. Raw material acquisition routes into the Scioto River valley of central Ohio, ca. 1800 cal yr bp (image courtesy J. Christopher Gillam, from Anderson et al. 2007).

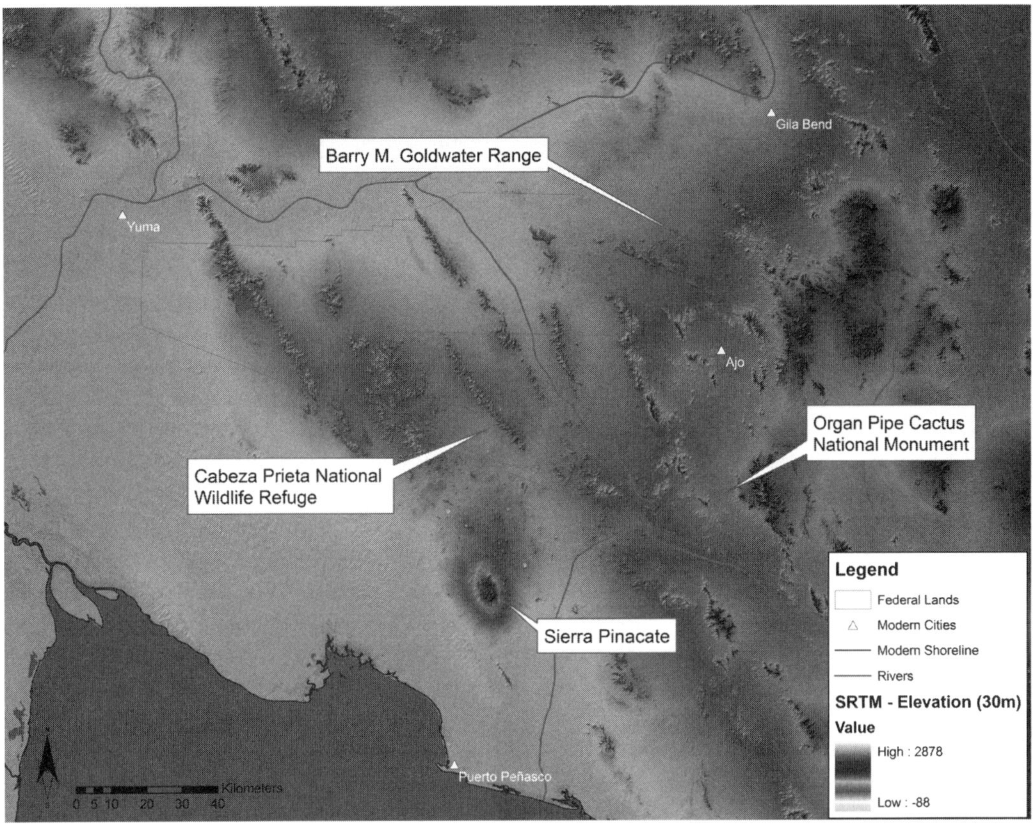

FIGURE 11.1. Map of the Western Papaguería study area, which extended from Yuma, Arizona, in the west to Gila Bend, Arizona, in the east and from the Gila River in the north to Puerto Peñasco, Sonora, Mexico, in the south.

tential for generating a significant amount of new information on the subjects of regional integration, regional travel, and interactions with and uses of landscapes.

However, only recently has the detection and interpretation of footpaths using GIS and remotely sensed data become a research topic in Arizona archaeology. The more traditional focus has been on remnants of prehistoric irrigation canals. For the most part, there has been a heavy reliance on black-and-white aerial photography and pedestrian survey, both of which are somewhat limited in thc types of information they can provide. In this region, therefore, opportunities exist to exploit advanced geospatial processing techniques and more complex forms of remotely sensed data—especially that collected by modern spaceborne and airborne platforms—in order to answer questions about regional integration and interaction. These methods offer, among many other advantages, (1) consistent data sources, (2) large-scale coverage, and (3) the ability to carry out sophisticated quantitative analyses.

This chapter, an extension of my study of the trail networks present in the Western Papaguería (White 2007), focuses on the Barry M. Goldwater Air Force Range (BMGR) in southern Arizona and the Sierra Pinacate Biosphere Reserve in northwest Mexico (Figure 11.1), which represent a significant portion of the Sonoran Desert along the border with Mexico—approximately 6.5 million acres. This area is traditionally regarded as an important trade corridor between the Gulf of California and the Hohokam regional system, which is centered on the Phoenix Basin. For centuries, high volumes of shell and salt came through this region and were distributed widely among the prehistoric peoples of the Southwest. Until fairly recently, it was thought that there were no permanent settlements here because the climate was too harsh to support agriculture. However, the results of several large archaeological projects that have been conducted by various Cultural Resource Management firms within the BMGR as part of its Section 106 compliance requirements have all found semi-permanent to permanent agricultural settlements. Some of these settlements, such as Verbena Village and Lost City, are quite large and appear to be major hubs for trade and social interaction. The largest settlements appear to be about a day's walk apart from one another along major trade routes through the region, but there are smaller outlying settlements scattered across the landscape (Hill and Bruder 2000; Tucker 2000a, 2000b, 2000c; Slaughter et al. 2000; Lyon 2000a, 2000b; Roberts et al. 2000; Ahlstrom and Roberts 2002). It is also clear that this region was in continuous use by native peoples from the early pre-Paleoindian through historic time periods.

This chapter explores the roles and meanings of trail networks and related landscapes for prehistoric cultures within the Western Papaguería. In general, it investigates how footpaths acted as mechanisms for interregional transportation, multiscalar community and regional integration, and ideological facilitation both synchronically and diachronically through the use of multispectral and hyperspectral remotely sensed data, pedestrian survey, and GIS-based predictive modeling. Additionally, it focuses on the development of efficient methods for the detection, identification, and recording of trail networks and associated archaeological features. In particular, this chapter explores three lines of inquiry concerning the roles and meanings of trails, networks, and landscapes in the Western Papaguería:

1. How effectively can both large-scale prehistoric footpath networks and what they tie together be found and interpreted using remote sensing, GIS, and traditional field methods?
2. What kinds of footpath networks operated in the Sonoran Desert region? What can their morphologies tell us about their functions? How do their functions compare and contrast across the region?
3. What can the functions of these footpath networks tell us about how past societies that moved through and lived in the region used the various landscapes at their disposal?

11.2. Technical Approach

Locating, documenting, and understanding the functions of prehistoric trail networks in as large an area as the Western Papaguería is no small task. Ordinarily, prehistoric trail network studies are conducted on a much smaller and far less daunting scale. The primary limitation of research on prehistoric trails and roads is the availability of reasonably unmodified prehistoric cultural landscapes. Another major limitation revolves around the practicality of surveying large landscapes using traditional pedestrian methods. To effectively and thoroughly survey a large prehistoric landscape in an effort to find prehistoric trails and related sites could take decades—depending on budget and crew size. In recent years, archaeologists have started to use high spatial resolution aerial photography and satellite imagery as a way to augment and streamline pedestrian survey, but obtaining the necessary imagery can be an expensive and time-consuming process—if the imagery exists at all. Additionally, as the size of the landscape increases, so does the overall cost of analysis on every level: financial, mental, physical, temporal, and computational. At the outset of this project, though, it was clear that many of these obstacles to conducting large-scale prehistoric trails research had been removed. The Sonoran Desert region is relatively pristine. The Air Force took care of the project's imagery needs. Accessing the necessary software and hardware to analyze such a large volume of imagery was generally not a problem. The locations of a large number of potentially important starting and ending points for travel (mainly archaeological sites and reliable water sources) within the two primary research zones of the region (BMGR and Sierra Pinacate) were known from various databases and literature searches. As a result, pedestrian survey could be kept to a minimum. So the question became one of how to best approach and carry out such a large-scale study.

Addressing the first line of inquiry presented above involved locating and documenting as many prehistoric trails as possible within the region through the process of examining large volumes of high spatial resolution digital imagery, recording the locations of potential trails spotted in the imagery, and then ground truthing those locations via pedestrian survey. Addressing the second line of inquiry necessitated building a GIS database that contained the accurate map locations of as many archaeological sites, sacred sites, water sources, and potential natural resources as possible for the entire region in order to have a world of potentially important starting and ending points for travel along located trails. Addressing the third line of inquiry required finding ways to connect subsets of the aforementioned starting and ending points (sites to sites, sites to water sources, sites to natural resources, sites to sacred places, through-travel to procure exotic materials, etc.) in the most efficient manner possible and comparing these modeled routes with the locations of confirmed prehistoric trails. The goal was to see if predicted routes matched up to trails verified on the ground within a particular landscape. If so, and if the starting and ending points were solely sites, the general conclusion would be that the major focus of landscape use was sociopolitical. If so, and if the points were a mixture of sites and natural resources, the general conclusion would be that the major focus of landscape use was economic. If the predicted routes for supplied starting and ending points did not match up with archaeological reality, then the general conclusion would revolve around ideological landscape use (if sacred sites and resources were involved) or would rely on further evidence from available ethnographic sources related to the movements of historically known native groups from the region. In some cases, trail function remained undetermined.

11.2.1. Locating and Interpreting Prehistoric Trails: A Hybrid Approach

When one is presented with a study area roughly 6.5 million acres in size, it is difficult to know where to begin with respect to locating and interpreting prehistoric trails accurately and efficiently. In the end, a hybrid approach consisting of locating trails via airborne and spaceborne imagery followed up by pedestrian ground truthing and in-field documentation was selected. The decision to use this method was based on its successful application by the Proyecto Prehistórico Arenal in Costa Rica (Butler 2003; Sever et al. 2003; Sheets et al. 1991; Sheets and Sever 1991; Sheets and McKee 1994; Weller 2003; White 2003); however, several modifications to the method had to be made so it would be suitable for research in the Sonoran Desert region.

The hybrid method, as it was used by the Proyecto Prehistórico Arenal, consists of manually examining high spatial resolution imagery for faint linear anomalies that do not conform to modern elements on the landscape (roads, fence lines, etc.), noting their locations, finding the anomalies in the field, and trenching across them to study their stratigraphy (Butler 2003; Sever et al. 2003; Sheets et al. 1991; Sheets and Sever 1991; Sheets and McKee 1994; Weller 2003; White 2003). If the stratigraphy was consistent with known formation processes for trails in the Arenal region, its status as prehistoric was confirmed and its extent, morphology, and age were fully documented when possible (Sheets and Sever 1991; Sheets and McKee 1994; Weller 2003; White 2003). Finding diagnostic artifacts on, in, or near trails was rare, so dating was generally limited to the presence or absence of diagnostic volcanic ash layers in direct contact with or in close proximity to the surface of the anomaly and for which ^{14}C dates were available (Sheets and Sever 1991; Sheets and McKee 1994; Weller 2003; White 2003). Occasionally, while one anomaly was being ground truthed, another was discovered that did not show up on the available imagery. In those cases, the anomaly was subjected to the same ground truthing process. The hybrid method was scaled up significantly to meet the spatial requirements of this project, and certain parts of it had to be modified to fit with the archaeological reality of the Sonoran Desert region, but overall the approach worked quite well. The main modification was the removal of the trenching component. Unlike in Costa Rica, Sonoran Desert trails exist as geologically stable entities right at the modern ground surface. They do not have vertical stratigraphy, but many show up as highly visible lines on the landscape (Figure 11.2). Also, any artifacts present near trails are visible on the surface as well, due to the unique formation processes of this region — something Hayden referred to as "fragile-pattern areas" (Hayden 1965).

The hybrid approach, as applied to the project study area, allowed for the rapid survey of large

FIGURE 11.2. Oblique three-dimensional example view of the high spatial resolution imagery used to locate trails on the BMGR, overlaid on a digital elevation model. Arrows highlight a detected trail segment (barely visible at the tip of each arrow), and insets show the actual confirmed trail segment found during subsequent ground truthing and a schematic representation of its relatively high contrast ratio (visibility).

areas and the identification of hundreds of potential trail segments. More detailed information on its application, as well as an in-depth discussion of the other criteria used to determine whether or not identified linear features were in fact trails, can be found in White (2007).

11.2.2. Understanding Trail Network Functions: Modeled Reality vs. Actual Reality

In order to discern trail network functionality and landscape use, I needed to find a way to reconstruct what a trail network would look like based on known important travel points (archaeological sites, water sources, raw material procurement loci, sacred sites, etc.) and then compare that reconstruction with the locations and morphologies of the trails verified in the field and recorded by previous surveys. The end goal was to create a reconstructed network that could account for all the verified trails, thereby situating them in a broader cultural context, but one that could also explain individual trail network functions with a reasonable degree of accuracy. The base assumption for this process was that if a modeled path matched up with an actual prehistoric path, then the factors that helped to determine the creation of the modeled path were similar to those that created the actual one. As the reader will see, this methodology worked out quite well for trails in the Western Papaguería.

The following subsections discuss the methodologies employed to create the modeled trail networks. The first step in this process was to systematically break up the study region into its two main surface types: desert pavements (easy to walk on, trails remain visible) and *bajadas* (loose soil, trails do not remain visible). With that complete, the next step was to create a way to accurately keep track of energy expenditure cost for travelers moving through the region across those two surface types. Once that was accomplished, a master cost trail network was created. This master network represented an approximation of all possible routes a traveler could take to get from one point in the region to any other point in the region. Locations on the network known to be important prehistorically were then highlighted in differing combinations, and the least costly direct way to connect all of them to each other was found — resulting in multiple theoretical networks on which verified trail segments could be overlaid to see if they matched up. Looking at where they matched up and differed shed significant light on the prehistoric uses of landscapes in the Sonoran Desert region.

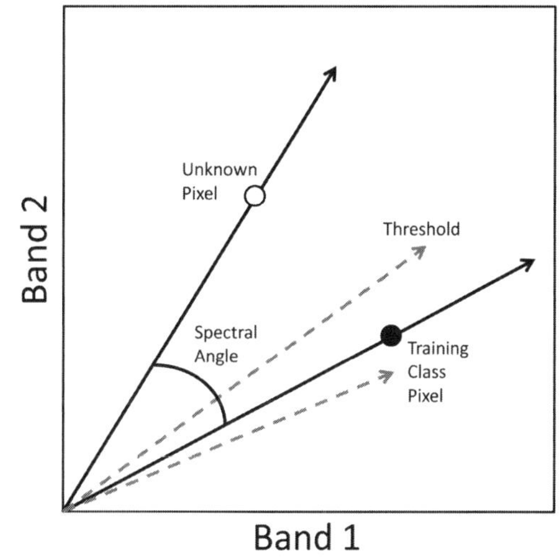

FIGURE 11.3. Spectral Angle Mapper classification, generalized.

11.2.3. Identifying Surfaces via Spectral Angle Mapper Classification

The Spectral Angle Mapper technique was used in combination with Landsat 7 ETM+ data in order to accurately and systematically locate desert pavements and *bajadas* in the study region. It was necessary to work with data from this sensor because it provides consistent measurements for large landscapes and records information in the portion of the electromagnetic spectrum where surfaces like desert pavement stand out against other geological materials.

The Spectral Angle Mapper (SAM) is a physically based spectral classification that uses an n-D angle to match pixels to reference spectra (Kruse et al. 1993). The algorithm determines the similarity between two spectra by calculating the angle between the spectra and treating them as vectors in a space with dimensionality equal to the number of bands (Figure 11.3). This technique, when used on calibrated reflectance data, is relatively insensitive to illumination and albedo effects because differences in intensities are simply values within the continuums mapped along the vectors used in the angular calculations. All SAM calculations and related processing steps were carried out in the Environment for Visualizing Images (ENVI).

The endmember spectrum used by SAM in this study to find desert pavement surfaces was extracted directly from a top-of-atmosphere reflectance Landsat 7 ETM+ mosaic made up of six scenes collected at approximately the same time. Several small regions known to contain desert pavements were provided via regions of interest (ROIs) as training sets, from which the SAM algorithm compared the angle between the endmember spectrum vector and each pixel vector in n-D space. Smaller angles represented closer matches to the reference spectrum. Pixels farther away than the specified maximum angle threshold in radians were not classified. The resulting image, using an angular distance threshold of 0.075 radians (narrow), highlighted the vast majority of areas of desert pavement terrain (and mountainous sources for the pavement) throughout the study region to a spatial accuracy of 30 m. Smaller areas of pavement likely remain unidentified in the image, grouped into the other image component: *bajadas*.

11.2.4. Energy Expenditure Estimation

In order to create a reasonably accurate estimation of a prehistoric trail network connecting known starting and ending points — based on limiting energy expenditure — I needed to find a reasonably accurate energy expenditure estimation method that could aid in building modeled trails. Pandolf, Givoni, and Goldman (1977:578) provide a good working formula for calculating energy expenditure based on the weight of the travelers, how much extra weight they are carrying, how fast they are walking, what type of terrain they are walking on, and terrain steepness. The formula was the result of several studies involving adult males walking on treadmills under varying conditions and was funded by the United States Army Research Institute of Environmental Medicine. A generalized version of their formula is as follows:

$$M_{watts} = 1.5W + 2.0(W+L)\left(\frac{L}{W}\right)^2 + \eta(W+L)(1.5V^2 + 0.35VG) \quad (1)$$

where
M_{watts} = metabolic rate, in watts
W = subject weight, in kilograms
L = external load, in kilograms
V = walking speed, in meters per second
G = slope, in percent
η = terrain coefficient

As Branting (2004:76) points out, though, the Pandolf equation breaks down when the traveler does anything besides walk on level ground or walk uphill. Walking downhill produces significantly negative cost estimates, which suggests that a traveler could recharge by walking in this direction—or at the very least not incur any cost at all. Since downhill walking still requires effort, the model is not entirely useful in this situation. It should also be noted that Rademaker et al. (this volume) highlight additional challenges with using the Pandolf equation in high-altitude mountainous areas, but that is not an issue in the Western Papaguería.

Santee et al. (2001) proposed an updated version of the Pandolf equation that takes into account downhill travel in a more accurate fashion. Their model assumes that the initial reduction in downhill walking energy cost is due to the negative work of gravity, but also that this is reduced by the eccentric action of muscles and joints. For slopes less than zero (negative), energy expenditure is calculated as:

$$M_{watts} = 1.5W + 2\left(\frac{L}{W}\right)^2 + \eta(W+L)(1.5V^2 + 0.35VG) - \eta\left(\frac{VG(W+L)}{3.5} - \frac{(W+L)(G+6)^2}{W} + (25 - V^2)\right) \quad (2)$$

where
M_{watts} = metabolic rate, in watts
W = subject weight, in kilograms
L = external load, in kilograms
V = walking speed, in meters per second
G = slope, expressed as a percentage
η = terrain coefficient

The addition of a component that allows for the accurate calculation of downhill travel makes the Pandolf equation very attractive from an analytical standpoint because it can account for terrain and load carriage. Thus Equation (1) was used in this study for positive slopes and Equation (2) was used for negative slopes. Walking speed was determined by using Tobler's hiking function (Tobler 1993; Kantner 1997, this volume) for a given slope:

$$W = 6e^{-3.5|S+0.05|} \quad (3)$$

where
W = walking speed, in kilometers per hour
S = slope, expressed as a fraction

Because the Pandolf-Santee formulas require walking speed in meters per second, returned values were multiplied by a conversion factor (1000/3600, or 0.277778). Output from Tobler's hiking function represents nominal speed attained over reasonable terrain. For the purposes of this chapter, it is assumed that the subject will continue to walk at the same speed, regardless of the terrain encountered, which will lead to variable energy expenditure rates.

A non-Western, prehistoric male/female subject weight of 60 kg was assumed (Eaton and Eaton 2003), as was an external load of 5 kg. The external load of 5 kg was chosen because of two main factors. First, each traveler would likely be carrying at least a gallon of water, which is the bare minimum the average human body needs to survive on a daily basis in a desert environment (Nelson et al. 1943:2). That gallon would weigh approximately 3.75 kg. Second, if we factor in clothes, tools, and other goods (carried for consumption or trade), reaching or exceeding a total load weight of 5 kg would be fairly easy. The load weight could have been increased, but the assumption was made that pedestrians would be traveling light. Load weight would increase significantly if large amounts of shell, salt, or obsidian were being transported. Terrain coefficients for the two major surface cover types were chosen based on those published by Soule and Goldman (1972). Archaeological sites, desert pavements, and related terrain received a coefficient value of 1.0 (no impedance), equivalent to a manmade blacktop surface. *Bajadas*, loose soil, and vegetated areas received a coefficient value of 1.2 (light brush, slight impedance). By comparison, loose sand (high impedance) has a coefficient value of 2.1.

11.2.5. The Least Cost Network Creation Process

Creating a master cost network for the study area involved applying the Pandolf-Santee formulas in a systematic way to every square meter of the surface contained within the area, in both the uphill and downhill directions to account for anisotropic movement (see Chapter 1 of this volume). Considering the size of the area covered by the available geospatial data (almost 65,000 km^2), this was not an easy task. With the use of remotely sensed data (Shuttle Radar Topography Mission), it became possible to obtain regularly spaced (30 m), consistently measured elevation values for the region, along with enough spectral information to carve up the landscape into different types of terrain (Landsat ETM+). As mentioned above, a Spectral Angle Mapper classification based on obtained spectral information identified the chosen terrain types (at 30 m intervals as well). A program written in IDL ingested both of these single-band datasets, registered them to one another geospatially (using nearest-neighbor resampling when transformations were required), combined both to produce a new co-registered multiband image where each pixel had both an elevation value and an associated terrain type, and then used the elevation and terrain types in conjunction with the Pandolf-Santee equations to build a cost-centric triangulated regular network (TRN) on which hypothetical pedestrians could travel.

The program built the TRN by first connecting the center of every pixel in the co-registered multiband image with the centers of its eight closest neighbors via a polyline vector segment. For each segment, the map locations and elevation values for the end points (pixel centers) were retrieved and used to calculate uphill and downhill slope between the two locations, along with the actual distance traveled (instead of assuming a constant horizontal value of 30 m for E-W and N-S travel and 42.42 m for NE-SW and NW-SE travel). The terrain type values for each end point were retrieved and appropriate terrain coefficients assigned (1.0 for pavement, 1.2 for *bajadas*). These values, subject weight, walking speed, extra load, uphill/downhill slopes, and actual walking distance were then used to calculate appropriate cost estimates in the uphill and downhill directions for travel between the two points. Four calculations were carried out. The first was the metabolic rate of walking uphill if the entire distance was covered by the first terrain type, and the second was the uphill rate if only the second terrain type was present. The third and fourth were the downhill rates if only the first or second terrain types were present, respectively. For the returned metabolic rates to be useful in this study, though, it was necessary to convert each calculated rate (watts) into kilocalories (Calories) used over the actual walking distance via the following formula:

$$C = (M_{watts}) \cdot \left(\frac{0.000239 \text{ kcal/sec}}{\text{walking speed in m/sec}} \right) \cdot 0.5 \cdot \text{Distance (in m)} \qquad (4)$$

The walking distance was halved before multiplication so that fractional cost estimates could be retrieved and more accurate overall cost estimates arrived at, since only half the walking distance in either direction is covered by each terrain type. The fractional costs for travel in each direction were then added together to create composite cost estimates. The inputted variables and the outputted composite cost estimates were then stored with the polyline segment in the database portion of an ESRI shapefile. Due to the size of the landscape used in this analysis, the resulting shapefile was several gigabytes in size. Figure 11.4 demonstrates the details of this workflow on a hypothetical dataset. A digital elevation model (1) and an associated terrain type classification image (2) are used to create a TRN (3). For a single hypothetical E-W-oriented line segment (4), travel between point A and point B is considered. Associated terrain coefficients and

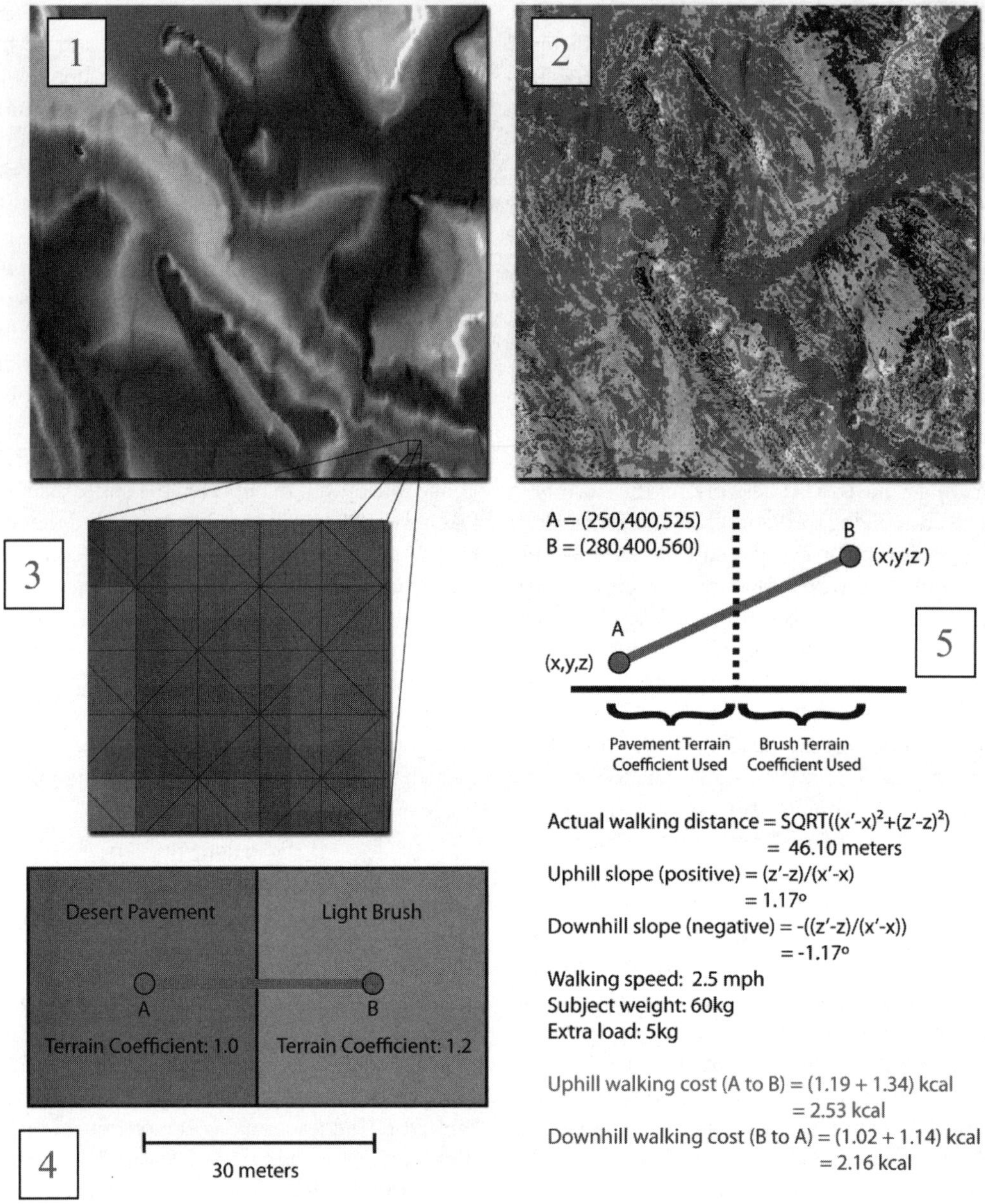

FIGURE 11.4. Creation of a cost-centric triangular regular network.

positional information (X,Y,Z) are used with derived distance and slope and traveler-specific information to calculate uphill and downhill caloric cost (5).

Two smaller zones within the generated TRN were chosen for detailed analysis. The first region was the Air Force side of the BMGR. It was chosen because a great deal of information was available with respect to the locations of archaeological sites, sacred places, water sources, and raw material procurement loci. Using the TRN and these archaeologically significant locations, we can build theoretical trail networks and compare them with known prehistoric trails in the region. The second region was the Sierra Pinacate, for which accurate information about prehistoric trail locations and water sources was known from an unpublished map drafted by Hayden (n.d.). Hayden (1967) hypothesized that water sources served as hubs in the trail network and that movement across the landscape was done in as efficient a manner as possible, which is something I was interested in testing.

To facilitate the generation of least cost routes that could connect particular starting and ending points, I converted the two TRN subsets into Network Datasets within ArcGIS, which turned each polyline segment into a link and each location where the polylines connected into a node on two vast networks. The Network Datasets were then used as input networks to multiple ArcGIS Network Analyst Shortest Distance analysis layers. Each analysis layer represented a different archaeological scenario for each chosen zone (e.g., traveling between sacred sites on the BMGR), and the impedances to travel were set to actual walking distance and caloric cost, with both uphill and downhill cost represented as separate variables to account for anisotropy (see Surface-Evans and White this volume). Discrete starting and ending points for travel associated with a particular scenario — in this study, archaeological sites, sacred places, water sources, and raw material procurement loci — were placed on top of the Network Datasets and fitted to the networks so that they now lay along an edge or at a node (whichever was closest). With the starting and ending points fitted, graph theory was employed via the ArcGIS Network Analyst's Solve process to find the shortest and least costly ways to travel between chosen points on the network via Dijkstra's algorithm (see Surface-Evans and White this volume). By providing both uphill and downhill cost estimates for each link, the network solving process could make more accurate choices about which directions to move, and by including a distance impedance, it ensured that route length was kept to the smallest value possible. This constraint essentially forced the solve process to consider more costly uphill travel when necessary (realistic) instead of always choosing to travel downhill in order to limit energy expenditure (unrealistic). Limitations can be placed on travel such that only the three closest end points to a particular starting point are found, or multiple starting points and only one end point are used. The possibilities are wide open.

Once recorded, the least cost paths determined by graph theory (see Gorenflo and Bell 1991 for an example) were turned into shapefiles and compared with shapefiles of known trails for both chosen zones. Relative accuracy of each modeled network was calculated by tabulating the number of times modeled trails came into contact with or closely mirrored the orientation and location of known prehistoric trails within each zone and dividing that tabulation by the total number of modeled trails for that zone. To avoid overdetermination within a modeled network that contained a large number of possible starting and ending locations in a small space — the process by which so many theoretical trails are generated that they come into contact with or mirror all actual trails in an area by pure statistical chance — I set the maximum number of provided locations that could be reached from any other provided location to ten. While ten appears to be an arbitrary number on the surface, it was determined through a long trial-and-error experimental process that involved modeling trails within several tightly clustered groups of archaeological sites on the BMGR. Setting the limit at ten often produced an overabundance of theoretical trails, though, so the limit was generally set to three, five, or seven for visual clarity and analytical integrity. The trap this project wanted to avoid was that of generating a large number of trails just so there were enough to account for all the known trails. That would be analytically useless and defeat the purpose of attempting to find efficient ways to get from one point to another. If the limit number is kept low, the software is forced to choose a very small number of the most efficient ways to connect locations to one another, which allows for a more valid comparison of modeled trails with actual ones. If some or all of the trails line up, that means something. If they do not line up, it is also meaningful. It should be noted that as the limit is increased, the trails that were modeled at a lower limit are still included in the new output — the current network is just expanded to a larger area that encompasses more possible destinations in more ways.

As discussed above, subsets of locations were chosen based on assumptions regarding potential landscape use (sites for an assumption of sociopolitical usage, sites and natural resources for an assumption of economic usage, sites and sacred places/resources for an assumption of ideological usage), and modeled networks that connected the locations were compared with actual trail locations. In some cases, the models worked quite well. In others, the models broke down and integrating information from other sources became necessary for the landscape use to be understood. The results of this process are discussed in the next section.

11.3. Results

Before discussing individual findings for the two regional landscapes that compose the study area, I want to set the stage by bringing in two complementary interpretive models. The first is that of Ahlstrom and Roberts (2002), who propose that regional interaction within the Western Papaguería can be described using one of four possible scenarios: (1) farmers as foragers, (2) expansion and contraction of the farming frontier, (3) the shell trade (including salt and obsidian procurement), and (4) interaction of hunter-gatherers and farmers. The potential patterns of human movement across the landscape that would reflect these scenarios are highlighted in Darling's ethnographically informed archaeological example for the Sonoran Desert (2006:11), which posits that prehistoric trails can be classified in four groups based on primary route function:

1. Local path: Connects one location to another (one starting and ending settlement, traveling within settlement areas)
2. Type I trail: Localized interaction/exchange, intraregional travel (multiple starting and ending settlements, traveling through hinterland areas)
3. Type II trail: Resource acquisition (one starting settlement, multiple ending points, traveling within settlement and hinterland areas)
4. Type III trail: Interregional travel, ritualized travel, long-distance exchange (multiple starting and ending points, bypasses settlement areas)

For the first scenario, one should expect to see a higher frequency of Type II trails since settlements would be fairly well fixed along the wetter areas, with occasional trips to the interior to procure particular resources. For the second scenario, there should be a higher frequency of Type I trails and local paths since multiple communities would have been established in the interior and would have interacted with another and with outsiders. For the third scenario, Type III trails would be dominant on the landscape since the related landscape is not being used for settlement or exploited for resources. For the fourth scenario, there should be more Type I and Type III trails since there would mainly be a population on the move (no settlements). The predictive modeling efforts discussed below were designed to use these complementary models.

For the BMGR, modeled networks were created to account for travel related to (1) finding reliable sources of water for travel through the region, (2) connecting individual habitation and activity sites to one another, (3) connecting sites with ceramics to one another, (4) connecting sites with metates to one another (suggesting semi-sedentary agricultural activities), and (5) connecting sacred sites to one another. When diagnostic pottery was available for a series of sites, more specific ceramic models were run in an attempt to capture cultural, geographical, and/or temporal variability. In most cases, the final interpretation used required fusing the prediction results of several models. For example, if a predictive model related to linking sites with agricultural components found the same trail as another model focused on linking sites with Hohokam ceramics, the final interpretation was that the trail connected sites with agricultural components, most likely during the Formative period, and the inhabitants of those sites interacted with the Hohokam on some level (directly or indirectly). When sites with agricultural components were connected, but no ceramics were present, the assigned time period for the trail was listed as both Archaic and Formative. When only undiagnostic ceramics were present at connected sites, the assigned time period for the trail was listed as both Formative and Historic.

Due to the fragmentary nature of information related to the trail system of the Sierra Pinacate and associated natural/cultural phenomena—in addition to the fact that ground truthing was not an option due to the remoteness of the area—it was not possible to carry out the above interpretive process on each identifiable trail segment present on Hayden's unpublished map (n.d.),

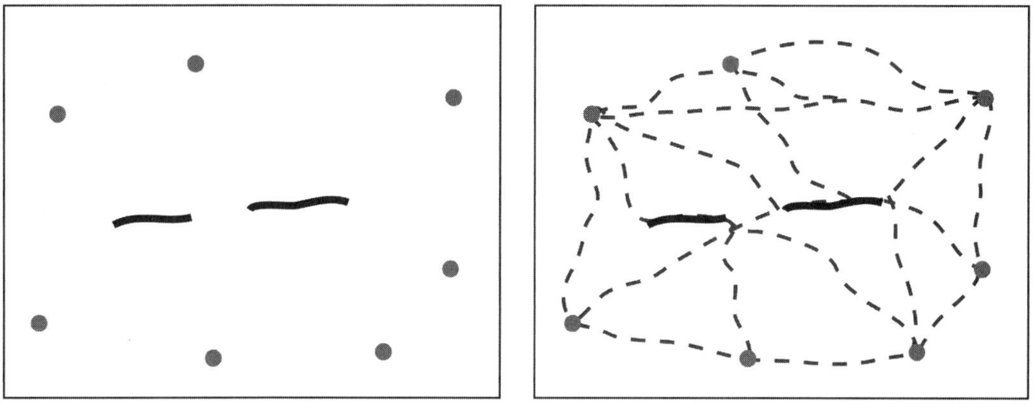

FIGURE 11.5. A hypothetical situation in which verified prehistoric trail segments (*solid lines*) appear to be floating among known archaeological sites (*dots*). This makes interpretation of trail function and related landscape use extremely difficult. However, when a least cost predictive model network designed to connect the sites to one another (*dashed lines*) is generated, it becomes possible to contextualize the segments.

which is shown in an abstracted form in Figure 11.5. Instead, a holistic approach was taken on the entire network, one that tested the assumption that the network primarily existed to move travelers between reliable sources of water (Hayden 1967:336). The results of that process are presented below in a comparative context with the results for the three landscapes of the BMGR.

Individually, the prehistoric trails of the Western Papaguería tell only a very small piece of the story. However, because the predictive modeling method used in this study builds *networks* of trails that simultaneously connect large groups of sites chosen for inclusion in specific models due to the presence or absence of particular similarities (sometimes hundreds or thousands), it becomes possible to speak about how individual trails that relate to those networks relate to entire landscapes and how people traveling along the routes may have used those landscapes. Figure 11.5 demonstrates this situation quite well. In many cases encountered during the analytical phase of this project, a confirmed trail composed of multiple segments (solid lines) would seemingly pass straight through a cluster of known habitation sites (dots). However, because of the trail's orientation, it was impossible to associate it with a single starting site and ending site for travel. A reasonable conclusion at this point would be to say that the trail's purpose is unknown since there are a multitude of ways to extend the end points of the segment so they connect to known sites. However, which set of connections is the right one? Alternatively, if a predictive trail network model is built that attempts to connect every site with its three closest neighbors (dashed lines), the orientation and position of the verified trail segments begin to make sense. The visible segments are part of a potentially much larger network that describes at least one of many ways that the local landscape was being used: intraregional integration of habitation sites.

Discussion of results for the predictive modeling process begins with three landscapes found inside the BMGR because more data were available for those areas. The three selected landscapes are referred to by their contemporary military designations (NTAC, STAC, ETAC), but their exact locations and the trails found on them, per request of the Air Force, cannot be specified here. Next there is a discussion of the Sierra Pinacate results, followed by a brief comparative discussion related to all four landscapes. Through examining the analytical results for individual trails, we begin to see patterns emerge with respect to when the trails within each landscape were in use (Table 11.1) and with whom the trail builders were affiliated (or at least interacted; Table 11.2). On NTAC, it appears that roughly two-thirds of the known trails were in use during the Formative (ceramic) period, most likely by groups with strong Patayan ties and weaker Hohokam ties, while fully one-third remain unexplained. On STAC, roughly half the known

TABLE 11.1. Temporal Assignment Relative Frequencies for Each Tactical Range Landscape, BMGR.

Time Period	NTAC	Freq	STAC	Freq	ETAC	Freq
Archaic	0	0.00	1	4.35	3	4.76
Formative	22	66.67	13	56.52	52	82.54
Historic	3	9.09	6	26.09	13	20.63
Unknown	11	33.33	6	26.09	11	17.46

Note: I determined the frequencies by taking the number of known trails located by predictive models focused on each broad time period, dividing it by the total number of known trails, then multiplying the resulting value by 100.

TABLE 11.2. Cultural Affiliation Relative Frequencies for Each Tactical Range Landscape, BMGR.

Cultural Affiliation	NTAC	Freq	STAC	Freq	ETAC	Freq
Archaic	0	0.00	1	4.35	3	4.76
Hohokam	8	24.24	9	39.13	30	47.62
Patayan	19	57.58	3	13.04	11	17.46
Historic	3	9.09	6	26.09	13	20.63

Note: I determined the frequencies by taking the number of known trails located by predictive models focused on each broad cultural entity, dividing it by the total number of known trails, then multiplying the resulting value by 100.

trails were in use during the Formative period by groups with stronger Hohokam ties and weaker Patayan ties, with another quarter in use during the Historic period and the last quarter unexplained. On ETAC, the overwhelming majority of known trails were in use during the Formative period by groups with the strongest Hohokam ties observed but that still had contact with Patayan groups. Residual percentages were equally split between the Historic period and remaining unknown. The presence of stronger Hohokam ties on ETAC is not unexpected, as the western edge of the Hohokam regional system extended to the eastern edge of the Western Papaguería, where ETAC is located.

Examples of modeled results for NTAC, STAC, and ETAC can be found in Plates 10 and 11. The same model was often run at several connectivity levels (connect each locus to its three closest neighbors, four closest neighbors, and so on) to test its effect on the prediction accuracy of the resulting trail network. In all cases, few additional trails were found and overall accuracy was significantly reduced when a connectivity value greater than five was used, so the results discussed below for each model focus on the output at a connectivity setting of five unless otherwise specified. Locations of prediction results for each model were then compared with the locations of actual prehistoric trails to see how good a job the prediction did at capturing the variability present in the archaeological record. Several evaluators were recorded: how many trail segments were found (not whole trails, so results will differ from what is presented in Tables 11.1 and 11.2), how many trail segments were missed, a percentage accuracy estimate of trails found versus total number of trails (relative frequency), the number of predicted trail segments (not whole trails) generated by the model, and a percentage accuracy estimate of trails found versus total number of predicted trail segments (absolute frequency). The absolute success metric is misleadingly low in all cases due to how ArcGIS generated the predicted trails. An individual predicted trail could have tens to hundreds of segments along its entire length if it spans a large distance, so the number used in the frequency denominator can be greatly inflated compared with what it would be if all relevant segments were combined to form coherent trails. Because of this multiplicative effect, relative frequencies are used in the discussion below.

As was noted earlier, and is borne out in the results shown above, increasing the connectivity level for a modeled network in an attempt to substantially increase the number of trails found does not work. This is an encouraging phenomenon since it suggests that models generated at the most fundamentally basic connectivity levels (five or less) have explanatory power.

The results of the predictive modeling activities and the delineation of the four basic trail groups indicate that the three different landscapes contained within the BMGR were being used in slightly different ways by their prehistoric occupants and travelers from outside the region. Approximately half the prehistoric trails on NTAC could be explained by the need to find reliable sources of water for travel through the region (Type III trails), but this need could not explain the presence of any of the trails on STAC and NTAC. Three-quarters of the trails on NTAC

TABLE 11.3. Importance of Each Potential Landscape Use for the BMGR.

Region	Finding Water	Linking All Sites	Ceramics/Exchange	Semi-sedentary	Sacred/Ritual
NTAC	Medium importance	Medium importance	High importance	Medium importance	High importance
STAC	Low importance	Extreme importance	Medium importance	Medium importance	Medium importance
ETAC	Low importance	Extreme importance	Medium importance	High importance	Low importance

could be explained as local paths or Type I trails that connected all types of habitation and activity sites to one another, while the percentages for STAC and ETAC were much higher. Half the NTAC trails could be explained by the linking of semi-sedentary sites (indicated by metates) to one another, while this explained only a third of STAC's trails and roughly two-thirds of ETAC's trails. Lastly, two-thirds of NTAC's trails could be explained by the linking of sacred sites to one another, while fewer than half of STAC's trails and fewer than a quarter of ETAC's trails could be explained this way. Table 11.3 shows these distributions in a more general way: from 0 percent to approximately 25 percent indicates that an activity was of potentially low importance to travelers, 25 percent to 50 percent indicates medium importance, 50 percent to 75 percent indicates high importance, and 75 percent to 100 percent indicates extreme importance.

At a broad level, then, it appears that sacred concerns and possibly intraregional exchange outweighed all other factors when it came to landscape use on NTAC, but linking communities to one another within settlement areas and facilitating interregional travel between the Gulf of California and the Phoenix Basin were important activities as well. This conclusion is in line with previous ones regarding NTAC that have suggested this is a locally important sacred landscape and a regionally important trade and travel corridor. On STAC, the focus appears to be on linking communities (intraregional organization), but there is evidence to suggest that exchange and sacred concerns played a role as well. Finding water was potentially not a paramount concern, as evidenced by the fact that large sites such as Verbena Village had an artificially constructed reservoir designed to collect runoff. On ETAC, linking communities appears to be far and away the major determining factor for trail generation, possibly followed by exchange. Sacred concerns and finding water appear to be deemphasized within this landscape. Water was likely not as much of an issue due to the fact that ETAC falls into an environmental zone with more rainfall.

TABLE 11.4. Documented Prehistoric Trail Segment Orientation on the BMGR.

Trail Segment Orientation	New	Known	Total	Total %
NS	2	28	30	25.42
EW	11	13	24	20.34
NW-SE	6	46	52	44.07
NE-SW	0	12	12	10.17

Note: Figures include trails documented by this project as well as by previous CRM fieldwork.

As was discussed above, one of the longest-standing professional archaeological beliefs about the prehistoric Western Papaguerían landscape is that its main use was tied directly to long-distance travel and exchange between the Gulf of California and the Phoenix Basin. While there is certainly evidence to indicate that Type III trails used by groups that are passing through exist on the various landscapes of the BMGR, the above data strongly suggest that other activities carried out by more permanent inhabitants were the main motivators behind trail formation. This finding lends further support to Bayman's recent (2007) thinking on the Papaguería, which he sees as a common-pool resource used by both local and extralocal populations. One additional piece of evidence that bolsters this conclusion comes in the form of trail orientation. If efficient travel between the Gulf of California and the Phoenix Basin was the primary reason trails existed in the region, one might expect that the majority of documented trails would be oriented N-S or NE-SW, which would be the most direct ways to get from the Hohokam or Patayan areas to the coast. As can be seen in Table 11.4, though, fewer than

TABLE 11.5. Prehistoric Trail Location Predictive Model Results for the Sierra Pinacate, Organized by Connectivity Setting.

	Sierra Pinacate Predicted Trails					
Network Model	**# Loci Used**	**Found**	**Missed**	**Rel %**	**Lines**	**Abs %**
Water sources—3 closest	52	77	177	31.05	110	70.00
Water sources—4 closest	52	104	144	41.94	165	63.03
Water sources—5 closest	52	115	133	46.37	220	52.27
Water sources—7 closest	52	130	118	52.42	330	39.39
Water sources—10 closest	52	136	112	54.84	495	27.47

Note: Travel between reliable water sources was the only model used for generating predicted trails. Relative accuracy, highlighted in grey, slightly exceeds 50 percent only when the connectivity level is set to an unreasonably high value.

half of all documented prehistoric trails within the BMGR are oriented N-S or NE-SW. The majority are oriented in directions that do not line up with that type of long-distance travel and exchange (E-W or NW-SE).

Another interesting item to note is that all but 21 of the documented trails on the BMGR (82 percent) fall within zones of high mammal richness around mountain ranges. This suggests that hunting activities (planned or opportunistic) may have played a role in trail placement and formation as well.

The trails of the Pinacate, documented by Hayden over several decades, are generally assumed to revolve around transporting people from one reliable water source to another in an efficient manner (Hayden 1967). A map of the trails (Hayden n.d.) tells a potentially different story. While many of the trails appear to connect water sources to one another, at first glance there are many trails (braided ones in particular) that appear to be functioning under a completely different set of rules (Plate 12).

This conclusion is strengthened when we run predictive models centered on linking water sources at various connectivity levels (Table 11.5). Even when the connectivity level is set very high (10), linking water sources (Type I trails) at best explains 55 percent of the documented trails in the Sierra Pinacate.

The unique braided structure of many of the trails missed by the predictive models, along with their placement far from known reliable water sources, suggests that even in this extremely harsh, unforgiving landscape, sacred concerns—not speed and efficiency—were extremely important motivations for travelers (Type III trails). In fact, the harshness of the environment and the placement of trails far from reliable water may speak to a need to intentionally make sacred travel physically demanding (see Darling 2006). Why these trails are braided remains a mystery. One possible reason could be that repeated efforts were made to overwrite earlier travel routes with newer ones that deviate slightly in their directionality and purpose. Another could be that pairs of travelers are meant to walk along the trails and crisscross each other at regular intervals as part of a ceremony. Both of these reasons are at best speculation.

Last but not least, it is important to consider similarities and differences between the three landscapes of the BMGR and the landscape of the Sierra Pinacate with respect to the four-scenario regional interaction model of Ahlstrom and Roberts (2002) combined with the four-type regional travel model of Darling (2006), discussed above. For the "farmers as foragers" scenario, one would expect mainly Type II trails to exist in the region since travelers are primarily moving between settlements and natural resources. For the "expansion and contraction of the farming frontier" scenario, one would expect local paths and Type I trails since many local settlements would have popped up within the region in relatively close proximity and their inhabitants would likely have interacted on a regular basis. For the shell trade scenario, one would expect a large number of Type III trails since the focus for travel would have been on long-distance transport of exotic materials. Lastly, for the "interaction of hunter-gatherers and farmers" scenario, one would expect a large number of Type I and Type III trails since there would be a significant amount of intraregional and interregional movement due to the fact that the local population would not be sedentary.

Judging by the trail network configurations and uses for the four landscapes, it is clear that some fit within the framework proposed by Ahlstrom and Roberts while others do not. The abundance of Type III trails on NTAC, in combination with the presence of many sacred sites, suggests that this particular landscape is a combination of their third and fourth scenarios. There is evidence that this landscape is an important long-term travel corridor for more complex groups; however, many of the identified trails link sacred sites or move in extremely inefficient ways across the landscape, indicating that a more mobile local population may have been engaged in ideological journeys. These findings are supported by Dooley's (2006) observations for the same landscape based on a study of desert pavement surfaces of different ages and their related artifact assemblages. STAC most clearly fits within the Ahlstrom and Roberts framework, sitting squarely within the second scenario. This finding is bolstered by the fact that the large site of Verbena Village had a sizable artificially constructed reservoir and that there were numerous agricultural fields to the east of the site (Rankin and Doyel 2007). ETAC, straddling the border between the Eastern and Western Papaguería, doesn't fit neatly into any one scenario. It has the regional integration aspects of the second scenario, but the area was less arid and therefore less risky to live in. The Sierra Pinacate is best characterized by the fourth scenario. This landscape was apparently home to a highly mobile population that was concerned about both practical and sacred matters.

11.4. Conclusion

Trails play an important role in how people, places, and things are connected to one another across space and time during prehistory. In many instances, footpaths are the only evidence available that demonstrates the existence of community ties within or even between geographic regions that bound middle-range societies. The nature of this connectedness, and hence the nature of regional integration, is often indeterminate if one looks at the footpaths or the end points and waypoints individually but not both at the same time in the context of an entire network; the two entities are part of a greater whole related to how people are organizing themselves across the landscape.

At its heart, the project discussed in this chapter is about uncovering and presenting archaeologically meaningful explanations of prehistoric human landscape use within the Sonoran Desert region. While there are several potential avenues that a researcher can take in order to understand prehistoric landscape use, the focus here is on trails and related networks. Trails connect people to one another and to places that are important to them across both time and space. Examining what trails connect—and how they do it—can provide a wealth of information about how a prehistoric society used a particular landscape.

When the archaeological reality for the Sonoran Desert region is compared with the trail network models generated from the above hypotheses, especially in light of what is known ethnographically, it becomes clear that trail network functions, roles, and meanings were not uniform or unilateral but instead consisted of varying combinations of many hypothetical scenarios across the four major prehistoric landscapes that make up the study area for the project. Three of these landscapes are currently in use as tactical bombing and gunnery training ranges by the U.S. Air Force (NTAC, STAC, ETAC), while the fourth consists of the Sierra Pinacate region of northwest Mexico.

If the trail network functions for each landscape are taken at face value, they indeed provide glimpses into the cultural landscapes of the prehistoric peoples that traveled through and inhabited the Western Papaguería. The results presented here strongly suggest that the social landscape of this region was not a geographic and functional monolith, used in only a set number of ways by all its permanent, semi-permanent, and transient inhabitants throughout prehistory, but instead exhibited a high degree of subregional variation motivated by strong local intracommunity ties and sacred/ritual concerns in addition to energy-efficient, long-distance travel motivated by exchange. Instead of one landscape used by all, there were at least four smaller landscapes used in ways that best fit the needs of the local populations—of which many or most were indigenous groups that interacted with more complex groups outside the region to varying degrees.

Acknowledgments

I thank the dedicated personnel of the United States Air Force, 56th Range Management Office (Luke AFB, Arizona), for their generous support of my research. In particular, James Uken, Adrianne Rankin, David Doyel, Maj. Daniel Garcia, and Chris Black made it possible for me to conduct fieldwork on the Barry M. Goldwater Air Force Range by arranging access, providing equipment and housing, and supplying data and technical expertise. I am also deeply grateful for the support of the National Geospatial-Intelligence Agency and the National Science Foundation. The former provided the vast majority of the remotely sensed data used in this study, and the latter provided travel funds for fieldwork through a Dissertation Improvement Grant (SBE 0437579). It is also important to highlight the invaluable contributions of the several graduate and undergraduate students from the University of Colorado (Tyr Fothergill, Sarah Case, and Joe Navarro) and the University of Denver (Kristin Safi) who volunteered their time to assist with the fieldwork portions of this project. Last, but certainly not least, I want to thank Steve Hayden, son of Julian Hayden, for providing access to his father's priceless notes and maps on the prehistoric and historic peoples of the Sierra Pinacate. The incredibly detailed information I gleaned from studying that collection made it possible to include the region in my analyses, which in turn greatly enriched the results of the entire project.

References

Ahlstrom, R. V. N., and H. Roberts
2002 Interpreting Formative Period Settlement in the Western Papagueria. In *Culture and Environment in the American Southwest: Essays in Honor of Robert C. Euler*, edited by D. A. Phillips Jr. and J. A. Ware, pp. 113–146. Anthropological Research Paper No. 8. SWCA Environmental Consultants, Phoenix.

Bayman, J. M.
2007 Papaguerian Perspectives on Economy and Society in the Sonoran Desert. In *Hinterlands and Regional Dynamics in the Ancient Southwest*, edited by A. Sullivan and J. M. Bayman, pp. 109–124. University of Arizona Press, Tucson.

Becker, K. M., and J. H. Altschul
2008 Path Finding: The Archaeology of Trails and Trail Systems. In *Fragile Patterns: The Archaeology of the Western Papaguería*, edited by Jeffrey H. Altschul and Adrianne G. Rankin, pp. 419–446. SRI Press, Tucson.

Branting, S. A.
2004 Iron Age Pedestrians at Kerkenes Dag: An Archaeological GIS-T Approach to Movement and Transportation. Unpublished Ph.D. dissertation, State University of New York, Buffalo.

Butler, M.
2003 El Sitio Poma, G-725PM, Proyecto Prehistórico Arenal, 2002–2003. *Vínculos* 28(1–2):61–76.

Darling, J. A.
2006 Pima Song and the Archaeology of Space. Paper presented at the University of Pennsylvania Museum of Archaeology and Anthropology International Research Conference, "Landscapes of Movement: Paths, Trails, and Roads in Anthropological Perspective," Philadelphia.

Darling, J. A., and B. S. Eiselt
2003 Trails Research in the Gila Bend. In *Trails, Rock Features and Homesteading in the Gila Bend Area: A Report on the State Route 85, Gila Bend to Buckeye, Archaeological Project*, edited by J. C. Czarzasty, K. Peterson, and G. Rice. Office of Cultural Resource Management, Department of Anthropology, Arizona State University, Tempe.

Dooley, M. A.
2006 Multi-Temporal Land-Use Patterning in the Western Papagueria: A Geoarchaeological Analysis of Pre-Columbian Cultural Landscapes. Unpublished Ph.D. dissertation, University of Nebraska, Lincoln.

Eaton, S. Boyd, and Stanley B. Eaton
2003 An Evolutionary Perspective on Human Physical Activity: Implications for Health. *Comparative Biochemistry and Physiology Part A* 136: 153–159.

Gorenflo, L. J., and T. L. Bell
1991 Network Analysis and the Study of Past Regional Organization. In *Ancient Road Networks and Settlement Hierarchies in the New World*, edited by C. D. Trombold, pp. 80–98. Cambridge University Press, Cambridge.

Hayden, J.
1965 Fragile-Pattern Areas. *American Antiquity* 31: 272–276.
1967 A Summary Prehistory and History of the Sierra Pinacate, Sonora. *American Antiquity* 32(3):335–344.
n.d. Trails and Tinajas of the Sierra Pinacate. Unpublished map.

Hill, M. E., Jr., and J. S. Bruder
2000 *Farmers and Foragers or Cultivators and Collectors? Opportunistic Occupation in the Interior Desert of the Western Papagueria*. Intermountain Cultural Resource Services Research Paper No. 57. Dames & Moore, Phoenix.

Kantner, J.
1997 Ancient Roads, Modern Mapping: Evaluating Chaco Anasazi Roadways Using GIS Technology. *Expedition* 39(3):49–61.

Kruse, F. A., A. B. Lefkoff, J. B. Boardman, K. B. Heidebrecht, A. T. Shapiro, P. J. Barloon, and A. F. H. Goetz
1993 The Spectral Image Processing System (SIPS)—Interactive Visualization and Analysis of Imaging spectrometer Data. *Remote Sensing of the Environment* 44:145–163.

Lumholtz, C.
1912 *New Trails in Mexico: An Account of One Year's Exploration in North-Western Sonora, Mexico, and South-Western Arizona, 1909–1910*. Charles Scribner's Sons, New York.

Lyon, J. D.
2000a *Procurement, Processing, and Passing Through: An Archaeological Survey of 14,390 Acres in the Central Sauceda Valley (East TAC) on the Barry M. Goldwater Range in Southwestern Arizona* I. SWCA Environmental Consultants, Tucson.
2000b *Procurement, Processing, and Passing Through: An Archaeological Survey of 14,390 Acres in the Central Sauceda Valley (East TAC) on the Barry M. Goldwater Range in Southwestern Arizona* II. SWCA Environmental Consultants, Tucson.

Nelson, N., L. W. Eichna, and W. B. Bean
1943 *Determination of Water and Salt Requirements for Desert Operations*. Army Medical Research Lab, Fort Knox, Kentucky.

Pandolf, K. B., G. B. Givoni, and R. F. Goldman
1977 Predicting Energy Expenditure with Loads While Standing or Walking Very Slowly. *Journal of Applied Physiology* 43(4):577–581.

Rankin, A. G., and D. E. Doyel
2007 NTAC 01 Culture History. Manuscript on file, 56 RMO/ESM, Luke AFB, Arizona.

Riley, C. L., and J. L. Manson
1991 The Sonoran Connection: Road and Trail Networks in the Protohistoric Period. In *Ancient Road Networks and Settlement Hierarchies in the New World*, edited by C. D. Trombold, pp. 132–144. Cambridge University Press, Cambridge.

Roberts, H., R. V. N. Ahlstrom, and D. B. Tucker
2000 *East TAC: An Archaeological Survey of 7,792 Acres on the Barry M. Goldwater Range in Southwestern Arizona*. SWCA Environmental Consultants, Tucson.

Rogers, M.
1945 An Outline of Yuman Prehistory. *Southwestern Journal of Anthropology* 1:67–198.

Santee, W. R., W. F. Allison, L. A. Blanchard, and M. G. Small
2001 A Proposed Model for Load Carriage on Sloped Terrain. *Aviation, Space and Environmental Medicine* 72(6):562–566.

Sever, T. L., P. Sheets, and D. Irwin
2003 Satellite and Airborne Remote Sensing Analysis for the Detection of Ancient Footpaths in Costa Rica. *Vínculos* 28(1–2):1–8.

Sheets, P. D., J. Hoopes, W. Melson, B. R. McKee, T. L. Sever, M. Mueller, M. L. Chenault, and J. Bradley
1991 Prehistory and Volcanism in the Arenal Area, Costa Rica. *Journal of Field Archaeology* 18(4): 445–465.

Sheets, P. D., and B. R. McKee
1994 *Archaeology, Volcanism, and Remote Sensing in the Arenal Region, Costa Rica*. University of Texas Press, Austin.

Sheets, P. D., and T. L. Sever
1991 Prehistoric Footpaths in Costa Rica: Transportation and Communication in a Tropical Rainforest. In *Ancient Road Networks and Settlement Hierarchies in the New World*, edited by C. D. Trombold, pp. 53–65. Cambridge University Press, Cambridge.

Slaughter, M. C., D. B. Tucker, and A. Lascaux
2000 *Trade Corridors and Ethnic Boundaries: An Archaeological Survey of 12,089 Acres along the Growler and San Cristobal Washes on the Barry M. Goldwater Range in Southwestern Arizona*. SWCA Environmental Consultants, Tucson.

Soule, R. G., and R. F. Goldman
1972 Terrain Coefficients for Energy Cost Prediction. *Journal of Applied Physiology* 32(5):706–708.

Tobler, W.
1993 *Three Presentations on Geographical Analysis and Modeling*. Technical Report 93-1. National Center for Geographic Information and Analysis, Santa Barbara, California.

Tucker, D. B. (editor)
2000a *Footsteps on the Bajada: An Archaeological survey of 15,813 Acres on the North Tactical Range of the Barry M. Goldwater Air Force Range in Southwestern Arizona* I. Cultural Resource

Report No. 99-140. SWCA Environmental Consultants, Tucson.
2000b *Footsteps on the Bajada: An Archaeological Survey of 15,813 Acres on the North Tactical Range of the Barry M. Goldwater Air Force Range in Southwestern Arizona* II. SWCA Environmental Consultants, Tucson.
2000c *South TAC II: An Archaeological Survey of 7,083 Acres in the Northeastern Growler Valley on the Barry M. Goldwater Air Force Range in Southwestern Arizona.* SWCA Environmental Consultants, Tucson.

Weller, E.
2003 Trenching in Search of Ancient Silencio Phase Footpaths. *Vínculos* 28(1–2):43–54.

White, D. A.
2003 Survey and Trenching in the Fincas of Castrillo and Mandela. *Vínculos* 28(1–2):27–42.
2007 Transportation, Integration, Facilitation: Prehistoric Trail Networks of the Western Papaguería. Unpublished Ph.D. dissertation, University of Colorado, Boulder.

PART 4

Constructive Criticisms and Theoretical Discussions

CHAPTER 12

Seven Solutions for Seven Problems with Least Cost Pathways

Scott Branting

The application of least cost pathway analysis as a method to address aspects of theoretical arguments within and beyond archaeology has seen wide and growing use. There are, however, a number of recognized problems with the approach, in both conception and implementation. This chapter takes Δa look at seven key problems in using least cost pathways (LCPs). It outlines what the problems are and discusses some solutions, based primarily on lessons I learned from developing network-based least cost transportation models and simulations as part of the Kerkenes Dağ and SHULGI research projects. These seven problems also form a convenient framework in which to highlight different aspects of the case studies found in this book.

12.1. Importance of Least Cost Pathways in Archaeology

Least cost pathway analysis has been used by archaeologists for decades now. Scholars such as Irwin-Williams (1977), Plog (1977), Ebert and Hitchcock (1980), Bell et al. (1988), Gaffney and Stančič (1991), Machovina (1996), Kantner (1996), Aldenderfer (1998), Llobera (2000), Bell and Lock (2000), Jennings and Craig (2001), Bell et al. (2002), Branting (2004), and White (2007) have all used these methods for a variety of purposes in a variety of different ways. While its employment started well before the advent of Geographic Information Systems (GIS) software on personal computers, GIS has played a major role in its increasing utilization in more recent years. The range of archaeological tasks that LCP analysis has been or could be used for includes:

- determining routes of movement between different settlements
- determining movement within a settlement or the catchment of a settlement
- determining herding and more nomadic routes beyond settlements
- determining the visibility of people moving through the landscape
- determining the visibility of features and structures to those moving through the landscape
- quantifying the economics and energetics of trade
- quantifying the logistics of past military campaigns and actions
- locating unknown settlements
- suggesting the importance of certain settlements or features
- parsing reconstructed settlement systems
- locating features directly related to movement within the landscape (such as roads, hollow ways, or trails)
- planning efficient transects for archaeological surveys

The reason for this growing popularity is multifaceted. However, the availability of GIS software and its increasing usage within archaeology since the late 1980s has played a major role.

The algorithms for LCP analysis seeing the most use are, unsurprisingly, those that have been the stock algorithms found in popular GIS software packages. Why develop your own, as did Livingood (this volume, Chapter 10), when there are already-developed solutions that can be used right out of the box? While this question has been illustrated in previous chapters and is discussed further below, there is no doubt that the ease of using preexisting least cost algorithms within a known GIS software environment are strong factors in the adoption of LCP analysis among a wider archaeological audience.

Going hand in hand with increasingly easy access to software and algorithms for undertaking LCP analysis has been the growing availability of prestructured geospatial data. To acquire topographic data, for example, by far the most widely used data for determining LCPs, used to involve laborious methods such as digitizing from contours or spot heights off paper maps on a digitizing tablet, acquiring heights from manual inspection of stereo-pair photography, or surveying the landscape by hand in the field. All these methods were relatively time-consuming tasks and required expertise in a range of ancillary GIS or surveying methods. They also entailed considerable expense, whether for acquiring the needed equipment, procuring maps, or taking the photographs. Digital elevation models (DEM), also known as digital terrain models (DTM), although in use since the 1950s, have become ever more readily available and more easily accessible over the past three decades. With a single 11-day mission of the Space Shuttle *Endeavour* in February 2000, a nearly global DEM was able to be constructed (http://www2.jpl.nasa.gov/srtm/). Such a process would have taken countless people countless years to produce by traditional means. In fact, with the advent of the Shuttle Radar Topography Mission (SRTM) in 2000 and the recent ASTER Global Digital Elevation Model (GDEM) dataset released in 2009 (http://asterweb.jpl.nasa.gov/gdem.asp), the coverage area of DEMs across the world has been massively expanded. They now seamlessly reach most corners of the global landmass in increasingly higher spatial resolutions. All this wealth of data is a few mouse clicks away from being ready to use within any standard GIS application. The availability of such pre-rectified and regularized data, which can quickly be incorporated into a GIS project, has been a second major factor in the increase of LCP analysis.

A third reason for the proliferation of LCP analysis is that there is a fundamental need in archaeology to be able to deal with more dynamic phenomena in the landscape, phenomena such as transportation. All archaeological evidence to some degree is spatially defined (Clarke 1977). Most archaeological work has focused on static features such as sites, systems of sites, or regions. Less work has been done with equally important features for understanding past behavior, features such as pathways and roads, which are the fossils of the spatially and temporally dynamic process of transportation (Trombold 1991; Ingold 1993; Tilley 1994; Snead 2002; Branting 2004; Miller 2006; Snead et al. 2009). Least cost pathways provide tools that can start to address this need. Yet tools like this for exploring and modeling more dynamic processes are often found wanting, regardless of whether they are used with more qualitative or quantitative data (Harris 2000). Part of the problem is limitations in preserved archaeological data (Kelly 1992; Close 2000). Equally problematic, however, is that few archaeologists have explored even a small portion of their potential.

Before turning to look at some of these problems, we should remember that LCP analysis and other transportation modeling methods were developed and are used much more widely than just within the discipline of archaeology. As with so many of the tools used by archaeologists, LCP analysis was developed for purposes and by researchers outside archaeology. This is not to say that it doesn't also have wide applicability within archaeology; however, sometimes a reevaluation of a tool in light of the needs and requirements of archaeology is necessary. In the case of transportation, there are areas critical for archaeological research that are poorly developed within more generalized analytical techniques. This includes, significantly, the modeling of traditional modes of transportation, such as animals, carts, canoes, and even pedestrians (Batty 2005:264; Zhou 2008; Crooks 2008; Livingood, Chapter 10). Archaeology is understandably often focused precisely on modes of transportation such as these, which

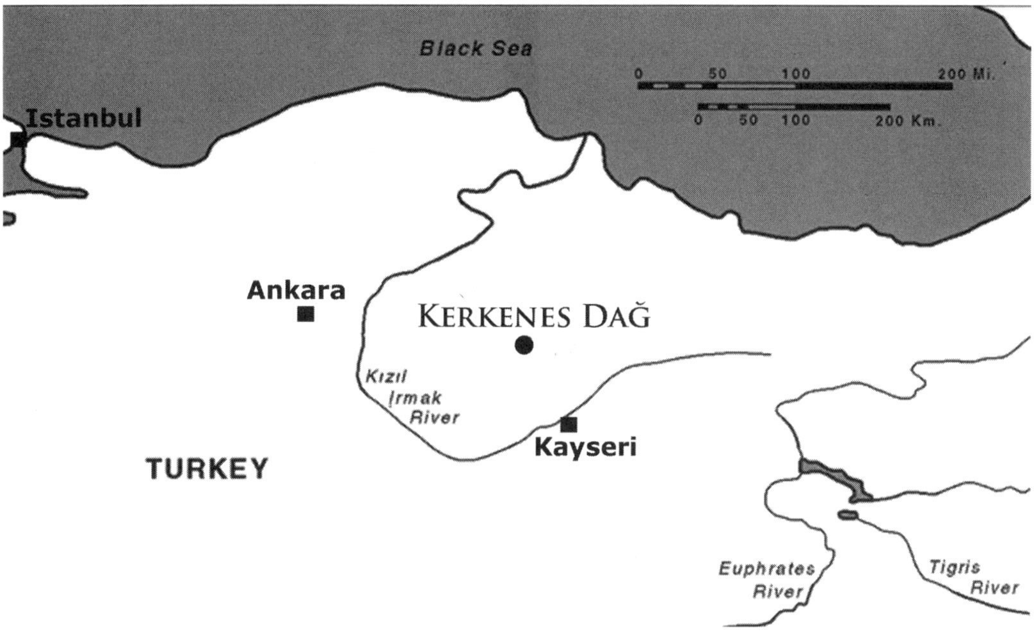

FIGURE 12.1. Location of Kerkenes Dağ, Turkey (basemap courtesy OI Map Series).

were heavily used before the Industrial Revolution. However, many of these modes remain critical in the modern world, especially pedestrian traffic in urban locations (Batty 2005:263–318). Therefore, archaeology can potentially make a strong contribution to understandings of both the past and the present by pursing research in these areas of common interest.

12.2. Kerkenes Dağ and SHULGI Research Projects

The Kerkenes Dağ Project provided my own entry into the utilization of LCPs. I briefly introduce this project and the SHULGI project, an agent-based modeling project developed with Argonne National Laboratory, so as to expose some of my own biases, to situate the critique that follows, and to provide key references for the work. The Kerkenes Dağ Project, directed since 1993 by Geoffrey and Françoise Summers, is investigating a large sixth-century BC city at Kerkenes Dağ in central Turkey (Figure 12.1) (http://www.kerkenes.metu.edu.tr/). While my previous survey work had looked at how flows of movement were taking place through time in an adjacent region and settlement system in Turkey (Branting 1996), Kerkenes Dağ provided an ideal opportunity for me to use and improve LCP analysis, particularly on the level of intra-urban transportation within an large ancient city. This was because of a combination of unique aspects of this city and innovative techniques that were being applied by the project.

Aspects of the city at Kerkenes Dağ provide constraints on a number of variables that can impede an effective implementation of LCP analysis. First, the time between the foundation and destruction of the city was relatively brief, around 40 to 80 years. This provides an important limit on the amount of reuse and remodeling of the urban network of streets, buildings, and spaces. Second, after the destruction of the city, very little overburden developed on top of the original surfaces, in most places not more than a meter, whether from erosion or anthropogenic factors. Only in a few restricted areas was there later settlement that partially or completely obscured the Iron Age urban network. This allowed for both the broad horizontal recovery of the plan of the city and relative ease in recovering archaeological material from its streets and surfaces. Third, the city covers a vast area by ancient standards, 271 hectares. This creates a significant urban environment paralleling in size smaller cities that exist today. While none of these factors is a necessary precondition for undertaking LCP analysis, they are all advantageous for trying out and testing new and existing formulations of such analysis at a large but subregional scale.

The archaeological advantages provided by the city at Kerkenes Dağ were complemented by the employment of a range of innovative methodologies undertaken by the archaeological project, starting in 1993. These began with techniques such as complete-coverage aerial photography survey and then expanded on that basis with a multidecade remote sensing program that has made particular use of complete-coverage magnetometry and ongoing large-area resistivity surveys (Summers and Summers 2010). This work has allowed most of the extent of the urban landscape to be revealed (Branting 2004; Kvamme 2003). In addition, intensive microtopographic GPS survey provided a three-dimensional complement to the reconstructed city plan, offering approximations of the paleosurfaces at less than a quarter of a meter resolution (Branting and Summers 2002).

Kerkenes Dağ, therefore, provides a unique opportunity for us to test LCP models against the plan of the city. Results of tests using basic LCP analysis and data of different resolutions proved particularly illuminating of potential problems, discussed in the next section (Branting 2007). In addition, more complex network-based LCP models were developed using a Transportation Geographic Information System (GIS-T) framework. These models, termed PGIS-T for pedestrian GIS-T, allow the viewing of aggregates of tens of thousands of virtual trips through the city by pedestrians with different demographic characteristics and potential reasons for moving about within the city (Branting 2004). The application of basic LCP analysis at Kerkenes Dağ and the development and testing of these PGIS-T models, both at Kerkenes Dağ and in the modern city center of Cambridge in England, revealed many of the seven problems that follow.

The SHULGI project, developed in partnership with Mark Altaweel of Argonne National Laboratory, took these PGIS-T models into the realm of agent-based modeling (Branting et al. 2007). Agent-based modeling, as seen in Ullah and Bergin's contribution to this volume (Chapter 9), was selected in large part because it provides an effective platform on which to expand into modeling interactions between different agents, a weakness of the aggregate models of the PGIS-T approach and traditional least coast analysis. SHULGI ported the equations underlying the PGIS-T models into Repast Simphony (http://repast.sourceforge.net/) to create an agent-based framework within Repast that could be applied to any network of streets or pathways. While initial tests focused on the Kerkenes Dağ street network in order to compare the agent-based modeling results with the results from the PGIS-T models, other tests applied SHULGI at larger regional scales. Elements of SHULGI have since been used for studying emergent riot behavior in differently formed urban contexts (Wise 2009) and are currently being used to help model the construction of the city wall at the archaeological site of Zincirli in southeastern Turkey.

12.3. Problems with Least Cost Pathways and Potential Solutions

With the potential that LCPs have for investigating the past and developing the future, it is important to reflect on problems with the methodology and its various implementations. To frame this discussion, I have chosen seven key problems that I have encountered in the years of working with aggregate and agent-based LCP methods. This is not intended to be an exhaustive list, but one that can point out some common difficulties, suggest possible solutions, and point to new directions for future research.

12.3.1. Problem 1: Least cost pathways are idealized patterns and are not necessarily based on any real paths used in the past

As has been mentioned several times already in this book, it is important to remember that any model is a simplification and an abstraction from reality. Models derived from LCP analysis of the locations of routes running through an area are no exception. They represent idealized situations that may or may not represent any kind of real usage of paths and roads in the area under study. However, while this is always an assumption that needs to be made explicit within research, it is not an impassable problem. An idealized path can always be used as a baseline against which to compare observations of known routes, as is seen in this volume in Rademaker et al. (Chapter 3), Phillips and Leckman (Chapter 4), White (Chapter 11), and Kantner (Chapter 13). Indeed, these deviations of both the routes themselves as well as their origin and destination points are often key to attempts to understand more complex cultural and social phenomena underlying the spatial distributions. While modern studies of transportation can rely on direct observation of individuals or travel diaries and the like, archaeology will always have to deal with proxy measures for "observing" movement since the people who once used these routes are usually dead. Even the traces left of these people's movement may have eroded away beneath the irresistible forces of wind and water. Yet this does not mean that the ideal models returned from LCP analysis are without utility in archaeology. Comparisons of ideal pathways with known tracks, settlements, or artifacts can be a very productive line of research, as is seen in almost every chapter in this volume, providing insights into a wider range of social, economic, or cognitive processes and practice (e.g., Kantner 1996; White, Chapter 11; Richards-Rissetto, Chapter 7).

The identification and testing of past routes of transportation, when present for comparison to LCPs, can yield surprising results. The first portion of this process is the identification of segments of pathways. This can be accomplished in any number of ways, including field surveys, local interviews, textual analysis, site distributions, distributions of artifacts, or remote sensing and other forms of geospatial data (e.g., Rademaker et al., Chapter 3; Phillips and Leckman, Chapter 4; Hudson, Chapter 6; White, Chapter 11). The role of geospatial data is growing increasingly useful in this regard as higher-resolution data, with more area coverage, is becoming easier and cheaper to acquire and utilize. When combined with data such as geo-rectified versions of earlier air photographs or spy satellite images, these present a powerful tool for initially locating segments of routes that have often been missed on the ground (Obenauf 1991; Sever and Wagner 1991; Wilkinson 1993; McKee et al. 1994; Lepper 1995; Hughes 1999; Erickson 2001; Ur 2003; Heckenberger et al. 2003; Sneed et al. 2009). At Kerkenes Dağ, the process of identifying ancient streets within the city was undertaken with the variety of types of geospatial data collected by the project, described above, combined with surface surveys and excavation (Branting 2004).

In certain situations, however, testing can be taken a step further. A variety of excavation and recording techniques have been and are being developed that can be used to tie the results of LCP analysis along these paths and roads to real-world data taken from them. This includes work undertaken by the Arenal Project in Costa Rica (McKee et al. 1994), by the Kerkenes Project in Turkey (Summers et al. 2005; Branting 2007, 2009), around Tell Brak in northern Syria (Wilkinson et al. 2010), and at Pompeii (Poehler 2009). These efforts have shown how excavations across past pathways and streets can collect valuable data that are necessary for testing and calibrating the results of models generated by LCP analysis. At Kerkenes, through work carried on in collaboration with C. A. I. French and Steve Boreham at the University of Cambridge, laser-particle-size analysis on loose soil samples taken from the different streets has so far provided excellent correlations to predicted levels of traffic from the modeling efforts. Additional techniques that show promise for corroborating these results include micromorphological analysis of samples from the streets, loss-on-ignition analysis to produce percentage calcium carbonate values, and electrochemical measurement of redox potential. Ongoing excavations, across the streets of Kerkenes or elsewhere, may soon provide additional methods for independent verification of the results of models generated by LCP analysis and potentially even for estimating traffic frequencies directly from ancient roads and paths.

12.3.2. Problem 2: Least cost pathways assume that those moving from place to place have perfect knowledge of the landscape from which to decide which routes to take

Assumptions of perfect knowledge are not merely a problem for LCP analysis. This point is raised in a wide range of economic and social modeling frameworks. As laid out in the introduction to this volume, in LCP analysis most of the algorithms used to determine route assignment start by assessing all the possible routes across the dataset and then apply this knowledge to find the least cost ones within the confines of the specific origin and destination pairs (start points and end points) that the user defines. In essence, from the viewpoint of a simulated individual starting from point A and moving to point B, that individual would know not only all pertinent information about the cost factors of moving along that route but also all other possible routes that are available within the landscape. This is obviously a large departure from the situation that a real-world person would face. It is also an issue that has been raised by a number of authors in this volume, including Rissetto (Chapter 2), Rademaker et al. (Chapter 3), Surface-Evans (Chapter 8), Ullah and Bergin (Chapter 9), and Livingood (Chapter 10).

There are several ways this problem could be at least partially circumvented. One would be to incorporate models of perception and concomitant action as a filter on the knowledge these virtual people would have at their disposal to make their decisions. A number of studies of cognitive processes at work in wayfaring and navigation have been undertaken in a range of fields such as psychology, geography, and neuroscience (Golledge 1999; Allen 2007; cf. Livingood, Chapter 10). Some of these studies focus on the physical apparatus of the brain and its impacts on navigation (Amorim 1999; Berthöz et al. 1999). Other studies focus on the interplay of senses and perception with movement (Okabe et al. 1986). More integrative approaches, like that of Stern and Portugali (1999), who study urban navigation using a combination of interrepresentational networks and decision field theory, provide a good starting point from which sophisticated models of how movement follows from cognition could be specified and tested. One important point to remember, however, is that over time, through trial and error, route selection can come to approximate optimal least cost routes even without the individuals realizing or knowing the totality of the system. Just because a route is least cost doesn't necessarily mean that it was consciously selected as such.

Agent-based modeling provides a framework within which imperfect knowledge on the part of an agent could be more effectively modeled within the decision-making process (Ullah and Bergin, Chapter 9). SHULGI, the agent-based modeling framework that has developed out of the work at Kerkenes Dağ, has so far focused mainly on incorporating geo-referenced data, network models, and detailed pedestrian movement algorithms for the agents (Branting et al. 2007). Consequently, most of the work on routing has remained based on relatively simple least cost algorithms such as A* (Hart et al. 1968) and Dijkstra (Dijkstra 1959), which were discussed in the first chapter of this book (Surface-Evans and White). Such algorithms function with global knowledge of the entire system, including the costs of all the pathways through the system, and decisions on route selection are made based on that knowledge. However, while SHULGI has so far not ventured into the incorporation of more complex decision making, the Repast framework within which SHULGI operates could incorporate them just as easily as other agent-based modeling systems. This would allow for more localized decision-making criteria and subsets of knowledge, as is suggested by Ullah and Bergin (Chapter 9), perhaps based on where the agent has been previously in that simulation or created from the start as a unique attribute of the agent, to play a key role in the decisions agents make. It would have the effect of not only allowing agents to have limited knowledge but also enhancing some of the interaction variables discussed below.

12.3.3. Problem 3: Least cost pathways often assume that a person would choose the optimal path rather than a merely satisficing or even nonsatisficing one

People, even when they have significant knowledge of part or all of the system, tend to make decisions on where to go and what to do in less than optimal ways. This can sometimes be explained by the use of alternative benefits or costs in people's decision-making processes. It could also be caused by the influence of external activities or institutional projects on how an activity is carried out, including elements of transportation between locations. It might be a secondary factor of interactions or expected interactions, including those based on misinformation as to who might be where, between the individual and people moving along or through that route at the same time. Or it could simply be that someone wants to rebel against conformity or introduce an element of randomness into his or her routine. With transportation decisions in particular, the mode of transportation can also play a role in the use of less than optimal routes. If animals are involved as a mode of transportation, as briefly mentioned by Ullah and Bergin (Chapter 9), they might make choices that differ from those of the people driving them. Certain types of transportation will also have requirements that may be met only by specific routes, such as Livingood's canoes (Chapter 10). For any of these reasons, and countless more, people may choose to deviate from what is viewed as the optimal path.

An important distinction should be made in least cost analysis between optimizing, or finding the absolute best path according to the criteria employed, and satisficing, or finding a path that would meet the need(s) behind the movement of the individual. The one undertaking the action rarely needs to choose an optimizing path if several alternative satisficing ones are present. The presence of such alternate routes allows greater freedom to satisfice not only a primary factor but other, secondary factors as well. For example, say it is 2 PM and I am required to be at a meeting at 3 PM at a location at least 20 minutes away. I have the flexibility to choose not only the optimal path but also a wide range of alternative paths that would suffice to get me to the meeting on time. I could choose to take a route 25 minutes in length that passes by my favorite coffee shop so I can stop and get a cup of coffee that I have been craving. This would allow me to meet the need of being at the meeting on time, while also allowing me to address the desire for a particular type of coffee before the meeting. I could choose to use a 30-minute path that takes me by a bank that I would need to visit at some point that day. This would allow me to fulfill two tasks with the same trip, both doing the banking that was required and making it to the meeting on time. I could even choose to take a 55-minute path through a local tree-filled park, allowing me to arrive at the meeting not only on time but also in a relaxed and refreshed state. However, I could not choose a route that took me to a place 90 minutes away, at least if I was still planning on arriving at the meeting on time. Alternative satisficing routes, while limited by travel time and factors within the broader decision calculus, nonetheless open up the process of route selection to a much wider range of choices than merely the optimal route.

Travel route assignment routines in most transportation GIS (GIS-T) or modeling packages have ways to stochastically assign routes. In other words, an element of randomness can be introduced to select among a set number of satisficing or even nonsatisficing routes rather than just the optimal one. This can be directly tied to a certain percentage variation from the optimal route, such as, say, the top 5 percent of most optimal routes, or it can take into account other variables in the decision calculus for selecting the appropriate sample. Surface-Evans's cost corridors reach a similar result in some ways by bracketing optimal cost values on the surface (Chapter 8). The use of stochasticity is certainly nothing new, for popular stochastic assignment routines such as like the STOCH method existed well before the advent of GIS (Dial 1971). The use of stochastic methods expands the range of routes returned by LCP analysis by allowing trips to deviate from the use of optimal paths. This could be used not only to better reflect satisficing choices but also to address a person's imperfect knowledge of the network or the potential for differential preferences. Particularly in archaeological applications, where there is less certainty about the start and end points of the particular trip, stochasticity could be used to choose from among a range of possible origin and destination points, and subsequent runs of either stochastic or optimal least cost route assignment could be used to identify potential corridors of travel between the constellation of points. Stochasticity can also be introduced within a Monte Carlo framework, as Ullah and Bergin (Chapter 9) do, utilizing multiple runs of stochastic simulations to reveal tendencies and convergences.

Stochastic routines for route assignment were used as checks against the initial least cost runs of the PGIS-T, in both the modern Cambridge City Center dataset and the street network from Kerkenes Dağ. Stochastic variations of up to 10 percent from optimal did not yield radical changes in the models developed of transportation through the street system of these cities. At the 10 percent threshold, some variation was noted at Cambridge, which could be interpreted based on observations of pedestrians and businesses (Branting 2004:109–110). However, with relatively little variation, and also only a small amount of information so far on land use within the city at Kerkenes Dağ or testing of the soils along the city streets, additional work with the PGIS-T and SHULGI systems there focused exclusively on least cost routines for route assignment. A return to implementations of stochastic route assignment methods is expected once more testing has been completed and more data have been generated against which to compare the results of the least cost and stochastically generated pathways.

12.3.4. Problem 4: Alternate spatial forms for least cost pathway analyses are often underutilized even when they could yield substantial advantages in a particular situation

Least cost pathway models, as well as the data used to populate their variables, take different types of spatial forms. Thill (2000) categorizes three types of models that have been used in GIScience for representing and analyzing transportation: raster models, network models, and discrete models. Each of these models has different problems and benefits. However, the reasons behind the adoption of a given model usually have less to do with evaluating these costs and benefits than with determining which is easiest to implement within a particular GIS software package and which format of data is more readily available. These have been key factors in the widespread adoption of raster and discrete models and the meager development of network models. In this book, all the chapters make use of raster models. Rissetto (Chapter 2), Surface-Evans (Chapter 8), and Livingood (Chapter 10) also include discrete models for comparative purposes, while only White (Chapter 11) incorporates network models. Certainly one type of model might be quite applicable to some situations yet entirely wrong for others, so to better balance the situation, I discuss each one in turn.

Raster models and raster data, as described in greater depth in the introductory chapter of this book, divide up space into a series of equally shaped and sized cells. In nearly all computer-based implementations, these cells take the form of a square, and the entire area is laid out in a grid of these squares much like a chessboard. Each cell has one or more values assigned to it. For transportation modeling, these values are some measure of part of the costs of moving through that cell. The costs can be determined by either quantitative or qualitative means and can reflect any number of potential factors, thoughts, concerns, or expenditures involved in moving around within the space (Branting 2004:67–71). Kantner (Chapter 13) provides a breakdown of the two primary costs, the "currencies" of time and energy, as he terms them, and details their application, while Surface-Evans and White (Chapter 1) go into more detail on the broader range of potential costs. Use of a raster framework movement is modeled by cumulatively merging whichever costs are selected from cell to adjacent cell, as if a king were being moved around a chessboard. The total costs of each different possible path through this grid from a given cell "A" to a given cell "B" can then be determined and the results directly compared.

The key benefits of using a raster model are the ease of its implementation and its computational speed, which is significantly faster than other types of models (Johnston and de la Barra 2000:135). It is also, as argued by Ullah and Bergin (Chapter 9), seen as a better way in many cases to model the underlying natural landscape, particularly over large regions. When implementing a raster, one need only decide on cell size, the raster's boundaries, and what values to assign. It requires no a priori knowledge about where roads or paths once ran, though as noted under Problem 1 they can be compared after the generation of the LCPs. This can save significant amounts of time and effort when one is setting up the model. It can also avoid problems of mistakenly identifying the course of particular roads or paths and thereby skewing the results of the analysis.

There are, however, drawbacks to using a raster model for analyzing ancient transportation. Foremost among them is the fact that people and things rarely move around the landscape like chess pieces, hopping from square to square within a grid. Instead, we walk, ride, or drive in a more linear fashion. Neither do we move exactly the same distance with each step, as is assumed by modeling movement through a grid that possesses uniform cell sizes. Instead, we take smaller steps when going down steep slopes and longer strides when rushing to get somewhere (Sun et al. 1996; Kawamura et al. 1991; Sato and Ishizu 1990). We can also move through a given square-shaped area in any number of ways. We could cut across a corner of the square just as easily as walking straight across the middle. Each of these different possible paths can have very different costs. This is in addition to the issue of isotropic vs. anisotropic, or bidirectional, movement through the cells, as discussed in Surface-Evans and White (Chapter 1) and in more depth by Kantner (Chapter 13) and Plog (1977:130). While there are important ways to work around some of these issues, such as anisotropicity, as described by Bell and Lock (2000:89–91), Bell et al. (2002), and, in this volume, Ullah and Bergin (Chapter 9), Livingood (Chapter 10), White (Chapter 11), and Kantner (Chapter 13), the workarounds can't completely address the fundamental geometric mismatch.

Care also needs to be taken, as discussed by Anderson (Chapter 14), in selecting an appropriate size of cell to use for raster models and data. Issues of scale are critical. As noted by Kvamme (1990), if the pixel size is too large, it can mask variations taking place within the aggregated pixel that significantly alter the output of the analysis. This was demonstrated at Kerkenes Dağ when I tried to match LCPs generated by a basic raster model and DEMs of varying cell sizes to the known segments of the urban street network (Branting 2007:103–105). The results of the correspondence analysis, particularly at raster cell size greater than 1 m, were striking. LCP analysis at the level of 30-m or 90-m pixel sizes ran over significant features of the landscape such as the later Byzantine castle perched on a massive rise in the city. Clearly, with larger cell sizes, significant topographic features that would have affected movement were lost within the area of these large pixels, as Kvamme warned.

Least cost models can take another very useful form: network models, which are used in this volume only by White (Chapter 11). Unlike raster models, which are based on a grid, network models and data are formed from a lattice of interconnecting linear segments joined to one another at specific intersection points called nodes. These segments, also known as links, can bend and curve throughout the area just as a road or path would do. They can be of any length, and different segments can vary quite widely in their respective lengths within the lattice. The only requirement for the network model is that the topology of the interconnecting segments within the lattice must be explicitly defined. Costs, just as with cells in the raster model, can be applied to each segment and node within the network. And the total cost of using different routes through the network can then be compared through the cumulative merging of the costs of movement over each consecutive segment between given points "A" and "B."

The key benefits of network models are their flexibility and the fact that they are a better fit for modeling the linearity of movement. Since people and things move in a linear fashion, not by hopping about from place to place without moving through the space in between, using a linear lattice of interconnecting segments better reflects how movement is undertaken. At the same time, the network model is not bound by the limitations of uniformity in cell size or structure. Segments can be placed anywhere in space, can themselves be of any length, can vary in length from segment to segment in the lattice, and can even overlie one another. The network models at Kerkenes Dağ, for example, could be broken into segments corresponding to the ca. 50-cm-long stride of the individual being modeled (Branting 2004:95). This allows for potentially much better modeling of modes of transportation such as walking, in which factors such as the slope of the ground can affect the length of each step and lead to non-uniformity. At the same time, the explicit topology underlying the lattice allows for easy modeling of varying bidirectional movement over any segment. These two primary benefits of the network model — flexibility and greater fidelity to how actual movement takes place — make it much better suited to handle the modeling of a variety of the nuances of traditional modes of transportation.

However, network models also have several drawbacks. Foremost among these are the time and effort that need to go into the design and construction of the network before modeling, as well as the additional processing time it can take during the modeling process. It is critical to accurately locate and define the pathways represented by each of the segments within the lattice. A quick or haphazard drawing of them will likely yield incorrect or highly uninformative results. The results from the network modeling of movement are closely tied to the structure and form of the network. If one does not know where the streets or paths once ran, the results will be without meaning. Yet luckily, as mentioned above in Problem 1, a growing number of studies, including several in this book, have demonstrated how entire lattices of ancient roads and pathways can be elucidated and mapped through remote sensing and traditional fieldwork.

The third category of models, discrete models, are little more than graphical representations of generalized measures of distance. Site catchments, buffers, and Thiessen polygons are all examples of discrete models. They are based on a measure of movement, such as a half-day's journey from the central village or the midpoint in traveling between villages, but they don't take into account the role that the landscape or other localized factors might play in defining that boundary. In this volume, Rissetto (Chapter 2), Surface-Evans (Chapter 8), and Livingood (Chapter 10) each contrast their own raster models with standard or existing discrete models to demonstrate the additional utility that might be gained by using more sophisticated models.

Finally, these three models are not mutually exclusive but can be combined to create hybrid models that merge aspects of two or more of them. One example is STREETS, an early agent-based system that combined network models with raster models, creating "walkability surfaces" in local areas linked by networks for routing (Schelhorn et al. 1999).

12.3.5. Problem 5: Many least cost pathway analyses use simplistic cost algorithms

LCP studies in archaeology have almost always used very simplistic algorithms to describe how things move from place to place. In this regard, this book is a welcome departure from the norm (cf. Kantner, Chapter 13). More sophisticated algorithms can yield impressive results, as preliminary results from the testing of the models at Kerkenes Dağ suggested, even with the inevitable incompleteness of archaeological data. Yet there are certainly many instances, including applications over large areas or where it is unnecessary for the research design, in which more simplified algorithms are not only appropriate but in some cases essential.

Five different levels of cost algorithms have been used in archaeology to model travel costs of individuals. Specific details on many of these algorithms as well as their application in this book are discussed in greater detail in Kantner (Chapter 13) and so are only briefly summarized here. The five levels are isotropic models, rule-of-thumb approaches, topographic-only models, physiological models, and finally the expansion of the physiological models through the application of kinematic variables. Isotropic models are those that deal with space as if it were a featureless plain without any topographic or other impedimental barriers. They most often are used with the discrete models described above. Rule-of-thumb approaches are similar to those found in hiking manuals for roughly estimating how long it will take a hiker to traverse particular types of terrain. Rule-of-thumb approaches also can include trial walks across a portion of the area to extrapolate estimates across the wider area. Topographic-only models are those that look at the upward and downward slopes of the terrain as the only variable in the model (e.g., Imhof 1950; Tobler 1993).

Physiological models are often based on regressions of test subject data or other such observational studies in the realms of biomechanics, engineering, ergonomics, and physiology (e.g., McDonald 1961; Murray et al. 1964; Murray et al. 1969; Finley et al. 1969; Minetti et al. 1993; Pandolf et al. 1977; Minetti 1995; Santee et al. 2001). The models can be quite complex and can incorporate a wide range of potential internal and external variables beyond just topographic slope to characterize pedestrian movement. They include variables such as terrain roughness, slickness, hardness and penetrability, loads carried, how loads are carried, footwear worn, the weight of the individual, and his or her age and sex. These physiological models can be further enhanced by the incorporation of values for variables in the models, drawn from studies looking at cognitive or social factors also involved in how people move (e.g., Lee 1970; Briggs 1973; Okabe et al. 1986; Sato and Ishizu 1990; Hanyu and Itsukushima 1995; Amorim 1999; Allen 2007). Using variables specified by different subpopulations allows a researcher to break down the physiological models in new ways that might be more applicable to a given research design.

The advantages of using more sophisticated models, when appropriate, have been explored in several archaeological studies. However, only a few studies have actually made use of the upper two levels of algorithms (e.g., Brannan 1992; Machovina 1996; Llobera 1999, 2000; Branting 2004; Branting et al. 2007; White 2007; and in this volume Rademaker et al., Chapter 3; White, Chapter 11; Kantner, Chapter 13). It is to be hoped that the growing number of studies using these more sophisticated models represents the start of a much more widespread usage. Both SHULGI and the PGIS-T systems utilized physiological models disaggregated by various kinematic variables, especially variables pertaining to the age and sex of the individual.

12.3.6. Problem 6: Least cost pathway analysis is often lacking in interaction variables

Transportation can be a very social activity, in terms of both why and how it is undertaken. While it is clear that the reasons people go to work or play has a great deal to do with interactions with other people and groups, it is also true that as they move about, they are also interacting with those around them. Studies have found that walking with, or at least in close proximity to, other people can dramatically alter how individuals within those groups walk (Sato and Ishizu 1990). This includes people who move at higher walking speeds and who decrease their speeds within a group, and those with slower walking speeds who increase their speeds in order to keep up with the group. Or individuals may slow down to have a conversation. All these interaction factors have significant effects on both the speed and the energy expenditures of individuals within that group. Walking with a group is but one example of a situation in which the interactions between individuals can have a significant effect on how individuals move. These interaction variables have yet to be adequately incorporated into any studies of movement within archaeology, including those in this book.

The development of SHULGI is primarily a consequence of the desire to start to address interaction variables during movement (Branting et al. 2007). An agent-based platform, such as Repast Simphony, on which SHULGI is based, is much more robust in dealing with the interactions of individual agents within a larger system. This was an aspect of movement that was completely lacking in the initial PGIS-T framework or in basic LCP analysis. By addressing this in an agent-based rather than aggregate simulation framework, we can code different agents with different models or different decision criteria. The interaction between individuals and groups of agents, both similar and different, can then be explored not only at the origins and destinations of the trip but in the movements between them. While SHULGI is just beginning to develop such interaction variables, this area of future expansion holds enormous potential for simulating higher-level interactions such as social ranking, economic exchanges, and resistance strategies.

12.3.7. Problem 7: Archaeological least cost pathway analysis often encounters problems in the determination of contemporaneity

Unlike the preceding problems, the problem of determining contemporaneity is much more specific to archaeological or historical research than it is to modern transportation research. The reconstruction of the particular placement of elements in settlement or nomadic systems, as well as the pathways linking them together, is often incomplete or fraught with missing data within the preserved archaeological record. This is an aspect that has been discussed earlier in facets of Problems 1, 4, and 5. However, even when abundant segments of pathways or large numbers of settlements are found, determining what was in use concurrently in any given year or even century can be problematic. The issue of contemporaneity has been raised already in this volume by Rademaker et al. (Chapter 3), Phillips and Leckman (Chapter 4), Nolan and Cook (Chapter 5), Hudson (Chapter 6), and White (Chapter 11).

The solution to this problem is not computational in nature. Instead, it can only be answered, if it can be answered at all in a specific archaeological context, by explorations in these different locations. This is not the only case in which well-devised archaeological strategies are a necessity for using LCP analysis. As noted above in Problem 1, results of such modeling and simulations should always find ways to lead back to methods for testing them in the field. There is sometimes an overreliance on building the incredible frameworks of conjecture that computational methods allow, without finding ways to turn these into testable hypotheses within the limitations of archaeological data. Not all archaeological problems can be solved through new and creative applications of computerized models, including those within the realm of LCP analysis. At Kerkenes Dağ, this problem of contemporaneity has been minimized through the short-lived nature of the city. Streets and roads were largely all contemporaneous during the short period the city was inhabited. Selection of specific sites with constrained periods of occupation is therefore one way that this problem can be approached. Another well-used approach is to date roads and trackways based on settlements or artifacts found along or within them (e.g., Phillips and Leckman, Chapter 4; White, Chapter 11). Textual evidence can also provide a solution to this problem, as well as offer valuable insights into the times and experiences of past travelers along these routes.

12.4. Conclusion

This discussion has raised seven prominent problems encountered in the use of LCP analysis with archaeology. Solutions for each, based in part on experience gained from developing network-based transportation simulations as part of the Kerkenes Dağ and SHULGI projects, were also explored. Areas explored included finding and dating pathways from the past to test the results of the models generated from analysis, choosing appropriate models in both form and cost algorithms to meet modeling needs, and adding new elements to LCP analysis such as interaction variables, decisions systems, and stochastic methods. Not all these ideas have yet been implemented in archaeological research, but those that have are yielding promising results. The incorporation of new forms of simulation and analysis will always offer the possibility to change the potentials and drawbacks of existing methods. The development of hybrid raster-network and raster-discrete models is but one example of how new methods can significantly change problems and benefits encountered in current approaches. Continuously reexamining and questioning issues such as those identified here can help guide the development of new methods that directly address these problems and provide illuminating new directions for future research.

References

Aldenderfer, Mark
1998 *Montane Foragers: Asana and the South-Central Andean Archaic.* University of Iowa Press, Iowa City.

Allen, Gary L. (editor)
2007 *Applied Spatial Cognition: From Research to Cognitive Technology.* Lawrence Erlbaum Associates, Mahwah, New Jersey.

Amorim, Michel-Ange
1999 A Neurocognitive Approach to Human Navigation. In *Wayfinding Behavior: Cognitive Mapping and Other Spatial Processes*, edited by Reginald G. Golledge, pp. 152–167. Johns Hopkins University Press, Baltimore.

Batty, Michael
2005 *Cities and Complexity: Understanding Cities with Cellular Automata, Agent-Based Models, and Fractals.* MIT Press, Cambridge, Massachusetts.

Bell, T. L., R. L. Church, and L. J. Gorenflo
1988 Late Horizon Regional Efficiency in the Northeastern Basin of Mexico: A Location-Allocation Perspective. *Journal of Anthropological Archaeology* 7:163–202.

Bell, T., and G. Lock
2000 Topographic and Cultural Influences on Walking the Ridgeway in Later Prehistoric Times. In *Beyond the Map: Archaeology and Spatial Technologies*, Vol. 321, edited by G. R. Lock, pp. 85–100. IOS Press, Amsterdam.

Bell, Tyler, Andrew Wilson, and Andrew Wickham
2002 Tracking the Samnites: Landscape and Com-

munications Routes in the Sangro Valley, Italy. *American Journal of Archaeology* 106(2):169–186.

Berthöz, Alain, Michel-Ange Amorim, Stephan Glasauer, Renato Grasso, Yasuiko Takei, and Isabelle Viaud-Delmon

1999 Dissociation between Distance and Direction during Locomotor Navigation. In *Wayfinding Behavior: Cognitive Mapping and Other Spatial Processes*, edited by Reginald G. Golledge, pp. 328–348. Johns Hopkins University Press, Baltimore.

Brannan, James A.

1992 On Modeling Resource Transport Costs: Suggested Refinements. *Current Anthropology* 33(1):56–59.

Branting, Scott

1996 The Alisar Regional Survey, 1993–1994: A Preliminary Report. *Anatolica* 22:145–158.

2004 Iron Age Pedestrians at Kerkenes Dağ: An Archaeological GIS-T Approach to Movement and Transportation. Unpublished Ph.D. dissertation, State University of New York, Buffalo.

2007 Using an Urban Street Network and a PGIS-T Approach to Analyze Ancient Movement. In *Digital Discovery: Exploring New Frontiers in Human Heritage; CAA 2006; Fargo, United States*, edited by Jeffrey T. Clark and Emily M. Hagemeister, pp. 99–108. Archaeolingua, Budapest.

2009 Kerkenes Dağ Project. In *The Oriental Institute Annual Reports, 2008–2009*, edited by G. J. Stein, pp. 88–95. Oriental Institute Press, Chicago.

Branting, Scott, and Geoffrey D. Summers

2002 Modelling Terrain: The Global Positioning System (GPS) Survey at Kerkenes Dağ, Turkey. *Antiquity* 76:639–640.

Branting, S., Y. Wu, R. Srikrishnan, and M. R. Altaweel

2007 SHULGI: A Geospatial Tool for Modeling Human Movement and Interaction. In *Proceedings of the Agent 2007 Conference on Complex Interaction and Social Emergence, Evanston, IL*, edited by M. North, D. Sallach, and C. Macal, pp. 475–487. Argonne National Laboratory, Argonne, Illinois.

Briggs, Ronald

1973 Urban Cognitive Distance. In *Image and Environment: Cognitive Mapping and Spatial Behavior*, edited by Roger M. Downs and David Stea, pp. 361–388. Aldine, Chicago.

Clarke, David L.

1977 Spatial Information in Archaeology. In *Spatial Archaeology*, edited by David L. Clarke, pp. 1–32. Academic Press, London.

Close, Angela E.

2000 Reconstructing Movement in Prehistory. *Journal of Archaeological Method and Theory* 7(1): 49–77.

Crooks, Andrew T.

2008 *Constructing and Implementing an Agent-Based Model of Residential Segregation through Vector GIS*. Working Paper Series 133. UCL Centre for Advanced Spatial Analysis, London.

Dial, R. B.

1971 A Probabilistic Multipath Traffic Assignment Model Which Obviates Path Enumeration. *Transportation Research* 5(2):83–111.

Dijkstra, E. W.

1959 A Note on Two Problems in Connexion with Graphs. *Numerische Mathematik* 1:269–271.

Ebert, J. I., and R. K. Hitchcock

1980 Locational Modeling in the Analysis of the Prehistoric Roadway System at and around Chaco Canyon, New Mexico. In *Cultural Resources Remote Sensing*, edited by T. R. Lyons and F. J. Mathien, pp. 167–207. National Park Service, Washington, D.C.

Erickson, Clark L.

2001 Pre-Columbian Roads of the Amazon. *Expedition* 43(2):21–31.

Etienne, Ariane S., Roland Maurer, Josephine Georgakopoulos, and Andrea Griffin

1999 Dead Reckoning (Path Integration), Landmarks, and Representation of Space in a Comparative Perspective. In *Wayfinding Behavior: Cognitive Mapping and Other Spatial Processes*, edited by Reginald G. Golledge, pp. 197–228. Johns Hopkins University Press, Baltimore.

Finley, F. R., K. A. Cody, and R. V. Finize

1969 Locomotive Patterns in Elderly Women. *Archives of Physical Medicine and Rehabilitation* 50:140–146.

Gaffney, Vincent, and Zoran Stančič

1991 *GIS Approaches to Regional Analysis: A Case Study of the Island of Hvar*. Znanstveni Institut Filozofske Fakultete, Ljubljana, Slovenia.

Golledge, Reginald G. (editor)

1999 *Wayfinding Behavior: Cognitive Mapping and Other Spatial Processes*. Johns Hopkins University Press, Baltimore.

Hanyu, Kazunori, and Yukio Itsukushima

1995 Cognitive Distance of Stairways: Distance, Traversal Time, and Mental Walking Time Estimations. *Environment and Behavior* 27(4):579–591.

Harris, Trevor
2000 Moving GIS: Exploring Movement within Prehistoric Cultural Landscapes Using GIS. In *Beyond the Map: Archaeology and Spatial Technologies*, Vol. 321, edited by G. R. Lock, pp. 116–123. IOS Press, Amsterdam.
Hart, P., N. J. Nilsson, and R. Bertram
1968 A Formal Basis for the Heuristic Determination of the Minimum Cost Paths. *IEEE Transaction on Systems Science and Cybernetics SSC* 4(2):100–107.
Heckenberger, Michael J., Afukaka Kuikuro, Urissapá K. Tabata, J. R. Christian, Morgan Schmidt, Carlos Fausto, and Bruna Franchetto
2003 Amazonia 1492: Pristine Forest or Cultural Parkland? *Science* 301:1710–1714.
Hughes, Karel J.
1999 Persistent Features from a Palaeo-Landscape: The Ancient Tracks of the Maltese Islands. *Geographical Journal* 165(1):62–79.
Imhof, E.
1950 *Gelände und Karte*. Eugen Rentsch, Zürich.
Ingold, T.
1993 The Temporality of the Landscape. *World Archaeology* 25(2):152–174.
Irwin-Williams, Cynthia
1977 A Network Model for the Analysis of Prehistoric Trade. In *Exchange Systems in Prehistory*, edited by T. Earle and J. Ericson, pp. 141–151. Academic Press, New York.
Jennings, Justin, and Nathan Craig
2001 Politywide Analysis and Imperial Political Economy: The Relationship between Valley Political Complexity and Administrative Centers in the Wari Empire of the Central Andes. *Journal of Anthropological Archaeology* 20: 479–502.
Johnston, R. A., and T. de la Barra
2000 Comprehensive Regional Modeling for Long-Range Planning: Linking Integrated Urban Models and Geographic Information Systems. *Transportation Research Part A* 34:125–136.
Kantner, J.
1996 An Evaluation of Chaco Anasazi Roadways. Paper presented at the 61st Annual Meeting of the Society for American Archaeology, New Orleans.
Kawamura, K., A. Tokuhiro, and H. Tahechi
1991 Gait Analysis of Slope Walking: A Study on Step Length, Stride Width, Time Factors and Deviation in the Centre of Pressure. *Acta Medica Okayama* 45:179–184.
Kelly, Robert L.
1992 Mobility/Sedentism: Concepts, Archaeological Measures, and Effects. *Annual Review of Anthropology* 21:43–66.
Kvamme, Kenneth L.
1990 GIS Algorithms and Their Effects on Regional Archaeological Analysis. In *Interpreting Space: GIS and Archaeology*, edited by Kathleen M. Allen, Stanton W. Green, and Ezra B. W. Zubrow, pp. 112–126. Taylor and Francis, New York.
2003 Geophysical Surveys as Landscape Archaeology. *American Antiquity* 68(3):435–457.
Lee, Terence
1970 Perceived Distance as a Function of Direction in the City. *Environment and Behavior* 2:40–51.
Lepper, Bradley T.
1995 Tracking Ohio's Great Hopewell Road: Aerial Reconnaissance and Infrared Photography Help Scholars Map a Sacred Roadway Nearly 60 Miles Long. *Archaeology* 48(6):52–57.
Llobera, Marcos
1999 Landscapes of Experiences in Stone: Notes on a Humanistic Use of a Geographic Information System to Study Ancient Landscapes. Unpublished Ph.D. dissertation, Oxford University, Oxford.
2000 Understanding Movement: A Pilot Model towards the Sociology of Movement. In *Beyond the Map: Archaeology and Spatial Technologies*, Vol. 321, edited by G. R. Lock, pp. 65–84. IOS Press, Amsterdam.
Machovina, Brett J.
1996 *Modeling of Pedestrian Mobility across a Natural Landscape Using a Geographic Information System*. Final report, submitted under National Park Service Contract 6115-4-8031 (Part II), Ohio State University, Columbus.
McDonald, Ian
1961 Statistical Studies of Recorded Energy Expenditure of Man: Expenditure on Walking Related to Weight, Sex, Age, Height, Speed, and Gradient. *Nutrition Abstracts and Reviews* 31(3):739–762.
McKee, Brian R., Thomas L. Sever, and Payson D. Sheets
1994 Prehistoric Footpaths in Costa Rica: Remote Sensing and Field Verification. In *Archaeology, Volcanism, and Remote Sensing in the Arenal Region, Costa Rica*, edited by Payson D. Sheets and Brian R. McKee, pp. 142–157. University of Texas Press, Austin.
Miller, Heather M.
2006 Comparing Landscapes of Transportation: Riverine-Oriented and Land-Oriented Systems in the Indus Civilization and the Mughal Em-

pire. In *Space and Spatial Analysis in Archaeology*, edited by Elizabeth C. Robertson, Jeffrey D. Seibert, Deepika C. Fernandez, and Marc U. Zender, pp. 281–291. University of Calgary Press, Calgary, Alberta.

Minetti, A. E.
1995 Optimum Gradient of Mountain Paths. *Journal of Applied Physiology* 79(5):1698–1703.

Minetti, A. E., L. P. Ardigò, and F. Saibene
1993 Mechanical Determinants of Gradient Walking. *Journal of Physiology* 471:725–735.

Murray, M. P., A. B. Drought, and R. C. Kory
1964 Walking Pattern of Normal Men. *Journal of Bone and Joint Surgery* 46A:335–360.

Murray, M. P., R. C. Kory, and B. H. Clarkson
1969 Walking Patterns in Healthy Old Men. *Journal of Gerontology* 24:169–178.

Obenauf, Margaret S.
1991 Photointerpretation of Chacoan Roads. In *Ancient Road Networks and Settlement Hierarchies in the New World*, edited by Charles D. Trombold, pp. 34–41. Cambridge University Press, Cambridge.

Okabe, A., K. Aoki, and W. Hamamoto
1986 Distance and Direction Judgment in a Large-Scale Natural Environment: Effects of a Slope and Winding Trail. *Environment and Behavior* 18(6):755–772.

Pandolf, Kent B., B. Givoni, and R. F. Goldman
1977 Predicting Energy Expenditure with Loads While Standing or Walking Very Slowly. *Journal of Applied Physiology* 43:577–581.

Plog, Fred
1977 Modeling Economic Exchange. In *Exchange Systems in Prehistory*, edited by T. Earle and J. Ericson, pp. 127–140. Academic Press, New York.

Poehler, Eric
2009 The Organization of Pompeii's System of Traffic: An Analysis of the Evidence and Its Impact on the Infrastructure, Economy and Urbanism of the Ancient City. Unpublished Ph.D. dissertation, University of Virginia, Charlottesville.

Santee, W. R., W. F. Allison, L. A. Blanchard, and M. G. Small
2001 A Proposed Model for Load Carriage on Sloped Terrain. *Aviation, Space and Environmental Medicine* 72(6):562–566.

Sato, H., and K. Ishizu
1990 Gait Patterns of Japanese Pedestrians. *Journal of Human Ergonomics* 19:13–22.

Schelhorn, T., D. O'Sullivan, M. Haklay, and M. Thurstain-Goodwin
1999 *STREETS: An Agent-Based Pedestrian Model.* Working Paper Series No. 9. Centre for Advanced Spatial Analysis. University College London, London.

Sever, Thomas L., and David W. Wagner
1991 Analysis of Prehistoric Roadways in Chaco Canyon Using Remotely Sensed Digital Data. In *Ancient Road Networks and Settlement Hierarchies in the New World*, edited by Charles D. Trombold, pp. 42–52. Cambridge University Press, Cambridge.

Snead, James E.
2002 Ancestral Pueblo Trails and the Cultural Landscape of the Pajarito Plateau, New Mexico. *Antiquity* 76(293):756–765.

Snead, James E., Clark L. Erickson, and J. A. Darling (editors)
2009 *Landscapes of Movement: Trails, Paths, and Roads in Anthropological Perspective.* University of Pennsylvania Museum of Archaeology and Anthropology, Philadelphia.

Stern, Eliahu, and Juval Portugali
1999 Environmental Cognition and Decision Making in Urban Navigation. In *Wayfinding Behavior: Cognitive Mapping and Other Spatial Processes*, edited by Reginald G. Golledge, pp. 99–119. Johns Hopkins University Press, Baltimore.

Summers, Geoffrey D., and Françoise Summers
2010 From Picks to Pixels: Eighty Years of Development in the Tools of Archaeological Exploration and Interpretation, 1927–2007, at Kerkenes Dağ in Central Turkey. In *Proceedings of the 6th International Congress on the Archaeology of the Ancient Near East*, edited by Paolo Matthiae, Frances Pinnock, Lorenzo Nigro, and Nicolò Lorenzo, pp. 669–683. Harrassowitz, Wiesbaden, Germany.

Summers, Geoffrey, Françoise Summers, and Scott Branting
2005 *Kerkenes News 7, 2004—Kerkenes Haberler 7, 2004.* METU Press, Ankara, Turkey.

Sun, Jie, Megan Walters, Noel Svensson, and David Lloyd
1996 The Influence of Surface Slope on Human Gait Characteristics: A Study of Urban Pedestrians Walking on an Inclined Surface. *Ergonomics* 39(4):677–692.

Thill, Jean-Claude
2000 Geographic Information Systems for Transportation in Perspective. *Transportation Research Part C* 8:3–12.

Tilley, Christopher Y.
1994 *A Phenomenology of Landscape: Places, Paths, and Monuments.* Berg, Oxford.

Tobler, W.
1993 *Three Presentations on Geographical Analysis and Modeling*. Technical Report 93-1. University of California, Santa Barbara.

Trombold, Charles D. (editor)
1991 *Ancient Road Networks and Settlement Hierarchies in the New World*. Cambridge University Press, Cambridge.

Ur, Jason A.
2003 CORONA Satellite Photography and Ancient Road Networks: A Northern Mesopotamian Case Study. *Antiquity* 77:102–115.

White, Devin A.
2007 Transportation, Integration, Facilitation: Prehistoric Trail Networks of the Western Papaguería. Unpublished Ph.D. dissertation, University of Colorado, Boulder.

Wilkinson, T. J.
1993 Linear Hollows in the Jazira, Upper Mesopotamia. *Antiquity* 67:548–562.

Wilkinson, Tony J., Charles French, Jason Ur, and Miranda Semple
2010 The Geoarchaeology of Route Systems in Northern Syria. *Geoarchaeology* 25(6):745–771.

Wise, Sarah
2009 RiotMonger: An Agent-based Simulation of Emergent Riot Behavior in Urban Contexts. B.S. honors thesis, University of Chicago, Chicago.

Zhou, Yuhong
2008 Agent-Based Modeling and Simulation for Pedestrian Movement Behaviors in Space: A Review of Applications and GIS Issues. In *Geoinformatics 2008 and Joint Conference on GIS and Built Environment: Geo-Simulation and Virtual GIS Environments*, edited by Lin Liu, Xia Li, Kai Liu, Xinchang Zhang, and Aijun Chen. *Proceedings of SPIE* 7143(1):71431I-1–71431I-9.

CHAPTER 13

Realism, Reality, and Routes

Evaluating Cost-Surface and Cost-Path Algorithms

John Kantner

The case studies in this book demonstrate how informative cost-path analysis can be for understanding the past, especially when creatively deployed to address specific questions whose answers require an understanding of human movement. (My use of "cost path" in this chapter is identical to "least cost path," or "LCP," found elsewhere in this volume.) Livingood's refinement of Hally's Mississippian polity boundaries in Chapter 10 provides a good example. By replacing arbitrary linear measurements to create hypothesized boundaries with distances measured by travel time, Livingood is able to account for regional geography, including waterways. Similarly, Richards-Rissetto's configurational analysis in Chapter 7 employs cost-distance considerations to expand classic space syntax methods to three rather than just two dimensions, with more realistic cost-of-movement values replacing the integration values typically used in space syntax. And in Chapter 9, Ullah and Bergin's consideration of a priori knowledge of least cost movement in a dynamically changing landscape provides a realistic augmentation of agent-based modeling in which agents no longer move across a landscape as if it is foreign to them no matter how often they experience it. The contributors to this volume join a growing number of archaeologists who are integrating cost-path analysis into projects that consider human use of space (e.g., Anderson and Gillam 2000; Byerly et al. 2005; Carballo and Pluckhahn 2007; Hare 2004; Howey 2007; Jennings and Craig 2001; Marín Arroyo 2009; Sakaguchi et al. 2010; Sherman et al. 2010; Siart et al. 2008; Taliaferro et al. 2010).

Despite increasing use of this approach, however, less clear is the degree to which the underlying methods and procedures of cost-path analysis are critically assessed before their implementation (Conolly and Lake 2006:252–256). Geographic Information System (GIS) software has become reasonably accessible, with user-friendly interfaces that obscure underlying procedures, the details of which are hidden away in online manuals and white papers. As a result, users can apply cost-path functions without full knowledge of the geographic information science on which the functions are based. Building on the stimulating chapters in this volume and the informative discussion in Branting's Chapter 12, this contribution considers the utility of the various cost-path approaches employed with a particular focus on the problem of algorithm selection (see also Kantner 2004).

13.1. Cost-Surface Algorithms

As described in Chapter 1, cost-path analysis requires four, and arguably five, components. First, a three-dimensional raster surface is needed that provides a representation of real-world topography (although see Nolan and Cook's demonstration in Chapter 5 that initial surfaces need not always be topographical). Digital elevation models (DEMs) typically are used, many of which are available online for free, although they of course come in different resolutions and accuracies that correspondingly affect the resolution and presumed accuracy of the resulting cost-path analysis, often significantly (Conolly and Lake 2006:252–253). Second, a set of objects—usually points, but they can also be vector objects—define the origin(s) and destination(s) between which cost paths will be generated. These objects are typically selected based on the larger research agenda that the cost-path analysis serves. Third, a cost surface (also called a "friction surface") consisting of a raster of cells, the values of which represent a cost of movement across the cell, is needed. It is typically generated with an algorithm that converts the topographic surface into some measure or proxy of cost. Fourth, a spreading algorithm generates a new surface in which each cell represents the *cumulative* costs from origin(s) to destination(s), typically using queen's or knight's moves to define possible movements from one cell to another. By tracking accumulating costs across the surface, this procedure can also identify the least costly path from origin(s) to destination(s). Finally, albeit rarely considered and poorly developed methodologically, some goodness-of-fit statistics or other independent tests would be helpful for assigning some level of confidence to the resulting cost paths.

Selection of the first component needed for cost-path analysis—the raw topographic surface—is fairly simple, the primary limitation being acquisition of a digital surface of sufficient resolution and accuracy for one to feel confident in its representation of the real world as experienced by humans moving over the landscape. Only fifteen years ago, the greatest challenge for GIS practitioners was the difficulty of acquiring DEMs with resolutions under 30 m, which was generally deemed too coarse to adequately model human movement. And even if one were able to find, and to afford, high-resolution DEMs, limitations in computing power made processing of those surfaces difficult. Today, with the availability of free, high-resolution, and seamless topographic surfaces, such obstacles in cost-path analysis might be a problem of the past. On the other hand, issues of projection, accuracy, and perhaps computing power still remain, especially as larger real-world surfaces are modeled with incredible precision. Such precision, moreover, presents its own problems, as mentioned below.

Selecting a cost-surface algorithm is arguably the most important component of successful cost-path analysis and is thus the focus of this chapter. The problem is that numerous algorithms exist (e.g., Kantner 2004; van Leusen 1999; Wheatley and Gillings 2002:154–156), but virtually no studies have attempted to evaluate which ones best model real human movement—and, more importantly, which ones decidedly fail to represent such movement (but see Aldenderfer 1998:11–15). The chapters in this volume demonstrate the general absence of consensus regarding the comparative utilities of available cost-surface algorithms. Several authors—Surface-Evans, Rissetto, and Rademaker and his colleagues, for example—use slope as a proxy for the cost of movement across a raster cell. As Hudson explains in her chapter, slope-derived "pathways are used here as merely a proxy for accessibility. The actual cost of traveling those paths is of secondary importance." Other authors select more complex and arguably realistic algorithms for generating their cost surface, as in Livingood's use of Tobler's so-called hiking function (1993), Ullah and Bergin's implementation of GRASS's "r.walk" function, and White's augmentation of the metabolic rate formula presented in Pandolf et al. (1977).

Slope has the longest history of use in cost-path analysis, with most of the earliest GIS practitioners arguing that slope is a reasonable predictor of actual human movement costs and that therefore it can be productively used as the cost surface (e.g., Anderson and Gillam 2000; Madry and Crumley 1990; Madry and Rakos 1996; Savage 1990; Wheatley 1996). Many recently published cost-path analyses still focus on slope as a proxy of cost (e.g., Byerly et al. 2005; Sherman et al. 2010; Siart et al. 2008). The strength of the relationship between slope and actual movement cost, however, has never been empirically demonstrated. In fairly simple topographic settings characterized by consistently gentle terrain, slope might indeed be expected to be a good proxy of movement costs. But in more complex landscapes, especially those marked by very high and/or very low slope values, two issues become important. First, the direction of movement across a raster cell becomes critical, for no one would suggest that it costs the same to go down a 90-degree slope as it does to go up it. This issue—characterized as isotropy vs. anisotropy—is discussed in more detail later in this chapter.

The second significant issue is that flat surfaces have a slope of zero, making slope a terrible proxy of movement costs across flat or nearly flat terrain. The path-selecting algorithms used by most GIS software will unfailingly select those paths that cross level surfaces, because with a slope of zero, they don't cost anything to move over. As described by Surface-Evans and White (Chapter 1), path selection is usually implemented through some version of Dijkstra's algorithm (1959), which is a so-called greedy algorithm that introduces its own set of assumptions and issues, including favoring the *locally* optimal solution. As a result, cost paths generated in this way theoretically could travel forever over flat surfaces, because as far as the algorithms are concerned, there is no cumulative cost whatsoever to follow those paths. This is no doubt why so many cost paths based on slope cost surfaces like to travel along the edge of oceans (e.g., Rissetto, Chapter 2). Ironically, the increasing resolution of DEMs used in cost-path analysis is likely making use of slope as a proxy for cost a much greater problem than it used to be, because path-seeking algorithms whose guiding principle is to minimize cost have many more low- or no-slope cells to work with at, for example, a 5-m resolution than they do at a 30-m one. The central issue is that humans do not behave like water; we do not follow paths of least resistance, because unlike water, we have somewhere we want to get to quickly and efficiently. On the other hand, we don't have the same slope limitations that wheeled vehicles do, making algorithms derived from transportation science (e.g., Atkinson et al. 2005) equally problematic for most archaeological applications.

13.2. Currencies

The suspected flaws with using slope as the proxy for cost when one is creating cost surfaces raises the question of what currency *should* be used. Virtually all cost-surface algorithms rely on one of two currencies—energy or time—that the cost-path algorithms attempt to optimize. Those of us enculturated into busy Western lifestyles are inclined to consider time the most important currency. Tobler's well-known hiking function (1993), which is employed here by Livingood, White, and Phillips and Leckman, is based on marching data from the Swiss military as reported by Imhof (1950). In its original formulation, it optimizes velocity:

$$V = 6e^{-3.5|S+.05|} \qquad (1)$$

where
V = walking velocity in km/hr
e = the base of natural logarithms
S = the slope measured in vertical change over horizontal distance

Tobler's algorithm can readily be solved for time, as is done in the case study at the end of this chapter.

Similarly, GRASS's r.walk algorithm, which Ullah and Bergin use in their work (Chapter 9), is derived from the nineteenth-century Naismith's Rule for walking times (Langmuir 1984; Naismith 1892):

$$T = [a\Delta L] + [b\Delta U] + [c\Delta D] + [d\Delta S] \qquad (2)$$

where
T = time in seconds
a = walking speed
ΔL = distance covered
b = walking speed moving uphill
ΔU = the uphill altitude difference in meters
c = walking speed moving downhill on a moderate slope
ΔD = the downhill altitude difference
d = walking speed moving down a steep slope
ΔS = the altitude difference for steep slopes

The default values of *a*, *b*, *c*, and *d* are determined based on Langmuir's *Mountaincraft and Leadership* (1984), a well-known treatise on mountaineering. A similarly derived algorithm is discussed by Marín Arroyo (2009):

$$\begin{aligned} T &= 0.6D\left(\frac{S}{11}+1\right) S \geq 0 \\ T &= 0.6D\left(\frac{S}{23}+1\right) S < 0 \end{aligned} \qquad (3)$$

where
T = time in seconds
S = slope as a percentage
D = distance in meters

This is abstracted from the Método de Información de Excursiones (Excursion Information Method), used in many European countries to identify mountaineering and hiking difficulty levels.

In contrast, several algorithms are available that derive the energetic costs of movement. Duggan and Haisman (1992), for example, use one such algorithm that is based on laboratory evaluations of human movement created by Pandolf and his colleagues (1977):

$$M = 1.5W + 2.0(W + L)\left(\frac{L}{W}\right)^2 + n(W + L)(1.5V^2 + 0.35VS) \quad (4)$$

where

M = metabolic rate in watts (i.e., kilojoules/minute)

W = the weight of the unclothed person in kilograms

L = the load carried by the person in kilograms

n = a value reflecting the terrain across which the person is moving

V = the speed of walking in meters/second

S = the percentage grade

Recent studies by Santee et al. (2001) and Kramer (2010) have evaluated and expanded this algorithm.

Unfortunately, not many archaeological applications can take advantage of these more elaborate energy-focused algorithms due to the number of variables that would be unknowable; as can be seen in the formula above, holding the variables constant would undermine the utility of this particular algorithm. In Chapter 3 of this volume, however, Rademaker and his colleagues use an elaborated version of Pandolf's algorithm to explore the impact of varying the values used in L, W, and V—a potentially informative application for complex formulas. White (Chapter 11) also employs an enhanced version of the Pandolf algorithm in which he creatively uses Tobler's hiking function to calculate V. Additional examples of algorithms that emphasize the energetic costs of human movement can be found in the larger cost-path and human biology literature (Brannan 1992; Ericson and Goldstein 1980; Hare 2004; Herhahn and Hill 1998; Kramer 2010; Llobera and Sluckin 2007).

Cost-path studies in archaeology have expended little effort exploring which currency humans would most likely want to optimize in their movement across the landscape, and time and energy are often treated as if they are reasonably equivalent. The differences between the two are not inconsequential, however, and they often lead to quite different solutions. The work presented by Rademaker and his colleagues in Chapter 3 provides a good illustration of this point, for they use both a derivation of the Pandolf formula and a simple slope-based approach for generating their cost surfaces and subsequent cost paths to Andean obsidian sources. The results illustrate how slope-based cost paths seek out low-slope cells, creating very lengthy cost paths, as was discussed earlier. More importantly, the energy-based cost paths differ significantly both from the slope-based cost paths and from one another as the parameters of Pandolf's algorithm are changed.

In a consideration of the impact of differing currencies on cost-path analysis, Livingood's Chapter 10 discussion of "subjective distances" as first described in Montello (1997) is especially welcome. The geographical literature on human decision making regarding path selection is considerable (e.g., González et al. 2008; Song et al. 2010), and it tends to confirm Montello's argument that people consider a combination of environmental features, travel time, and travel effort, among other factors, when choosing routes, a point also made in Chapter 4 by Phillips and Leckman. This is reflected in a few recent "multicriteria" cost-surface analyses, such as Howey's (2007) integration of waterways, land cover, and slope for her Michigan study area (see also Atkinson et al. 2005); Richards-Rissetto's "urban DEM" approach (Chapter 7) similarly considers the built environment and its effect on travel routes, as does Livingood (Chapter 10) with waterways. People also are likely to choose quite different paths when traveling over unknown landscapes than they do when moving over familiar terrain, a point made by Rademaker and his colleagues (Chapter 3) when they invoke the notion of "sequentially optimal" path segments. Part of the issue is a consideration of temporal scale: human path selection over the short time frame might be expected to more heavily weigh time, acute environmental factors, and extremes in effort, while paths created over generations of human use might be more likely to meet expectations of human behavioral ecology and truly approach energetic optimality (e.g., Helbing et al. 1997).

13.3. Isotropic vs. Anisotropic Approaches

Cost-path algorithms differ according to whether they assume that the costs to travel across a given space should be isotropic—the same no matter in which direction the space is crossed—or anisotropic (Wheatley and Gillings 2002:152–153; also see Surface-Evans and White, Chapter 1). Most of the contributions to this volume use an isotropic approach, with the exception of Livingood (Chapter 10), Ullah and Bergin (Chapter 9), and White (Chapter 11), mirroring a trend for the majority of cost-path analyses in archaeology to implicitly or explicitly assume that travel cost is reasonably isotropic. Part of the reason for this assumption is that, until recently, most software packages did not readily accommodate anisotropic modeling.

Intuition suggests that the cost of traveling down a slope is less than trudging uphill, and a few anisotropic algorithms have been developed, most notably Marín Arroyo's formula (2009) and Tobler's walking function, described earlier and implemented successfully in several studies (e.g., Carballo and Pluckhahn 2007; Gorenflo and Gale 1990; Kantner 1997; Sakaguchi et al. 2010; Taliaferro et al. 2010). While these appear to more realistically represent the energetic costs of traversing variable topography, simply assigning different costs to uphill and downhill movement does not necessarily make the algorithm an accurate representation of human movement. Empirical observations have in fact suggested that the time it takes to move up and down slopes is perhaps more symmetrical than intuition would suggest—depending, of course, on the cost measure considered. This is especially true in topographically complex terrain, where a traveler could spend nearly as much time braking against a downhill slope as he or she would spend ascending that same slope, especially if burdened with a load (e.g., Marble 1996:Figure 1; Wheatley and Gillings 2002:Figure 7.4). GRASS's r.walk anisotropic algorithm, described earlier and employed by Ullah and Bergin (Chapter 9), recognizes that moving downhill is favorable up to a specific threshold, at which point it becomes more costly (see also Kramer 2010; Zhang et al. 2010). This captures what is almost certainly a nonlinear relationship between slope and movement, although claims for an exponential relationship are probably inaccurate (Brannan 1992). This asymmetrical anisotropy is the reason why it is incorrect to assume that round-trips along cost paths make directionality irrelevant and that therefore isotropic algorithms are reasonable approximations of human movement costs.

An additional challenge for implementing anisotropic algorithms lies in the GIS software used for the cost-path analysis. Simply entering an anisotropic formula is not sufficient, for the specific GIS function must be able to consider the direction of movement—that is, whether movement from one cell into a neighboring cell is going uphill or downhill (Collischonn and Pilar 2000). ESRI's ArcGIS, for example, includes at least two standard cost-surface functions, only one of which is designed to handle anisotropic movement, and then only if a "vertical factor table" of modifiers for each positive and negative slope degree is provided. This demonstrates how important it is to fully understand what the GIS functions do, a point made more critical by the increasing user-friendliness of GIS software, which hides detailed algorithms and complex procedures behind simple graphic user interfaces. This no doubt leads to some published studies that fail to provide details on how their cost surfaces and cost paths were generated (e.g., Fletcher 2008; Siart et al. 2008).

13.4. Case Study: Difference in Cost-Path Algorithms

As a demonstration of these issues, consider a brief example based on a study area in northwestern New Mexico for which I have conducted cost-path analyses (Kantner 1997, 2004; Kantner and Hobgood 2003). The specific area is a 2,500-km^2 region roughly centered on Hosta Butte, a prominent feature on top of Lobo Mesa (Figure 13.1). To the south is the Red Mesa Valley, which runs roughly east-west and is bisected by the Continental Divide; the headwaters of the Rio Puerco and Rio San Jose are found here. Even farther south is the Zuni Mountain uplift. North of Lobo

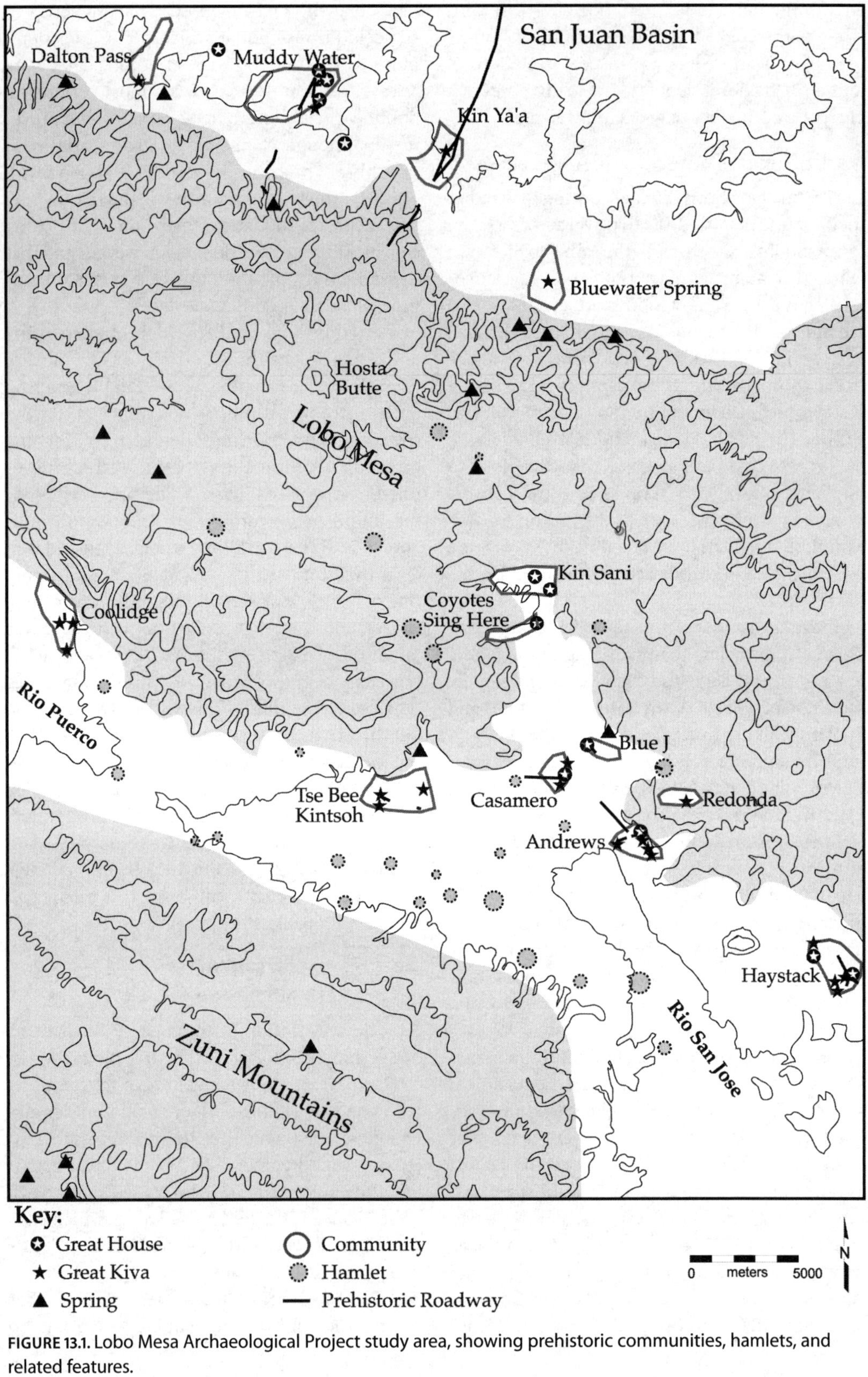

FIGURE 13.1. Lobo Mesa Archaeological Project study area, showing prehistoric communities, hamlets, and related features.

Mesa is the southern edge of the broad and comparatively flat San Juan Basin. I selected this region because it has been the focus of my archaeological investigations for many years, but it is also an ideal testing ground for cost-surface algorithms because it features considerable topographic complexity. There are also a number of prehispanic and contemporary communities and roadways to work with, most notably those associated with the eleventh-century Chaco Canyon pilgrimage center to the north. Here I generate a series of cost paths illustrating the points made earlier in this chapter about the importance of algorithm selection.

For the first cost-path analysis, I employed slope as the proxy for movement cost, the commonly used approach described earlier. With ESRI's ArcGIS software, I converted the National Elevation Data 30-m-resolution surface for the study area into percent slope values for each cell. I then used this as the input, along with distance, for ArcGIS's Costdistance function in order to generate a cumulative cost surface from the centrally located, eleventh-century community of Casamero to every other contemporaneous community in the study area. The results, which route least cost paths along the flattest topography (Figure 13.2), are not surprising. The slope function in ArcGIS determines the maximum rate of change for any given cell in the raster compared with all of its neighboring cells, which results in high costs for any topographic variability. On the other hand, any consistently flat areas are assigned very low costs. As a result, using slope as a cost surface, especially since this first analysis uses an isotropic implementation, identifies cost paths that minimize all vertical movement. In Figure 13.2, this can be seen in the Red Mesa Valley and the San Juan Basin, where the cost paths often follow long, circuitous routes.

To demonstrate the sensitivity of cell size for using slope as a proxy for movement costs, I did the same analysis on a DEM with a resolution of 10 m. As suggested earlier, increasing resolution is more likely to expose the flaws in using slope, and the resulting cost paths shown in Figure 13.2 are consistent with this suggestion. A 10-m surface has nine times as many cells as a 30-m surface, exposing a greater range of variation and providing the spreading algorithm many more near-zero cost options. The most extreme example in Figure 13.2 is the lengthy 10-m cost path that meanders across the northern third of the map.

I next used an implementation of Tobler's hiking function that employs travel time as the currency to optimize. This replicates what I did in my 1997 study, for which I selected Tobler's function because I reasoned that optimizing travel time was more important than alternative currencies, especially since travel from one community to another was over distances that would make for a long day across a high desert without many sources of water. I also appreciated the advantage of an anisotropic algorithm, although at the time I could not successfully implement the anisotropic part of Tobler's function before the implementation of vertical factor tables in ArcGIS (my 1997 work was in ESRI's earlier Arc/Info software). The algorithm employed was as follows:

$$T = \frac{D}{6e^{-3.5|S+.05|}} \qquad (5)$$

where
T = time in hours
D = distance in kilometers (converted from meters, the DEM's spatial unit)
e = the base of natural logarithms
S = the slope measured as vertical change over horizontal distance

To illustrate the importance of anisotropic vs. isotropic algorithms, I used both implementations of Tobler's hiking function. The results, shown in Figure 13.3, reveal that the least cost paths generated in this way are quite different from those identified simply through the use of slope, and tend to be much more direct between source and destination. Both the anisotropic and isotropic implementations favor shorter routes over rough terrain, forgoing some of the obvious, albeit more circuitous, canyon routes off Lobo Mesa that today are the conduits for paved roads; wheeled vehicles are limited by slope thresholds that do not as seriously impede human movement. The anisotropic cost paths tend to significantly diverge from the isotropic paths when going downhill, which is not surprising since the isotropic implementation regards all slopes as having positive values.

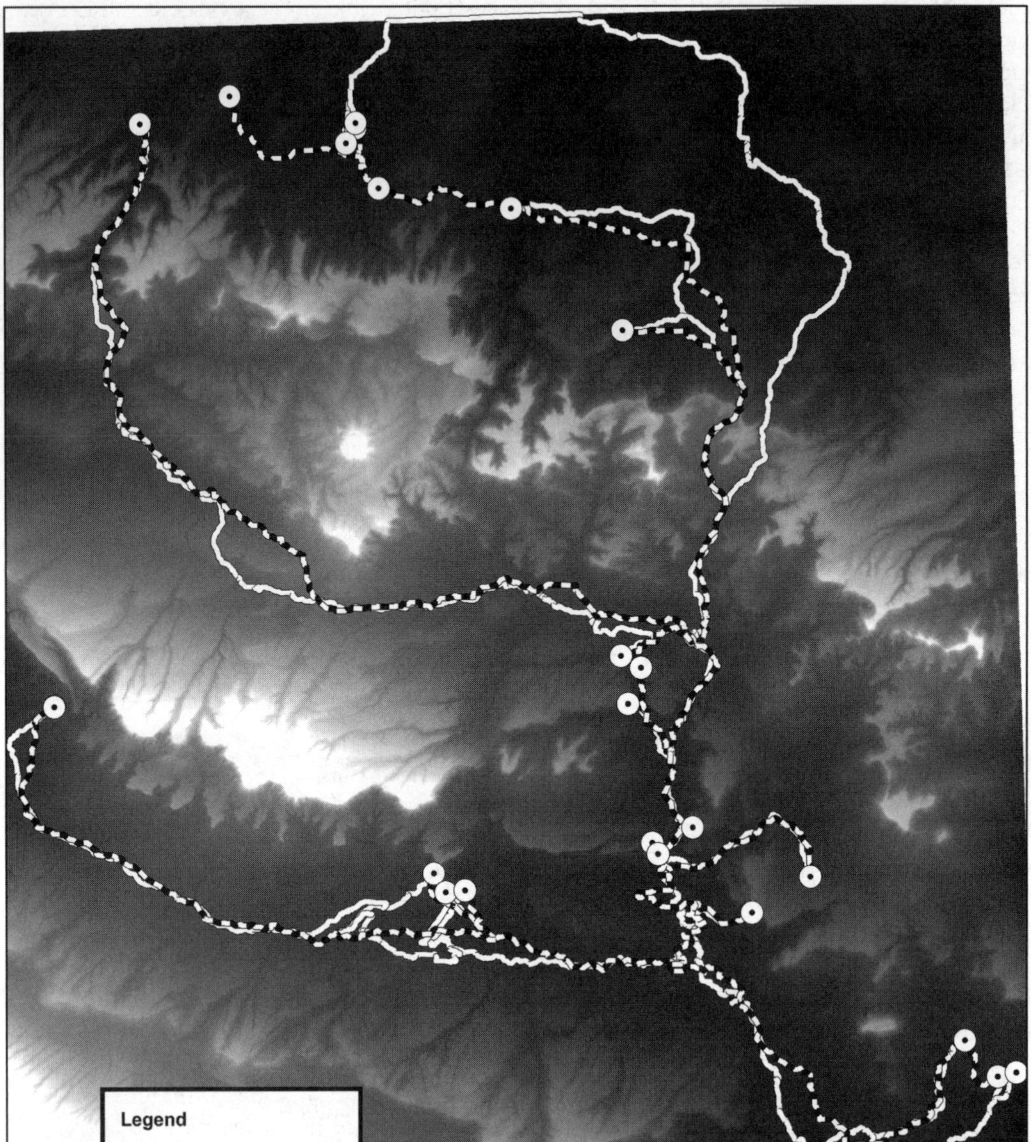

FIGURE 13.2. Cost paths from the prehistoric community of Casamero to all other contemporaneous communities in the study area. The cost paths were generated from cost surfaces that used slope as a proxy for movement cost. Slopes were themselves derived from digital elevation models with two different cell resolutions: 10 m and 30 m.

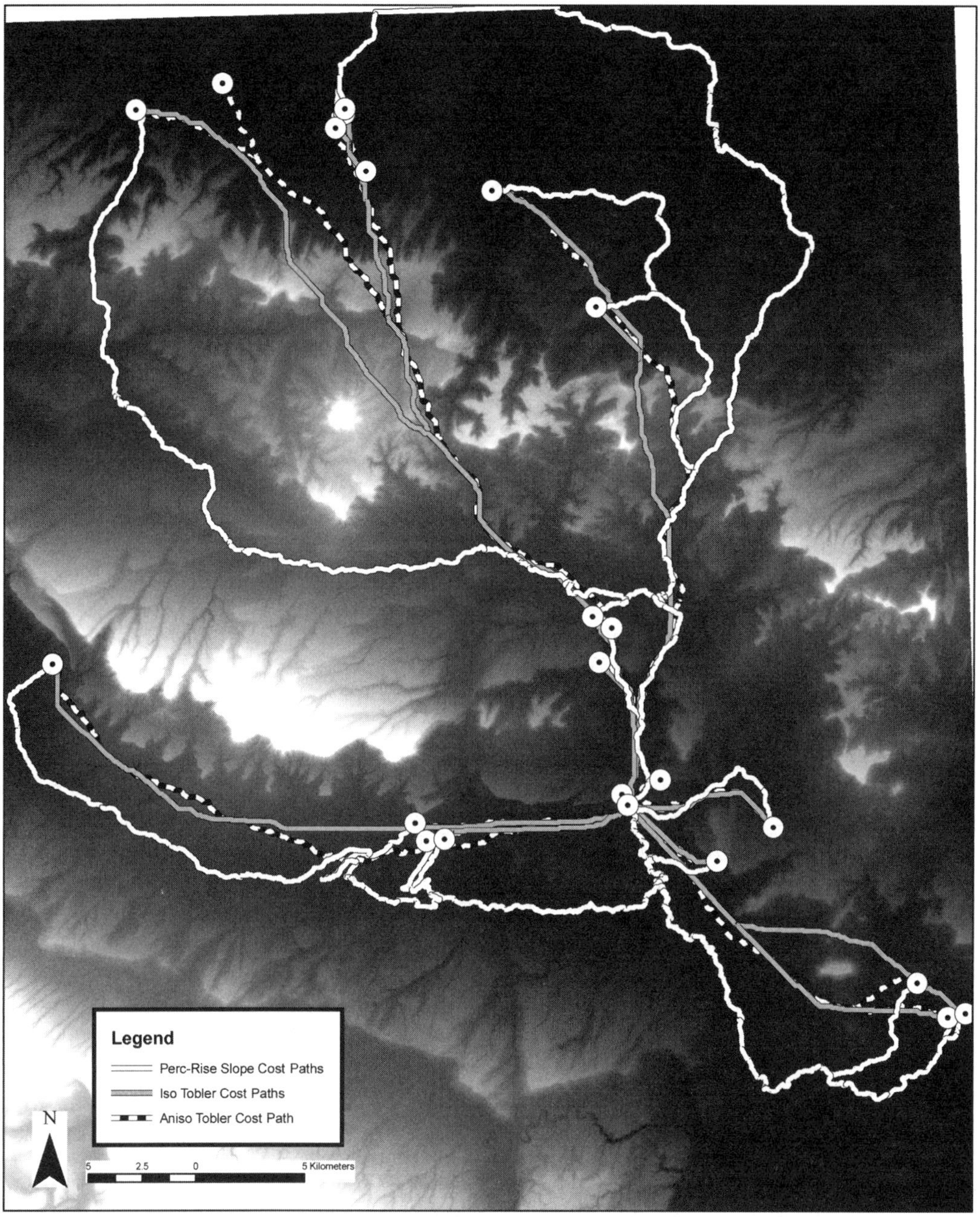

FIGURE 13.3. Cost paths from the prehistoric community of Casamero to all other contemporaneous communities in the study area. This figure compares the 10-m-slope cost paths with isotropic and anisotropic implementations of Tobler's hiking algorithm.

The anisotropic Tobler cost path between Casamero and the site of Dalton Pass, located in the extreme northwestern corner of the study area, suggests that travel time between those two points would be 10.4 hours, a reasonable estimate for a journey of approximately 48 km at 4.6 km/hr. The slope-derived cost path between those two points, however, is closer to 68 km, and extracting travel time from Tobler's cost surface for that route suggests a journey of 12.5 hours, or 5.4 km/hr; although longer in distance, it takes great pains to minimize slope change, allowing for a greater speed even though it takes 20 percent more time to get from Casamero to Dalton Pass. On the other hand, what is not evaluated is whether *less energy* would be expended to make the longer, slope-minimized journey. To evaluate this possibility, I used the algorithm employed in Rademaker et al. (Chapter 3), which is itself derived from the experimental work of Pandolf et al. (1977) detailed above, to generate another set of cost paths from Casamero; a 60-kg person with no load traveling 1.2 m/s was assumed. The results (Figure 13.4) suggest that Pandolf's energy-based algorithm yields results similar to, albeit not identical with, the cost paths generated by Tobler's time-based algorithm. The two solutions appear to diverge the most in relatively simple topographic settings, such as in the Red Mesa Valley.

13.5. Concluding Comments

The case study provided in this chapter briefly assesses the impact of cell size, anisotropy vs. isotropy, energy vs. time currencies, and topographic complexity on cost-path analysis. Because different algorithms yield markedly different results — even when restricted to waterways, as Livingood demonstrates in Chapter 10 — we need to expend considerable effort defending the methods we use. Using slope alone as a proxy for cost diverges most significantly from the other approaches evaluated here and therefore seems the most questionable. But ultimately, the issue of which approach best replicates real patterns of human mobility is an empirical one. We need either to base our methods on empirically derived algorithms or to empirically test our algorithms with contemporary examples of human movement before applying them to prehistoric cases. Or we can do both.

In reference to the former approach toward empirical evaluation, more attention needs to be paid to the currency that is ultimately being optimized. Tobler's hiking function and Livingood's canoeing rate optimize time, whereas others, like that of White and Rademaker and his colleagues, optimize energy in generating least cost paths. We also could imagine other optimization scenarios, such as minimizing time spent between water sources or minimizing some measure of risk for moving across each cell of a surface raster. In the 1997 case study, partially replicated above, I assumed that people in the past would have been more cognizant of how much time the trip took and perhaps less aware of the link between travel and the exact amount of energy (i.e., food) required to make the trip. On the other hand, human behavioral ecology would have us believe that energetic cost is the currency with which we should be concerned. These two approaches yield different results, and more empirical work clearly needs to be conducted on this issue, such as the research that Tripcevich (2008) did in the Andes, Malville (1999) did in Nepal, and geographers have conducted mostly in Western societies (e.g., González et al. 2008). Cross-culturally, what do people minimize when they travel? At the very least, we need to explicitly rationalize our use of particular optimization currencies.

Another area that merits discussion is the issue of simple vs. complex algorithms, which mirrors the debate over simple vs. complex models in general, as Branting notes in Chapter 12. There is a definite appeal to cost-path approaches that take into account terrain or land cover in addition to topography (e.g., Howey 2007), and those that consider traveler weight and load and sex and footwear and so forth (e.g., Kramer 2010; Pandolf et al. 1977). But as Livingood points out in reference to the effect of varying canoeing rates, and as I attempted to show here, cost-path studies are extremely sensitive to small changes in variables, especially when we're working with small regions or topographically diverse but less extreme landscapes. Sensitivity analyses of specific variables, especially those for which we arbitrarily assign cost factors, might lead us to seriously reconsider the utility of complex models.

Finally, let me end with a brief discussion of the related issues of suboptimality, stochasticity,

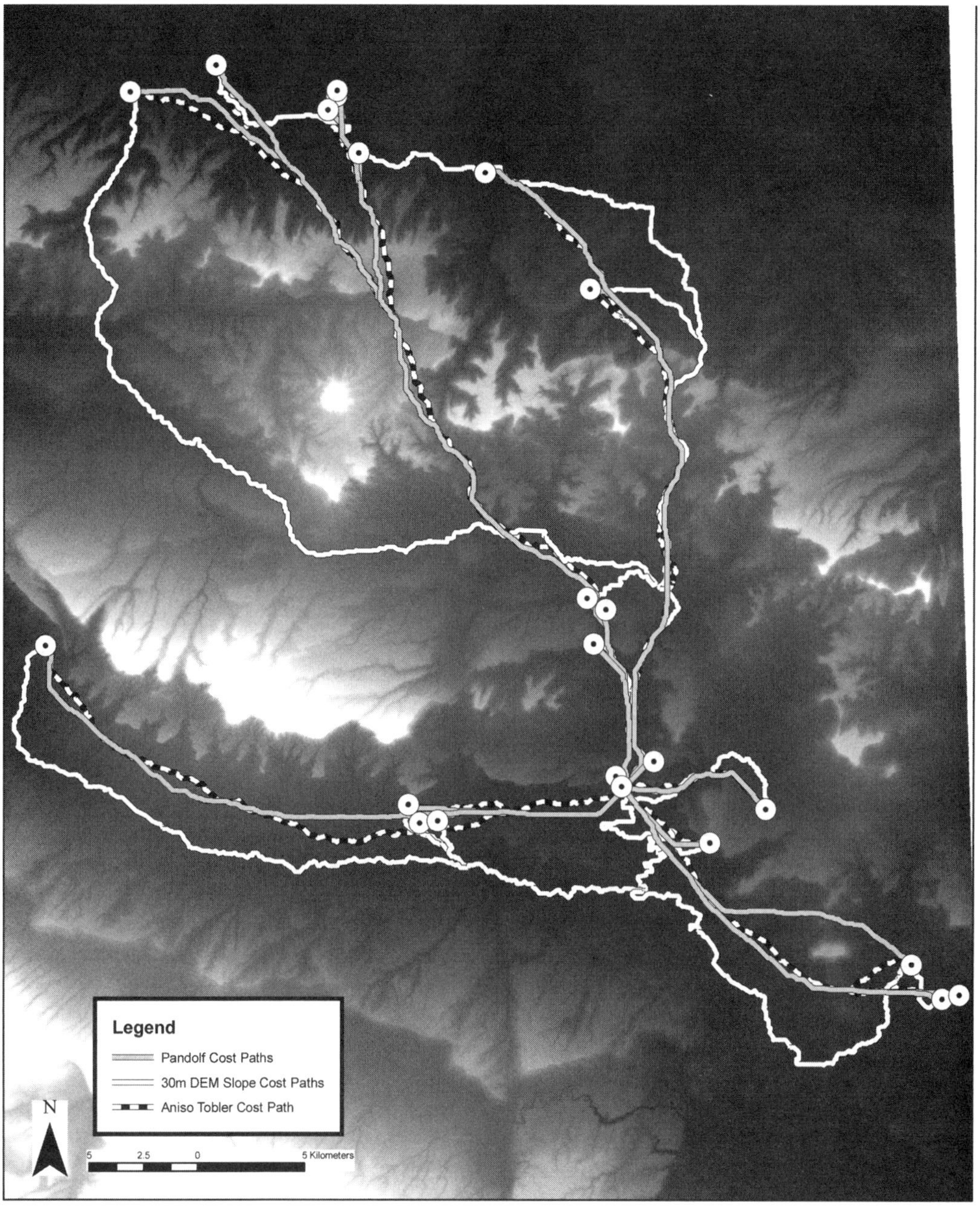

FIGURE 13.4. Cost paths from the prehistoric community of Casamero to all other contemporaneous communities in the study area. This figure compares cost paths generated by Pandolf's energy-based algorithm with cost paths derived from the 30-m slope surface and an anisotropic implementation of Tobler's hiking algorithm.

and statistical testing. One of the unresolved issues is how to accommodate the fact that people do not always make perfect decisions regarding the paths on which they choose to travel, a point also discussed by Branting in Chapter 12. For my 1997 study, this was an issue because I was comparing cost paths with actual paths, and I wanted some way to determine how close a fit was close enough, an issue also explored by Phillips and Leckman in Chapter 4. I considered adding a stochastic element and iteratively running the model so as to create something akin to confidence intervals, similar to Branting's suggestion, but the problem is that the stochasticity could lead to paths all over the place, especially on less complex topographies, making evaluation of modeled paths more of an art than a science. None of the case studies in this volume address the issue of statistical validation, but I think that going forward, this will be a significant challenge for cost-path applications. Work along these lines has advanced further in ecology than it has in archaeology (e.g., Pinto and Keitt 2009), and approaches for understanding and predicting animal movement across complex landscapes may be informative as archaeologists refine the tools of GIS to better re-create and understand the human past.

References

Aldenderfer, M. S.
1998 *Montane Foragers: Asana and the South-Central Andean Archaic*. University of Iowa Press, Iowa City.

Anderson, D. G., and J. C. Gillam
2000 Paleoindian Colonization of the Americas: Implications from an Examination of Physiography, Demography, and Artifact Distribution. *American Antiquity* 65(1):43–66.

Atkinson, D. M., P. Deadman, D. Dudycha, and S. Traynor
2005 Multi-criteria Evaluation and Least Cost Path Analysis for an Arctic All-Weather Road. *Applied Geography* 25:287–307.

Brannan, J. A.
1992 On Modeling Resource Transport Costs: Suggested Refinements. *Current Anthropology* 33(1):56–60.

Byerly, R. M., J. R. Cooper, D. J. Meltzer, M. E. Hill, and J. M. LaBelle
2005 On Bonfire Shelter (Texas) as a Paleoindian Bison Jump: An Assessment Using GIS and Zooarchaeology. *American Antiquity* 70(4): 595–629.

Carballo, D. M., and T. Pluckhahn
2007 Transportation Corridors and Political Evolution in Highland Mesoamerica: Settlement Analyses Incorporating GIS for Northern Tlaxcala, Mexico. *Journal of Anthropological Archaeology* 26(4):607–629.

Collischonn, W., and J. V. Pilar
2000 A Direction Dependent Least Cost Path Algorithm for Roads and Canals. *International Journal of Geographical Information Systems* 14:397–406.

Conolly, J., and M. Lake
2006 *Geographical Information Systems in Archaeology*. Cambridge Manuals in Archaeology. Cambridge University Press, Cambridge.

Dijkstra, E. W.
1959 A Note on Two Problems in Connexion with Graphs. *Numerische Mathematik* 1:269–271.

Duggan, A., and M. F. Haisman
1992 Prediction of the Metabolic Cost of Walking with and without Loads. *Ergonomics* 35(4): 417–426.

Ericson, J. E., and R. Goldstein
1980 Work Space: A New Approach to the Analysis of Energy Expenditure within Site Catchments. In *Catchment Analysis: Essays on Prehistoric Resource Space*, edited by F. J. Findlow and J. E. Ericson, pp. 21–30. Anthropology UCLA Vol. 10. Department of Anthropology, University of California, Los Angeles.

Fletcher, R.
2008 Some Spatial Analyses of Chalcolithic Settlement in Southern Israel. *Journal of Archaeological Science* 35:2048–2058.

González, M. C., C. A. Hidalgo, and A.-L. Barabási
2008 Understanding Individual Human Mobility Patterns. *Nature* 435:779–782.

Gorenflo, L. J., and N. Gale
1990 Mapping Regional Settlement in Information Space. *Journal of Anthropological Archaeology* 9:240–274.

Hare, T. S.
2004 Using Measures of Cost Distance in the Estimation of Polity Boundaries in the Postclassic Yautepec Valley, Mexico. *Journal of Archaeological Science* 31:799–814.

Helbing, D., J. Keltsch, and P. Molnár
1997 Modelling the Evolution of Human Trail Systems. *Nature* 388:47–49.
Herhahn, C. L., and J. B. Hill
1998 Modeling Agricultural Production Strategies in the Northern Rio Grande Valley, New Mexico. *Human Ecology* 26(3):469–487.
Howey, M. C. L.
2007 Using Multi-criteria Cost Surface Analysis to Explore Past Regional Landscapes: A Case Study of Ritual Activity and Social Interaction in Michigan, AD 1200–1600. *Journal of Archaeological Science* 34(11):1830–1846.
Imhof, E.
1950 *Gelände und Karte*. Rentsch, Zürich.
Jennings, J., and N. Craig
2001 Politywide Analysis and Imperial Political Economy: The Relationship between Valley Political Complexity and Administrative Centers in the Wari Empire of the Central Andes. *Journal of Anthropological Archaeology* 20(4): 479–502.
Kantner, J.
1997 Ancient Roads, Modern Mapping: Evaluating Prehistoric Chaco Anasazi Roadways Using GIS Technology. *Expedition Magazine* 39(3): 49–62.
2004 Geographical Approaches for Reconstructing Past Human Behavior from Prehistoric Roadways. In *Spatially Integrated Social Science: Examples in Best Practice*, edited by M. F. Goodchild and D. G. Janelle, pp. 323–344. Oxford University Press, Oxford.
Kantner, J., and R. Hobgood
2003 Digital Technologies and Prehistoric Landscapes in the American Southwest. In *The Reconstruction of Archaeological Landscapes through Digital Technologies*, edited by M. Forte, P. R. Williams, and J. Wiseman, pp. 117–123. BAR International Series 1151. Archaeopress, Oxford.
Kramer, P. A.
2010 The Effect on Energy Expenditure of Walking on Gradients or Carrying Burdens. *American Journal of Human Biology* 22:497–507.
Langmuir, E.
1984 *Mountaincraft and Leadership*. Scottish Sports Council, Glasgow.
Llobera, M., and T. J. Sluckin
2007 Zigzagging: Theoretical Insights on Climbing Strategies. *Journal of Theoretical Biology* 249: 206–217.
Madry, S. L. H., and C. L. Crumley
1990 An Application of Remote Sensing and GIS in a Regional Archaeological Settlement Pattern Analysis: The Arroux River Valley, Burgundy, France. In *Interpreting Space: GIS and Archaeology*, edited by K. M. S. Allen, S. W. Green, and E. B. W. Zubrow, pp. 364–380. Taylor and Francis, London.
Madry, S. L. H., and L. Rakos
1996 Line-of-Sight and Cost-Surface Techniques for Regional Research in the Arroux River Valley. In *New Methods, Old Problems: Geographic Information Systems in Modern Archaeological Research*, edited by H. D. G. Maschner, pp. 104–126. Occasional Paper 23. Center for Archaeological Investigations, Southern Illinois University, Carbondale.
Malville, N. J.
1999 Porters of the Eastern Hills of Nepal: Body Size and Load Weight. *American Journal of Human Biology* 11(1):1–11.
Marble, D. F.
1996 *The Human Effort Involved in Movement over Natural Terrain: A Working Bibliography*. Part I of the Final Report submitted under National Park Service Contract 6115-4-8031. Copies available from National Park Service Contract 6115-4-8031.
Marín Arroyo, A. B.
2009 The Use of Optimal Foraging Theory to Estimate Late Glacial Site Catchment Areas from a Central Place: The Case of Eastern Cantabria, Spain. *Journal of Anthropological Archaeology* 28(1):27–36.
Montello, D. R.
1997 The Perception and Cognition of Environmental Distance: Direct Sources of Information. In *Spatial Information Theory: A Theoretical Basis for GIS (1997 International Conference on Spatial Information Theory)*, edited by S. C. Hirtle and A. U. Frank, pp. 297–311. Lecture Notes in Computer Science. Springer, Berlin.
Naismith, W. W.
1892 Excursions: Cruach Ardran, Stobinian, and Ben More. *Scottish Mountaineering Club Journal* II:136.
Pandolf, K. B., R. L. Burse, and R. F. Goldman
1977 Role of Physical Fitness in Heat Acclimatization, Decay, and Reinduction. *Ergonomics* 20: 399–408.
Pinto, N., and T. H. Keitt
2009 Beyond the Least-Cost Path: Evaluating Corridor Redundancy Using a Graph-Theoretic Approach. *Landscape Ecology* 24(2):253–266.
Sakaguchi, T., J. Morin, and R. Dickie
2010 Defensibility of Large Prehistoric Sites in the

Mid-Fraser Region on the Canadian Plateau. *Journal of Archaeological Science* 37(6):1171–1185.

Santee, W. R., W. F. Allison, L. A. Blanchard, and M. G. Small
2001 A Proposed Model for Load Carriage on Sloped Terrain. *Aviation, Space, and Environmental Medicine* 72(6):562–566.

Savage, S. H.
1990 Modelling the Late Archaic Social Landscape. In *Interpreting Space: GIS and Archaeology*, edited by K. M. S. Allen, S. W. Green, and E. B. W. Zubrow, pp. 330–355. Taylor and Francis, New York.

Sherman, R. J., A. K. Balkansky, C. S. Spencer, and B. D. Nicholls
2010 Expansionary Dynamics of the Nascent Monte Albán State. *Journal of Anthropological Archaeology* 29(3):278–301.

Siart, C., B. Eitel, and D. Panagiotopoulos
2008 Investigation of Past Archaeological Landscapes Using Remote Sensing and GIS: A Multi-method Case Study from Mount Ida, Crete. *Journal of Archaeological Science* 35(11): 2918–2926.

Song, C., Z. Qu, N. Blumm, and A.-L. Barabási
2010 Limits of Predictability in Human Mobility. *Science* 327(5968):1018–1021.

Taliaferro, M. S., B. A. Shriever, and M. S. Shackley
2010 Obsidian Procurement, Least Cost Path Analysis, and Social Interaction in the Mimbres Area of Southwestern New Mexico. *Journal of Archaeological Science* 37:536–548.

Tobler, W.
1993 *Three Presentations on Geographical Analysis and Modeling*. Technical Report 93-1. University of California, Santa Barbara.

Tripcevich, N.
2008 Llama Caravan Transport: A Study of Mobility with a Contemporary Andean Salt Caravan. Paper presented at the 73rd Annual Meeting of the Society for American Archaeology, Vancouver, British Columbia.

van Leusen, M.
1999 Viewshed and Cost Surface Analysis Using GIS. In *New Techniques for Old Times: CAA98; Computer Applications and Quantitative Methods in Archaeology*, edited by J. A. Barcelo, I. Briz, and A. Vila, pp. 215–223. BAR International Series 757. British Archaeological Reports, Oxford.

Wheatley, D.
1996 The Use of GIS to Understand Regional Variation in Earlier Neolithic Wessex. In *New Methods, Old Problems: Geographic Information Systems in Modern Archaeological Research*, edited by H. D. G. Maschner, pp. 75–103. Occasional Paper 23. Center for Archaeological Investigations, Southern Illinois University, Carbondale.

Wheatley, D., and M. Gillings
2002 *Spatial Technology and Archaeology: The Archaeological Applications of GIS*. Taylor and Francis, New York.

Zhang, H., A. Bevan, D. Fuller, and Y. Fang
2010 Archaeobotanical and GIS-Based Approaches to Prehistoric Agriculture in the Upper Ying Valley, Henan, China. *Journal of Archaeological Science* 37:1480–1489.

CHAPTER 14

Least Cost Pathway Analysis in Archaeological Research

Approaches and Utility

David G. Anderson

14.1. Introduction

The chapters in this book demonstrate that least cost pathway analysis (hereafter LCPA) can be an important tool for exploring the past, and they illustrate various ways it can be used. While the approach has been known by archaeologists for some time, the method has not seen widespread use, something that this book should help to remedy. My first introduction to the procedure came in the mid-1990s, when my colleague Chris Gillam and I were discussing ways to examine how colonizing populations might have moved into the Americas. We were looking for a more quantitative and replicable method than the traditional intuitive approach of drawing arrows indicating possible movement pathways on maps, however well informed these efforts might have been by consideration of factors such as physiography, glacial geology, or sea-level change (e.g., Sauer 1944). Chris suggested that we try LCPA, which he had seen put to great effect by Fred Limp (1990) a few years earlier, in an attempt to locate where in Arkansas the DeSoto expedition may have traveled, an effort to reconcile historic accounts with local archaeology and physiography. We thought the method could prove useful, although the hemispherical scale of the project was daunting, something we simplified somewhat by conducting separate analyses for North and South America.

The results, in which possible movement pathways were compared with artifact distributions and tied to demographic reconstructions of possible population growth that may have occurred after entry, were presented at the 1997 meeting of the Society for American Archaeology in Nashville and later published in *American Antiquity* (Anderson and Gillam 1997, 2000), albeit not without stirring a little controversy (cf. Moore and Mosely 2001; Anderson and Gillam 2001). The paper has since, fortunately, been widely read and variously applied, sometimes with quite serendipitous results. A primary least cost pathway for human radiation in South America, for example, ran to the east of the Andes instead of down the Pacific coast, and it has subsequently been determined that this is an area of great linguistic diversity, suggesting it may have been a location of early settlement (Dahl et al. 2011). Likewise, the movement pathways generated in North America, calculated from entry points starting in the Pacific Northwest or passing through the ice-free corridor, to accommodate the principal routes now currently believed plausible, trended to the south and east. Some morphometric characteristics of Amerind populations exhibit similar trends, suggesting initial dispersal may well have occurred in these directions (Jantz et al. 2010). While we had assumed that the pathways would lead to the discovery of new archaeological sites, since some ran through regions previously given minimal consideration, these examples from linguistic and physical anthropological research illustrate how LCPA can sometimes lead to quite unexpected findings.

We have both continued to employ LCPA extensively in our research, Chris in South America and the Russian and Japanese Far East, and together examining Paleoindian site distributions and later Archaic, Woodland, and Mississippian period interaction networks in eastern North America (Anderson 2010:286–287; Anderson et al. 2007; Anderson, Miller et al. 2010; Anderson, Yerka et al. 2010; Banks et al. 2006; Gillam 2008, 2009; Gillam and Tabarev 2004; Gillam et al. 2007, 2008; Uchiyama et al. 2009). The work has led us to address a range of interesting questions and develop more appropriate mapping routines that take into account sea-level rise, pluvial and periglacial lake extent, and ice sheet margins in our Paleoindian research (all of which would have affected movement), and allowed us to better integrate compilations of primary data into other GIS-based applications (Anderson, Miller et al. 2010). As the chapters in this book impressively document, LCPA encompasses a wide range of possibilities, and the results of individual analyses invariably suggest new ones. In the pages that follow, I present my reactions to specific chapters and then offer a brief synopsis of some recent research on interaction networks in eastern North America.

14.2. Observations on the Chapters

A primary lesson of LCPA is that traditional site-catchment analysis (e.g., Vita-Finzi and Higgs 1970), based on the assumption that resource procurement can be monitored using movement directed to areas defined by concentric circles with radii of varying extent, does not accurately reflect the way humans made use of the landscape (e.g., Rissetto, Chapter 2; Surface-Evans and White, Chapter 1; Surface-Evans, Chapter 8). Humans are not able to fly directly between two points. They must walk and deal with terrain features such as slopes, vegetation, watercourses, and so on. Aspects of terrain such as the presence or absence of potable water or the availability of desired resources can profoundly affect the way people move over the landscape. Indeed, the analyses in this volume show that path-distance measures obtained from LCPA rarely correspond to straight-line distances; almost invariably, they are much greater, especially in areas of varied terrain (e.g., Rissetto 2009, Chapter 2). As Rissetto (Chapter 2) argues, furthermore, it is possible for our thinking about resource procurement to be dominated by assumptions about the size of catchments and the "likely" areas human groups are thought to exploit, when in reality, as his work and that of several other authors herein show, they make use of landscapes very differently. Human resource procurement "catchments" are typically anything but circular, but instead are highly irregular in shape and varied in size.

Another lesson from the contributions to this book is that LCPA can be applied to human groups in all times and places. Although Rissetto's (Chapter 2) examination focuses on resource procurement and mobility patterns associated with Magdalenian settlements in Cantabrian Spain, the artifact category employed in his and indeed several of the chapters — stone tools — are as old as humanity itself. LCPA, in conjunction with lithic raw material sourcing studies, can thus be used to examine human settlement and mobility patterns wherever such artifacts are found; similar analyses, of course, can be conducted using other resource or raw material types, meaning the approach is not limited to stone tools (e.g., Copé 2007; Gillam and Tabarev 2004; Krist and Brown 1994; Suárez and Gillam 2008).

Rissetto's analysis, dealing with late Pleistocene environments, also illustrates how changes over time in topography and resource structure and availability must be considered in LCPA. Human movement patterns were profoundly shaped by climate change over time, which at a broad scale affected sea levels, river courses, and the presence of glaciers and pluvial or periglacial lakes (Rademaker et al., Chapter 3), or at closer intervals affected crop yields, erosional patterns, and biotic resource distributions (e.g., Ullah and Bergin, Chapter 9). Another aspect of the environment to consider is how much it has been modified by humans since the period under examination, through drainage channelization, agricultural terracing, deforestation, urbanization, road building, and so on (e.g., Rademaker et al., Chapter 3; Ullah and Bergin, Chapter 9). Even within landscapes not too different from those of the present, seasonal differences should also be considered, such as in plant, animal, or potable water availability, or flooding patterns or ice cover. River floodplains, for example, although commonly shown in LCPA to be likely movement corridors, may have been avoided during periods when flooding occurred, at least as far as foot traffic was concerned. As Livingood (Chapter 10) argues, watercourses were not just walked alongside but also traversed with boats or canoes in many human cultures; when we conduct LCPA, water travel must be considered a possibility, one that likely assumed great importance in some times and places in the past (e.g., Erlandson 2002; Jodry 2005; see also Surface-Evans, Chapter 8).

Rissetto's research (Chapter 2) also demonstrates the importance of having reliable as well as precise information on raw material occurrence, and hence the critical importance of sourcing studies. Materials occurring at point sources or with a limited distribution are ideal for LCPA, while raw materials that occur widely are likely to be of little use in such studies. Lithic reduction analyses have often been used to inform on mobility, employing assumptions about the characteristics of materials obtained from greater or lesser distances (e.g., Andrefsky 2005, 2009; Goodyear 1979; Surovell 2009); assemblage analyses combined with sourcing and LCPA can provide multiple lines of evidence toward the resolution of mobility patterns. As Rissetto also argues, however, just because humans could have used certain routes doesn't mean they did; people don't always act in optimal or rational ways, and other considerations, such as the location of other people or resources or sacred sites, or even the need for a line of sight to their destination, or as a conscious strategy to reinforce status differences may have shaped movement (e.g., Phillips and Leckman, Chapter 4; Richards-Rissetto, Chapter 7; Gould and Saggers 1985).

Even given these assumptions, LCPA, by offering direct pathways where people *could* have moved, and from the perspectives of efficiency and rationality likely *did* move, allows us to explore specific locations for evidence of their presence. That is, LCPA offers a valuable means of predicting archaeological site locations, places on the landscape where people might have been present (e.g., Rademaker et al., Chapter 3; Surface-Evans, Chapter 8). Rather than regard every point within a specific catchment radii as a likely place of settlement, we can use LCPA to assign greater or lesser probabilities, or likelihoods, of sites being present at a specific location. LCPA can thus be used to refine more traditional predictive modeling efforts, by documenting where within zones having a high probability of site occurrence people likely moved.

At a larger scale, LCPA can help us get a better perspective on seasonal or annual range mobility or extent of trade and exchange networks. Likewise, by knowing what kinds and quality of stone is within what cost distances, we can better understand the uses it is being put to. Many North American Paleoindian groups, for example, are known to have traveled great distances to obtain high-quality stone (Goodyear 1979). LCPA may be able to document whether, for highly mobile foragers, such strategies were energetically as efficient as, or at least not much more costly than, the use of closely available but lower-quality stone sources. Even if high-quality sources were "expensive" to access (i.e., with a high associated travel cost), Rissetto's analysis shows that in such cases they were typically extensively exploited when visited, presumably precisely because they were so hard to reach. That is, given the difficulty of access, it would be important to come away with as much material as possible. Given large-scale procurement, the actual costs may be no more, or not appreciably more, than those involved with the use of more readily accessible materials.

Rissetto (Chapter 2), like many of the authors in the book, offers additional means by which his analyses can be elaborated on in the future. These include using different kinds of or higher-resolution data, employing other cost/currency measures such as nutritional calories, and varying the scaling of slope or caloric costs to terrain conditions. Many of the authors also emphasize that we must take into consideration social or natural features that constrain movement, such as other human settlements, vegetation and terrain conditions, mountain passes, fords, water bodies, or even uphill vs. downhill travel (e.g., Rademaker et al., Chapter 3; White 2007 and Chapter 11). Moreover, the routes revealed through the employment of one measure, such as slope, may be somewhat to very different from those based on other measures, such as nutritional calories/energy (e.g., Rademaker et al., Chapter 3). Output will also vary when differing factors affecting movement are employed, such as the use of watercraft as opposed to pedestrian travel, warranting care in the selection of weighting factors and the interpretation of analytical output. Indeed, another lesson from some of the chapters herein is that LCPA often yields uncertain or probabilistic results. That is, rather than consider movement as inevitably following narrow lines or pathways, we may find it more appropriate to present output in the form of buffers or corridors that encompass the likely range of variation in routes (Rademaker et al., Chapter 3; Surface-Evans, Chapter 8).

As a number of the chapters herein also demonstrate, the resolution in the landscape data (e.g., 5 m, 30 m, 90 m, 1 km) used in LCPA can be critically important (Ullah and Bergin, Chapter 9). Data at one level of resolution may be ideal for some kinds of analyses and totally inappropriate for others. For continental to hemispherical scale analyses that examine movement pathways over great distances, 1-km resolution would likely be fine, while for analyses of movement within smaller areas, such as those conducted in most of the studies reported in this book, such a scale would be far too coarse-grained. Another factor to consider when we examine movement patterns is that some intermediate locations were likely places to be visited or avoided and must be incorporated as such into LCPA. Raw materials may have been passed between major nodes in settlement hierarchies, for example, if procurement was controlled, or movement may have been designed to avoid such nodes if conflict (or smuggling) between groups was likely. Likewise, in complex societies, as Richards-Rissetto (Chapter 7) documents, movement in urban settings may have been deliberately constrained to ensure or deny access to some groups by others, as part of strategies to reinforce relationships of power, prestige, and social inequality.

Rademaker, Reid, and Bromley's examination of Paleoindian settlement and mobility patterns in southern Peru also aims to resolve trail networks that existed in the remote past. Their chapter documents the routes people may have used to bring obsidian from sources at some distance in the interior highlands to the coastal site of Quebrada Jaguay. Significantly, as in several of the other studies in this book, the authors used the LCPA results to look for archaeological sites presumably contemporaneous with Quebrada Jaguay. They located a number of previously unrecorded Paleoindian sites along the solution pathways, a process they appropriately described as "connecting the dots." Their field survey was also guided by expectations derived from optimal foraging as well as colonization and migration theory with regard to the desirability of specific settings, rather than the assumption that every place along a route would be ideal for settlement. Where the least cost routes intersected the theoretically predicted important places in the kind of hunter-gatherer settlement systems presumed to have been present, their fieldwork showed that these were indeed good places to find sites (see also Surface Evans, Chapter 8).

Rademaker et al.'s research also included ground truthing the movement pathways that their LCPA revealed, demonstrating that some were far less likely to have been used than others, due to difficult terrain features, however optimal they have looked analytically. Another important and indeed critical part of their analysis was determining the location, extent, and geochemical signature(s) of the Alca obsidian used as an archaeological marker of movement; this effort also required extensive field data collection (e.g., Rademaker 2006; Rademaker et al., Chapter 3). Lest anyone think archaeologists conducting GIS least cost analyses are sedentary types sitting in front of computers, it should be noted that the field efforts of several of the authors in this book involved checking routes and looking for sites in difficult and sometimes dangerous terrain. Fieldwork informed by theory and cutting-edge analytical tools such as LCPA, however, is a more accurate picture of the discipline than Hollywood fantasies involving treasure maps, high-speed vehicle chases, or encounters with space aliens.

Rademaker, Reid, and Bromley also make the important point that groups exploring and colonizing a landscape for the first time would likely traverse it quite differently than those who had lived within it for generations, specifically that initial exploration would be more likely to be risk-averse and conservative than optimal.

Landscape-learning theoretical arguments, they suggest, need to considered when examining the movements of the first peoples in an area (e.g., Anthony 1990; Barton, Schmich et al. 2004; Beaton 1991; Kelly 2003; Meltzer 2002; Steele and Rockman 2003). An important lesson, they note, is that colonizing routes may be non-optimal with respect to *accumulative* least cost, since colonizing populations, barring the use of scouts (sensu Anthony 1990:902–903), would have had no idea what lay very far ahead. Instead, *sequentially* optimal least cost paths, employing risk-averse behavior and segments involving efficient movement over short distances, are a more appropriate way to proceed with such analyses. In addition, while foragers likely ranged widely in their pursuit of resources and shelter, they also probably quickly learned which areas and routes to take and which to avoid. Rademaker et al.'s (Chapter 3) research, in fact, shows that some pathways into the interior never saw much if any use, since local terrain conditions were difficult and habitats for settlement or resource procurement along them were poor.

As Rademaker et al. (Chapter 3) also document, we must consider characteristics not only of the landscape in our analyses but also of the people traversing it. That is, individual movement is constrained by ability and walking speed as well as body and burden weight; likewise, group movement patterns are constrained by factors such as the presence and numbers of children or the elderly, as well as the rate of the slowest members (e.g., Gillam 2008, 2009; Pandolf et al. 1977). Additionally, just because a solution pathway may traverse one side of a valley, hunter-gathering populations may have used other parts of it, depending on where potable water, exploitable plants or animal populations, south-facing rockshelters, sacred sites, or a host of other variables occurred. Areas ideal for settlement and particularly for multiband aggregation among hunter-gatherers (e.g., Conkey 1980; Anderson 1990), as Rademaker et al. (Chapter 3) and others in this book argue, may thus have been selected for reasons other than their proximity to travel routes, although this was probably an important factor. All the chapters similarly highlight the importance of considering both cultural and environmental factors shaping movement.

Phillips and Leckman's (Chapter 4) analysis of travel routes, what they call footpaths or water transport trails in the Fort Bliss area of west Texas and eastern new Mexico, brought back fond memories of my own work on Jornada Mogollon lithic and ceramic scatters in the low desert near El Paso in the late 1970s (Anderson and Carter 1985). At the time, we were interested in finding and documenting sites, defined as concentrations of artifacts, with only minimal attention given to their relation to watercourses or larger settlements, or whether they were located along or helped to define trails. The use of GPS and GIS technology in recent decades has revolutionized how archaeologists examine past human use of the landscape and the accuracy with which we can record locational data. In particular, Phillips and Leckman argue that a nonsite or landscape archaeological survey approach, in which the locations of all encountered artifacts are precisely recorded, is far more effective at documenting past land use, including where trails were located, than the traditional archaeological focus on large, dense scatters of remains. In particular, they advocate looking for linear arrangements of sites and artifacts, particularly those leading out from larger settlements to important resources such as water, lithic or other useful raw materials, or sacred sites, as a means of reconstructing trail networks. As an aside, in my own survey work in the area, we too piece-plotted the location of all the artifacts we found, although in the absence of GPS technology the locations of the sites we worked on were only approximately determined, to within perhaps 50 m at best, and even that level of accuracy was only possible because we had aerial photographs showing the location of dirt roads and individual mesquite and creosote bushes. The artifactual data Phillips and Leckman worked with, in contrast, were at a maximum 15-m resolution, and were often much more precisely delimited.

Phillips and Leckman (Chapter 4) use archaeological as well as ethnographic evidence to argue that broken pottery often occurs along pathways leading to and from water. That is, if ceramics were used to carry or store water, broken vessels or vessel fragments would be expected along the routes to such sources, and leading away from them as well. Indeed, isolated "pot busts" are a fairly common feature in many parts of the Southeast, where I am most familiar with the archaeological record; while breakage during transport is commonly assumed, efforts at interpretation rarely go further than this. Such data could be used to infer settlement locations or trail networks, and may also indicate activity loci where other, more perishable remains were employed and are hence not visible archaeologically, such as some kinds of plant collecting or processing. The broken pots found in these areas were likely brought there to provide water for the task groups, for drinking or for use in the processing activity. In cases in which perishable containers other than pottery were in use, as during the preceramic Archaic and Paleoindian periods in the region, examination of lithic artifacts and scatters might prove a useful alternative method of analysis. Phillips and Leckman suggest such an approach might eventually prove productive in their study area, although they argue that lithics were too widespread to be as useful for delimiting trails compared to pottery, which is much less commonly found away from settlements. Any artifacts found along inferred trails, of course, must also be evaluated for evidence of contemporaneity, through analyses to determine when and how long these routes were in use.

Phillips and Leckman (Chapter 4) also observe that trails do not always follow an idealized least cost pathway, due to geomorphic factors, vegetation patterns such as dense underbrush, landform features, socially constituted group boundaries, or sacred features on the landscape. They make the very good point that having the ability to follow a convenient line of sight, such as a distant mountain peak or some other prominent terrain feature, may shape where trails develop in otherwise undifferentiated terrain. They explore this possibility using viewshed analysis between sources or starting points and destinations, a technique that is seeing increasing use in modern archaeology. One important conclusion they reach is that having visible landmarks is more important in determining trail locations than following optimal routes, especially in terrain where it is possible to lose sight of such landmarks. As they put it, "intervisibility between departure and destination locations plays an important role in path making" (see also Waldron and Abrams 1999). Indeed, where prominent terrain features are being examined that likely served as aggregation loci or sacred sites in prehistory, among the first questions that should be asked is, How were such features accessed? Such analyses in turn can help us better document where the peoples who came to these locations originated.

Trail networks are well documented in the historic period, and widely inferred in prehistory, although actual landscape features created by human movement (i.e., wear patterns on resistant material, sunken roadbeds, associated debris scatters) may not always or even commonly be evident. Archaeologists often find historic sites using old maps and carefully examining the terrain along the roads or trails delimited on them. Only rarely, however, do well-delimited trails survive from prehistory, such as the Chacoan and Incan road systems. Even the exact routes of many historic period Indian trails or Anglo/European expeditions, much less the routes of earlier trail networks or of initial colonizing populations, remain partially or largely unknown (Anderson and Gillam 2000; Hudson 1990; Kantner 1997; Limp 1990; Myer 1928; Snead 2002, 2008; Tanner 1989; Ware and Gumerman 1977; Whitley and Hicks 2003). When portions of obvious trails or routes are found, however, LCPA can be used to infer where remaining segments may have been located. Likewise, when we are attempting to locate trails mapped historically, LCPA can quickly indicate whether the mapping bears any relation to reality, at least if these routes are assumed to have approximated ideal least cost pathways; such analyses can also be used to suggest places on the landscape we can look for traces of these trails. Structures or shelters, cairns, worn or sunken pathways, rock art, and the occurrence of artifacts or other landscape features such as water sources or dramatic physiographic features can all be used to infer possible routes people may have regularly used, as well as their periods of use (e.g., Phillips and Leckman, Chapter 4).

Phillips and Leckman also note that the existence of known roads or trails may constrain where archaeologists look, under the assumption that sites are likely to be located along such presumably optimally placed corridors. Less charitably, I would suggest, this is also because some archaeologists (not anyone herein!) don't wander very far from their vehicles or venture into difficult terrain during field survey work. Indeed, given the kind of data already in many GIS systems today, we can explore such "observer" or "recorder" effects by comparing site locations and numbers with their distance from modern roadways by individual or organization. It very quickly becomes clear when conducting such analyses, as I have done on occasion out of curiosity or when overseeing cultural resource management (CRM) contracts for the National Park Service, whose survey coverage is more thorough and hence likely more reliable. LCPA can thus be used to examine the movement patterns of modern archaeologists as well as those of past populations.

What all the chapters in this book provide are clear examples of how to link LCPA with archaeological data. Indeed, some of the archaeological and locational datasets employed are extensive, such as the 14,000 15-×-15-m survey units in the Chihuahuan Desert of New Mexico used by Phillips and Leckman (Chapter 4), the nearly 600 Mayan households and larger building complexes at Copán examined by Richards-Rissetto (Chapter 7), and the 45 Mississippian mound centers in northern Georgia examined by Livingood (Chapter 10). LCPA can also be linked to other innovative approaches, such as the use of viewshed analyses by Phillips and Leckman and Richards-Rissetto (Chapters 4, 7), precipitation surfaces and other environmental data layers (Nolan and Cook, Chapter 5; Ullah and Bergin, Chapter 9), and lithic sourcing studies (Rissetto, Chapter 2; Rademaker et al., Chapter 3). One of the important things about LCPA, from a scientific perspective, is its replicability; analysts using similar methods and datasets should arrive at similar or identical mapped outcomes, something that lends confidence to interpretations based on them, as well as their application in other research. Finally, the ground truthing of output from LCPA is a standard aspect of much of this work, at least at smaller scales of analysis, further removing such research from categorization as an armchair method somehow apart from the real world of archaeology.

Kevin Nolan and Robert Cook (Chapter 5) used LCPA to determine likely trading relationships among late prehistoric Fort Ancient culture sites in the Ohio area, based on expectations from environmental modeling and theoretical assumptions about factors shaping interaction between groups. A baseline analysis employed dendroclimatologically based reconstructions of rainfall patterns at 50-year intervals from AD 800 to 1400 (after Cook et al. 2004) to determine potential crop yields and, from that, the likely political stability of agricultural populations in their study area. Such an approach was initially developed in the Southwest and has been conducted with some success in other parts of eastern North America in recent years (Anderson 1994; Anderson et al. 1995; Benson et al. 2009; Blitz and Lorenz 2006; Meeks and Anderson 2006, 2007). Employing arguments by Winterhalder (1996) and Kelly (1995) about conditions under which territorial behavior will occur, Nolan and Kelly maintain that interaction was likely to be greater between groups whose harvest returns were different, and that warfare and active boundary maintenance and defense, if not outright conflict, were more likely when harvest returns were highly variable and intercorrelated. They conducted LCPA from two major centers, SunWatch and Reinhardt, to outlying late prehistoric sites yielding certain types of remains (i.e., fauna, ceramics) and superimposed the output over the precipitation reconstructions for specific periods.

Significantly, Nolan and Cook also conducted systematic LCPA from the two principal sites to locations on the landscape at varying distances and at cardinal and intermediate intercardinal or subcardinal directions (i.e., N, NE, E, SE, S, SW, W, NW) purely as analytical controls, to show how movement may have occurred regardless of whether identified sites were present in a given direction. Use of such controlled comparisons should be standard in LCPA, and indeed in any analysis employing partial samples of either sites or landscapes, to determine whether outputs are being skewed by sample size or locations, extent of survey coverage, prevailing landscape characteristics, or some other parameter (Anderson and Smith 2003:187–188). The LCPA showed that SunWatch may have been more of a cultural nexus, or center for interaction, than Reinhardt, that is, a site whose location shaped the occurrence of other settlements and the interaction pathways between them. The results of Nolan and Cook's analyses, while preliminary and to a large extent exploratory, as well as constrained by questions of contemporaneity between the centers and outlying sites (which the authors acknowledge), illustrate how both climate and terrain features can help shape interaction extent and direction and hence regional political conditions. The authors' analyses were also tied to prosaic but long-studied aspects of the archaeological record, ceramic stylistic variability. Not surprisingly, during periods when greater interaction was inferred from climatic conditions, greater similarity in ceramic assemblages was also noted. Although they were originally classified as different foci by Griffin (1943), the LCPA indicated that Fort Ancient sites in the Miami and Scioto valleys interacted, reinforcing more recent interpretations that these areas were closely linked and similar culturally (Brady-Rawlins 2007; Cook 2008; Nolan and Cook 2010).

Erin Hudson (Chapter 6), like Richards-Rissetto (Chapter 7), uses space syntax analysis, or what is sometimes called social space theory—the relationships between the built environment, spacing of structures and settlements, and social organization—to examine whether segregation in architecture reflected segregation in other aspects of social life (e.g., Bustard 2003; Grøn 1991, Hillier and Hanson 1984; Van Dyke 1999). Community accessibility and integration, in this approach, is measured by degree of access, which ranged from open to restricted, in part through a determination of the number of ways a room or building could be reached or entered. Hudson expanded the analysis beyond building accessibility to employ LCPA as well as line-of-sight and viewshed analyses as proxies for measuring the openness and accessibility of entire communities, in this case the Pueblo II Lion Mountain site cluster and the late Pueblo III Gallinas Springs community in west-central New Mexico. In this her work is similar to what Richards-Rissetto (Chapter 7) calls configurational analysis, albeit the latter study was conducted within an urban environment, the late Classic period Mayan city of Copán. Correlates of communities characterized by open access, Hudson argues, include high visibility, line of sight between communities, large viewsheds, and numerous routes of access, suggesting fairly egalitarian relationships. Opposite patterning would be expected in sites with restricted access, something in turn suggesting greater social inequality. Important differences between the Pueblo II and Pueblo III occupations were observed in terms of ease of access and line-of-sight networks, factors that imply that great differences also existed in the way and extent to which these communities interacted with the rest of the world. The Lion Mountain peoples, the author observes, apparently wanted to see and be seen; their sites were highly visible on the landscape, and access between sites within the cluster was open. Yet at the same time, the overall cluster was located in a setting (primarily along ridgetops) that restricted access. They wanted the world to know where they were, and possibly visit, but had some restrictions on access. In this case, high visibility did not necessarily equate with ready access and presumably a warm welcome. The Gallinas Springs site, a large, ca. 500-room pueblo, in contrast to the Lion Mountain sites, was more hidden and inaccessible, and there was evidence for internal inequalities and increased sociopolitical complexity. To understand these findings (i.e., of space syntax and LCPA within a specific area), as Hudson's work demonstrates, we must sometimes evaluate the results over much larger regions, and over time as well. During Pueblo II and Pueblo III times, for example, dramatic changes were occurring in the eastern Southwest, and the differences between the Lion Mountain and Gallinas Springs settlements appear to be part of regional trends involving population migration, relocation, and nucleation, as well as increased warfare.

Differences in access and organization linked to site setting and community layout need to be explored beyond the Southwest or Mesoamerica, where some of the major space syntax studies conducted to date by archaeologists have occurred. In the Southeast, for example, some Mississippian sites are located prominently along major waterways, while others are located in the interior along minor drainages well away from main channels (e.g., Anderson 1994:250–253; Hally 1993, 1999; Smith 1978). Why this is the case appears linked to larger regional patterns of warfare and political organization, factors that may help to explain local situations. Likewise, some sites have multiple exterior palisade or internal fence lines, as well as clusters of buildings around discrete plazas, restricting visibility and access to some precincts, whereas other sites are more open, with single plazas and minimal internal partitioning (e.g., Anderson 1994, n.d.; Holley 1999; Pauketat 2007). Access to community buildings, such as structures atop mounds, may also be restricted, as documented ethnohistorically (e.g., Hudson 1990, 1997), although this is harder to demonstrate prehistorically. Intra- and intersite analyses of centers in the eastern Woodlands would thus appear to be another profitable area for space syntax, viewshed, line-of-sight, and (following Richards-Rissetto) configurational approaches, particularly as ever more complete community plans are appearing through excavation and remote sensing analyses (e.g., King et al. 2011).

Heather Richard-Rissetto (2010, Chapter 7) examined the distribution and arrangement of elite and commoner households at the late Classic period Mayan center of Copán in the development of an Urban Digital Elevation Model, to explore how and why the city was laid out as it was and possible patterns of movement and interaction by its inhabitants—what she calls the relationships between "site configuration and social connectivity" among people in different social classes and in different parts of the city. As the author notes, we have moved beyond simplistic analyses of such things as cardinality, the geographic orientations of the built environment, to exploring how site configuration not only enshrines cosmic beliefs but leads to the enactment of such beliefs through patterns of daily movement. We have also moved beyond simpler forms of space syntax analysis, which previously focused on architectural characteristics of individual or small numbers of buildings, often of a specific type (i.e., temples, palaces), to multiscalar analyses of architecture and terrain features to encompass entire communities and landscapes.

Configurational analyses subsume space syntax studies and include such things as examining pedestrian access to and visibility of different parts of a community (Hillier 1999; Hillier and Hanson 1984; Hillier et al. 1993; Preziosi 1979; Richards-Rissetto 2010). They have the advantage of considering space in three rather than two dimensions, the traditional axial space syntax approach, and, as used by Richards-Rissetto, the natural as well as the built environment. The layout of the Mayan city of Copán was examined, and "integration values," or movement costs, were calculated based on whether spaces were readily accessible and visible or not and whether people were more or less likely to walk to or through them. The analysis, significantly, also evaluated the impact even minor differences in topography had on movement as facilitators, barriers, or effectively neutral. Richards-Rissetto's research was directed to determining how differing social groups within Copán and over the surrounding landscape—including both presumed elites and commoners (whose social status was inferred in part based on the types of structures in which they lived)—were integrated within the society as a whole.

GIS technology is enabling since it allows travel costs between one location and all others to be quickly calculated, with terrain and cultural factors weighted as appropriate, providing a numerical measure or value of integration. Such analyses offer replicable measures of integration and accessibility, given as the average (or alternatively the total) travel costs between a specific building or location and whatever subset of other buildings or landscape features the researcher wishes to examine. Richards-Rissetto provides exceptional detail on her methods at Copán, offering a useful guide for researchers interested in conducting similar analyses in other locations, which could be any urban setting anywhere, or indeed any human community. Her study also employs multiple structure types and spatial scales of analysis from the urban core to the surrounding countryside, and generates insights and patterning not always seen at specific levels of resolution.

As Richards-Rissetto's work shows, the results of LCPA are not always as expected. Elites, for example, while living in places where access was spatially restricted, were nonetheless in settings that were among the most visible and had the greatest accessibility to all other members of society, in part because of their often central or strategic locations in the community. They were thus the most connected of the people within Copán, at least in terms of their ability to see or access all other residents, although it is doubtful whether they actually visited many commoner residents. Commoners, in contrast, rarely had ready access to many other members or classes of society, and while they might be able to access the center, their entry into elite precincts was restricted. The difference between commoners and elites was thus reflected, in part, by who could see and hence be seen by the most other members of society. The analyses also demonstrate that some structure/household types were likely misclassified by earlier researchers, at least in terms of who may have lived in them, given their locational characteristics as reflected in calculated integration values. Configurational analyses thus offer an independent means of evaluating site or building classifications, and can be used by themselves or in tandem with other measures to generate such classifications. They can also be used to determine whether social hierarchies were present, as well as document the relative amount of access or control subgroups may have had, based on whether specific ranges of integration values fell within discrete groupings. Such analyses can also be conducted with subsets of site types within different settings, as was done at Copán between the urban core and hinterland, to determine whether the degree of integration differed between the two areas.

Conducting such configurational analyses diachronically, Richards-Rissetto also suggests, would be one way to evaluate whether Copán society became more stratified over time, something indicated by the fact that more and more elites established residences in and near the urban core (Fash 2001). Too often research employing LCPA is synchronic in approach, which may not reflect how movement patterns or settlements themselves changed over time. Configurational analyses, Richards-Rissetto's research demonstrates, are thus useful for the exploration of a wide range of archaeological questions, from establishing site and structure classificatory systems, to evaluating whether social hierarchies were present (both overall and how they operated in different parts of the landscape), to investigating how measures of social hierarchy and integration changed through time. If ancient cities were generative arenas of change, or crucibles of interacting factions in which early state formation occurred, as scholars such as Yoffee (2005:62) have argued, configurational analyses offer a means by which such groups might be recognized and their actions documented.

Sarah Surface-Evans's contribution (Chapter 8) was directed to testing whether Shell Mound Archaic (SMA) sites in the Falls region of the Ohio River were "positioned" within the landscape to access and control key resources. The analysis, through a direct comparison of the two procedures, shows the value of LCPA/corridor cost-surface modeling as opposed to more traditional site-catchment analyses (see also Rademaker et al., Chapter 3). Using cost-weighted corridor analyses as opposed to direct or linear distance pathway analyses, Surface-Evans evaluated how movement was constrained by terrain and other environmental factors. Indeed, Surface-Evans, like others in this volume, as noted previously, provides an effective demonstration that the use of fixed catchment radii as a means of documenting how people accessed the landscape must be replaced by analyses considering the "cost" of access or movement over the landscape. Site-catchment analyses may still prove useful for providing first approximations of the kinds of resources and activities that may have occurred within a specific distance of a given point, but they can no longer be considered an effective way of examining how humans made use of landscapes. Indeed, all such fixed measures linking distance with resource procurement zones, what are sometimes called foraging radii, should be replaced with cost surfaces, or what Surface-Evans calls "cost catchments."

Reconstructing hypothetical travel pathways between shell mound sites and examining river travel and overland corridors, Surface-Evans demonstrates that the SMA sites in her sample were highly accessible to each other. This in turn suggests a high degree of interaction or affinity between the peoples creating these sites. An important part of her research, presented at the conference that led to this volume (Surface-Evans 2009a, 2010) but not reported here, was something also noted by other researchers in this book, that LCPA can be used to suggest areas where previously unknown and undocumented sites might be located on the landscape (see also Rademaker et al., Chapter 3). In particular, she suggests that sites were likely to occur in locations where travel paths converged, or intersected, what she calls "high traffic areas" (Surface-Evans 2009a, 2009b, 2010). The lesson for researchers using LCPA is not just to look along trails for new sites, or where trails pass by or near areas likely to facilitate interaction—such as settings rich in exploitable food resources or prominent physiographic features on the landscape—but also to examine areas where different trails may have intersected.

Isaac Ullah and Sean Bergin (Chapter 9) combine agent-based and landscape process modeling with LCPA to simulate human interactions with the environment, specifically at four hypothetical village locations in the Penaguila Valley of eastern Spain. Their analyses employ high-resolution 5-m DEM data, a fine-grained scale unimaginable even a few years ago, yet one now feasible given advances in computational capabilities. As the authors note, while much LCPA employs an accumulative, or what they call a "cellular automata movement model," in reality almost all human movement is based on some prior knowledge of what lies ahead, save for the very first representatives of colonizing populations entering an area. *Sequentially* optimal LCPA (Rademaker et al., Chapter 3) is thus preferred. Ullah and Bergin's analyses appear decidedly complicated, but as they argue, such a comprehensive approach is critical to realistic simulation; Occam's razor (*lex parsimoniae*) is not a principle that offers much comfort or utility when one is investigating complex phenomena. Significantly, the authors consider dynamic aspects of the landscape in their modeling, such as vegetation cover and soil fertility, erosion, and deposition, as well as changes over time in affected human populations. Explicit details about how the analyses were conducted are provided, with reference to computational algorithms as well as programs. Their analyses, as they explicitly state, are designed to assist archaeologists in understanding "the effects of village location on agropastoral economy, village population, and landscape change," serving as a "hypothesis generator against which to test data from the archaeological record." The procedures should prove useful for examining the establishment and impact of initial Neolithic farming populations in the region, about which much is already known (e.g., Barton, Schmich, and James 2004; Barton, Bernabeu, Aura, and Molina 2004). I look forward to analyses linking the modeling with the local archaeological record, and the new understanding such an approach will undoubtedly provide.

Patrick Livingood's (Chapter 10) reevaluation of David Hally's (1993, 1999, 2006) classic work on Mississippian mound center spacing in northern Georgia is an elegant demonstration of the value of applying new analytical procedures to old datasets and observations developed from them. Livingood uses LCPA to evaluate whether cost-distance measures should replace the straight-line geographic distance measures Hally arrived at twenty years ago linking the spacing and political organization of complex societies in the region. Hally (1993) documented two major groupings of sites, mound centers located within 18 km of one another that he argued were likely within the same polity, and centers much farther away (>32 km) that were assumed to be within different polities. Happily, Livingood's work, using travel time as opposed to distance, both supports and improves on the original model, revealing a clear bimodal grouping of centers comparable to that found by Hally. In Livingood's analysis, the break came between those centers within six hours' travel time of each other (one way) and those beyond this range. Livingood looked at both overland and waterborne travel, and examined factors influencing the rate of travel, such as the delays involved in crossing rivers and whether travel was on trail or off trail. Livingood's analyses, significantly, include measures derived from real people traversing the landscape, in this case from historic records of canoe trips down the Mississippi. He also argues the need to consider the occurrence of swamps, logjams, falls, and shoals; river ways were not the unimpeded routes we sometimes think they were, but were often more difficult to move along than overland routes. As he notes, we also need to consider movement by water as anisotropic in nature, more difficult in one direction than another, as anyone who has paddled a canoe with and against a current can testify. Water travel appears to have been important in North America as far back as the Paleoindian era (Jodry 2005) and is thus a form of movement that LCPA should incorporate wherever possible.

I can sympathize with the extensive computational time Livingood noted (in his original presentation) that it took to complete his analyses. The pathway analyses Chris Gillam and I conducted in the mid-1990s evaluating Paleoindian colonization routes over the Americas took massive amounts of time to complete with the technology then available—half a day or more, given the millions of 1-km grid cells we were employing for each continent—and problems such as power fluctuations often literally short-circuited analyses in midstream, requiring us to start over again many times. Happily, LCPA is now much more routine and, with improved computers and software, nowhere near as time-consuming as it once was. If we are to develop and evaluate ever more sophisticated models of past human behavior, however, we are going to need to continue to refine our theoretical approaches, our field data collection efforts, and our computational algorithms.

The final case study in the book, by Devin White (Chapter 11), examines prehistoric and historic trails in the Western Papaguería region of the U.S. Southwest, showing how they were both transportation networks and a means of local and regional integration. As in the analyses from the eastern part of the Southwest by Phillips and Leckman (Chapter 4), water, useful raw materials, and settlement locations strongly dictated trail network occurrence. White's study explores an array of research questions using multiple forms of analysis, including remote sensing, field survey, and GIS-based predictive modeling. Predictive modeling was used to generate efficient movement pathways between settlements and water sources, which were then compared with trail segments located using imagery followed by on-the-ground examination, or truthing. This proved a highly effective means of locating past trails and reconstructing larger networks from surviving segments. The analysis was facilitated by the presence of landscapes favoring trail preservation (i.e., desert pavement) coupled with fairly minimal recent human disturbance of the terrain in the study area. Finding archaeological signatures of trails in more densely vegetated environments, as some of the chapters in this book demonstrate (Nolan and Cook, Chapter 5; Surface-Evans, Chapter 8), is a major and still largely unresolved challenge facing archaeologists (Phillips and Leckman, Chapter 4). White's methods for examining imagery to resolve terrain types, and from there possible trail segments, should have widespread utility around the world, and ideally not just in sparsely vegetated environments.

White's use of movement coefficients weighted by terrain characteristics, such as whether one was walking on hard ground or loose soil (e.g., Pandolf et al. 1977; Santee et al. 2001), or upslope or downslope, or the weight of the load being carried—together with his consideration that different types of trails would be likely to occur given the type and scale of interaction and the kinds and dating of the sites and activities being connected—exemplifies the type of diachronic, multivariate, and multiscalar LCPA done today. Indeed, his research demonstrates that single variables, however important they may seem (i.e., the location of water sources in a desert environment), probably only rarely account for all the variation observed. As White, and indeed many of the contributors to this volume show, modeling interaction requires consideration of not just energetic efficiency but other factors such as "sacred/ritual" considerations that bind local communities; wherever possible, we should consider entire networks rather than focus on local segments or areas.

14.3. Conclusion

The chapters in this book offer innovative and thoughtful applications of LCPA, and the future looks bright for this form of analysis in archaeology. Indeed, I expect that the research examples presented here will help generate many new studies. In recent years, with Chris Gillam and other colleagues, I have continued to conduct LCPA, looking at changes in the location of trail networks over time in eastern North America, during both the prehistoric and early historic eras (Anderson et al. 2007) (Plates 13 and 14). These analyses, which look at the flow of materials into and out of major prehistoric centers such as Poverty Point in Louisiana, the Scioto River valley of southern Ohio, and Cahokia in Illinois, during the Late Archaic, Middle Woodland, and Mississippian periods, respectively, demonstrate rather conclusively that road or trail networks are shaped as much by regional political geography — the location, size, spacing, and influence of centers on the landscape, and their relationship to one another — as by physiography or natural resource occurrence. Aspects of the so-called natural environment are certainly important: transportation routes clearly make use of mountain passes and fords, follow river valleys, and often lead to locations where large numbers of people can make a living from the bounty of the surrounding landscape. But trail networks in eastern North America also changed over time. When the Poverty Point site was at its height ca. 3200 years ago, people apparently came and went from northeastern Louisiana in large numbers, probably in much greater numbers as a proportion of the regional population than do so today, although New Orleans, to the south, is a good example of a likely modern-day equivalent. In later times, movement corridors changed, as first Hopewellian and then Mississippian centers rose and fell on the landscape. Only rarely were centers or large numbers of people located in many of these areas previously. In the early historic era, in turn, many trading paths as well as Indian trails led to previously all-but-ignored places on the eastern seaboard such as Charlestown, St. Augustine, and Jamestown (Myer 1928; Tanner 1989). Modern railroad and interstate corridors in turn were built to tie new manufacturing and population centers together, sometimes following older routes, but sometimes not.

Formal roads might be replaced by less formal trails, or vice versa, when environmental or political conditions changed. These analyses of regional interaction provide important lessons: first, not all roads or trails were straight, even though that is perhaps the view permeating popular interpretation, and second, neither ideologically significant directionality nor transport efficiency were the sole explanations for the location of pathways. Third, as White (Chapter 11) demonstrates, while long-distance trail networks may have spanned an area, where sites were located often shaped local patterns of movement or routes that trumped efficiencies at larger scales. Fourth, even when a site was reduced in influence or abandoned, its presence (at least in some instances) may have continued to shape movement. In the Southeast, for example, major Mississippian period mound centers such as Etowah and Moundville underwent periods of reduced or changed use, from large towns to more vacant centers with smaller resident populations; both, however, were apparently still considered important by surrounding populations (Knight and Steponaitis 1998; King 2003). Some sites that were once major centers were, in fact, consciously avoided because there were prohibitions against visiting them. In the Southwest, the perception of Chaco Canyon among some contemporary Puebloan peoples, as a place where bad things happened, as Steven Lekson (1999, 2008:198–200) has suggested, may help explain why the site was avoided by later peoples. In the twelfth century, however, as Kantner (1997, 2004, Chapter 13) and others have classically documented, many very real roads led to Chaco.

A final major lesson these LCPA examples from prehistoric eastern North America provide, subsuming the others, is that interaction networks must be considered diachronically. That is, they are profoundly shaped by changes in regional political geography, which must be carefully monitored. All roads may have led to Chaco, or to Rome, or to Poverty Point or Cahokia when these centers were at their height, but when they declined, the roads leading to them often did so as well, as did the relationships between the peoples they connected. Interaction networks, like the centers themselves, apparently took appreciable time to create, maintain, or — once lost and if conditions permitted — reconstitute. Often this occurred in new locations, necessitating new trails. Portions of these new routes may have reappeared in the same places where older trails ran, but this was typically when movement was in the same direction or when it was constrained by physiographic features such as river courses, fords, or mountain passes. Population distributions and political conditions were more important in shaping movement, and these changed over time (Anderson 2010:286–287), something archaeologists need to consider when conducting LCPA.

LCPA can also evaluate the efficiency or lack thereof of the routes early explorers took across the landscape. One of the first least cost pathway analyses conducted in archaeology, by Fred Limp (1990), examined the route of a portion of

Hernando DeSoto's expedition in Arkansas in 1541. As noted, it inspired the work Chris Gillam and I did later in that decade examining possible colonization pathways in the remote past (Anderson and Gillam 2000) and offers a means by which more recent expeditions, military campaigns, or trade routes can be potentially reconstructed in the absence of precise historic documentation. Least cost pathway analysis, as the chapters in this book illustrate, has much to tell us about how, why, and where peoples move around on the landscape, both in the past and in the shaping of our current world.

Acknowledgments

I thank Sarah Surface-Evans and Devin White for inviting me to participate in the session and for their patience in awaiting this chapter. A special debt of thanks is owed to my good friend and colleague Chris Gillam, who has worked with me and guided me on least cost pathway analysis for many years and who provided final versions of the figures included herein.

References

Anderson, David G.

1990 The Paleoindian Colonization of Eastern North America: A View from the Southeastern United States. In *Early Paleoindian Economies of Eastern North America*, edited by Kenneth Tankersley and Barry Isaac, pp. 163–216. *Research in Economic Anthropology* Supplement 5. JAI Press, Greenwich, Connecticut.

1994 *The Savannah River Chiefdoms: Political Change in the Late Prehistoric Southeast.* University of Alabama Press, Tuscaloosa.

2010 The End of the Southeastern Archaic: Regional Interaction and Archaeological Interpretation. In *Trend, Tradition, and Turmoil: What Happened to the Southeastern Archaic?* edited by David Hurst Thomas and Matthew C. Sanger, pp. 273–302. Proceedings of the Third Caldwell Conference, St. Catherines Island, Georgia, May 9–11, 2008. Anthropological Papers Vol. 93. American Museum of Natural History, New York.

n.d. Monumentality in Eastern North America during the Mississippian Period. In *Early New World Monumentality*, edited by Richard L. Burger and Robert M. Rosenwig. University Press of Florida, Gainesville, in press.

Anderson, David G., and E. Suzanne Carter

1985 Archaeological Investigations at Four Open Air Sites in Eastern El Paso, Texas. In *Views of the Jornada Mogollon: Proceedings of the Second Jornada Mogollon Archaeology Conference*, edited by C. M. Beck, pp. 78–106. Contributions in Anthropology 12. Eastern New Mexico University, Portales.

Anderson, David G., and J. Christopher Gillam

1997 Paleoindian Colonization of the New World: Implications from an Examination of Physiography, Artifact Distributions, and Demography. Paper presented at the 62nd Annual Meeting of the Society for American Archaeology, Nashville, Tennessee.

2000 Paleoindian Colonization of the Americas: Implications from an Examination of Physiography, Demography, and Artifact Distribution. *American Antiquity* 65(1):43–66.

2001 Paleoindian Interaction and Mating Networks: Reply to Moore and Moseley. *American Antiquity* 66(3):530–535.

Anderson, David G., J. Christopher Gillam, Christopher Carr, Thomas E. Emerson, and Jon L. Gibson

2007 Resolving Interaction Networks in Eastern North America. Paper presented at the 72nd Annual Meeting of the Society for American Archaeology, Austin, Texas.

Anderson, David G., D. Shane Miller, Stephen J. Yerka, J. Christopher Gillam, Erik N. Johanson, Derek T. Anderson, Albert C. Goodyear, and Ashley M. Smallwood

2010 PIDBA (Paleoindian Database of the Americas) 2010: Current Status and Findings. *Archaeology of Eastern North America* 38:63–90.

Anderson, David G., and Steven D. Smith

2003 *Archaeology, History, and Predictive Modeling: Research on Fort Polk, 1972–2002.* University of Alabama Press, Tuscaloosa.

Anderson, David G., David W. Stahle, and Malcolm R. Cleaveland

1995 Paleoclimate and the Potential Food Reserves of Mississippian Societies: A Case Study from the Savannah River Valley. *American Antiquity* 60:258–286.

Anderson, David G., Stephen J. Yerka, and J. Christopher Gillam

2010 Employing High Resolution Bathymetric Data to Infer Possible Migration Routes of Pleistocene Populations. *Current Research in the Pleistocene* 27:60–64.

Andrefsky, William, Jr.
2005 *Lithics: Macroscopic Approaches to Analysis.* 2nd ed. Cambridge University Press, Cambridge.
2009 The Analysis of Stone Tool Procurement, Production, and Maintenance. *Journal of Archaeological Research* 17:65–103.
Anthony, David W.
1990 Migration in Archeology: The Baby and the Bathwater. *American Anthropologist* 92:895–914.
Banks, William E., Francesco d'Errico, Harold Dibble, Leonard Krishtalka, Dixie West, Deborah I. Olszewski, A. Townsend Peterson, David G. Anderson, J. Christopher Gillam, Anta Montet-White, Michel Crucifix, Curtis W. Marean, María-Fernanda Sánchez-Goñi, Barbara Wohlfarth, and Marian Vanhaeran
2006 Eco-Cultural Niche Modeling: New Tools for Reconstructing the Geography and Ecology of Past Human Populations. *PaleoAnthropology* 2006:68–83.
Barton, C. M., J. Bernabeu, J. Emili Aura, and L. Molina
2004 Historical Contingency, Nonlinearity, and the Neolithization of the Western Mediterranean. In *Current Issues in Mediterranean Landscape Archaeology*, edited by L. Wandsnider and E. Athanassopoulos, pp. 99–124. University of Pennsylvania Press, Philadelphia.
Barton, C. M., J. Bernabeu Auban, O. Garcia Puchol, S. Schmich, and L. Molina Balaguer
2004 Longterm Socioecology and Contingent Landscapes. *Journal of Archaeological Method and Theory* 11:253–395.
Barton, C. Michael, Steven Schmich, and Steven R. James
2004 The Ecology of Human Colonization in Pristine Landscapes. In *The Settlement of the American Continents: A Multidisciplinary Approach to Human Biogeography*, edited by C. Michael Barton, Geoffrey A. Clark, David R. Yesner, and Georges A. Pearson, pp. 138–161. University of Arizona Press, Tucson.
Beaton, J. M.
1991 Colonizing Continents: Some Problems from Australia and the Americas. In *The First Americans: Search and Research*, edited by T. D. Dillehay and D. J. Meltzer, pp. 209–230. CRC Press, Boca Raton, Florida.
Benson, Larry V., Timothy R. Pauketat, and Edward R. Cook
2009 Cahokia's Boom and Bust in the Context of Climate Change. *American Antiquity* 74: 467–483.
Blitz, John H., and Karl G. Lorenz
2006 *The Chattahoochee Chiefdoms.* University of Alabama Press, Tuscaloosa.
Brady-Rawlins, Kathleen
2007 The O.C. Voss Site: Reassessing What We Know about the Fort Ancient Occupation of the Central Scioto Drainage and Its Tributaries. Unpublished Ph.D. dissertation, Ohio State University, Columbus.
Bustard, Wendy
2003 Pueblo Bonito — When a House Is Not a Home. In *Pueblo Bonito: Center of the Chacoan World*, edited by Jill Neitzel, pp. 80–92. Smithsonian Institution Press, Washington, D.C.
Conkey, Margaret W.
1980 The Identification of Prehistoric Hunter-Gatherer Aggregation Sites: The Case of Altamira. *Current Anthropology* 21:609–630.
Cook, Edward R., Connie A. Woodhouse, C. Mark Eakin, David M. Meko, and David W. Stahle
2004 Long-Term Aridity Changes in the Western United States. *Science* 306:1015–1018.
Cook, Robert A.
2008 *Sunwatch: Fort Ancient Development in the Mississippian World.* University of Alabama Press, Tuscaloosa.
Cooper, Laurel M.
1995 Space Syntax Analysis of Chacoan Great Houses. Unpublished Ph.D. dissertation, University of Arizona, Tucson.
Copé, Silvia M.
2007 El uso de la arquitectura como artefacto en el estudio de paisajes arqueológicos del altiplano sur brazilieño, Río Grande do Sul. *Revista de Arqueología* 2:15–34.
Dahl, Östen, J. Christopher Gillam, David G. Anderson, José Iriarte, and Silvia M. Copé
2011 Linguistic Diversity Zones and Cartographic Modeling: GIS as a Method for Understanding the Prehistory of Lowland South America. In *Ethnicity in Ancient Amazonia: Reconstructing Past Identities from Archaeology, Linguistics, and Ethnohistory*, edited by Alf Hornborg and Jonathan D. Hill. University of Colorado Press, Boulder.
Erlandson, Jon M.
2002 Anatomically Modern Humans, Maritime Voyaging, and the Pleistocene Colonization of the Americas. In *The first Americans: The Pleistocene Colonization of the New World*, edited by Nina G. Jablonski, pp. 59–92. Memoirs No. 27, Wattis Symposium Series in

Anthropology. California Academy of Sciences, San Francisco.

Fash, William L.

2001 *Scribes, Warriors, and Kings: The City of Copán and the Ancient Maya.* Thames and Hudson, London.

Gillam, J. Christopher

2008 Prospectus on Archaeological Geographic Information System (GIS) Applications and Modeling in the Japanese Sea Basin. In *Neolithic and Neolithisation in the Japanese Sea Basin: Individual and the Historical Landscape*, pp. 57–63. Far East National University, Vladivostok, Russia.

2009 Empirical Modeling of Cultural Landscapes and Their Environmental Impact. In *Neolithisation and Landscape: NEOMAP International Workshop*, edited by Junzo Uchiyama, Kati Lindström, Carlos Zeballos, and Keisuke Makibayashi, pp. 25–44. Research Institute for Humanity and Nature, Kyoto, Japan.

Gillam, J. Christopher, David G. Anderson, and A. Townsend Peterson

2007 A Continental-Scale Perspective on the Peopling of the Americas: Modeling Geographic Distributions and Ecological Niches of Pleistocene Populations. *Current Research in the Pleistocene* 24:86–90.

Gillam, J. Christopher, David G. Anderson, Stephen J. Yerka, and Shane Miller

2006 Estimating Pleistocene Shorelines and Land Elevations for North America. *Current Research in the Pleistocene* 23:207–208.

Gillam, J. Christopher, and Andrei Tabarev

2004 On the Path of Upper Paleolithic Obsidians in the Russian Far East. *Current Research in the Pleistocene* 21:3–5.

Gillam, J. Christopher, Andrei V. Tabarev, Masami Izuho, Yuichi Nakazawa, Chen Quanjia, Batmunkh Tsogtbaatar, and Yongwook Yoo

2008 The Far East Archaeological Database (FEAD): A Maximum 1-Minute Resolution Database for Exploring the Big Picture. *Current Research in the Pleistocene* 25:197–200.

Goodyear, A. C.

1979 *A Hypothesis for the Use of Cryptocrystalline Raw Materials among Paleoindian Groups of North America.* Research Manuscript Series 156, South Carolina Institute of Archaeology and Anthropology. University of South Carolina, Columbia.

Gould, Richard, and Sherry Saggers

1985 Lithic Procurement in Central Australia: A Closer Look at Binford's Idea of Embeddedness in Archaeology. *American Antiquity* 50: 117–136.

Griffin, James B.

1943 *The Fort Ancient Aspect: Its Cultural and Chronological Position in Mississippi Valley Archaeology.* Museum Papers 20. University of Michigan, Ann Arbor.

Grøn, Ole

1991 A Method for Reconstruction of Social Organization in Prehistoric Societies and Examples of Practical Application. In *Social Space: Human Spatial Behavior in Dwellings and Settlements*, edited by Ole Grøn, Ericka Engelsted, and Inge Lindblom, pp. 100–117. Odense University Press, Odense, Denmark.

Hally, David J.

1993 The Territorial Size of Mississippian Chiefdoms. In *Archaeology of Eastern North America: Papers in Honor of Stephen Williams*, edited by James B. Stoltman, pp. 143–168. Mississippi Department of Archives and History, Jackson.

1999 The Settlement Pattern of Mississippian Chiefdoms in Northern Georgia. In *Settlement Pattern Studies in the Americas: Fifty years since Virú*, edited by Brian R. Billman and Gary M. Feinman, pp. 96–115. Smithsonian Institution Press, Washington, D.C.

2006 Nature of Mississippian Regional Systems. In *Light on the Path: The Anthropology and History of the Southeastern Indians*, edited by Thomas J. Pluckhahn and Robbie Ethridge, pp. 26–42. University of Alabama Press, Tuscaloosa.

Hillier, Bill

1999 *Space Is the Machine: A Configurational Theory of Architecture.* Cambridge University Press, Cambridge.

Hillier, Bill, and Julienne Hanson

1984 *The Social Logic of Space.* Cambridge University Press, Cambridge.

Hillier, Bill, A. Penn, J. Hanson, T. Grajewski, and J. Xu

1993 Natural Movement; or, Configuration and Attraction in Urban Pedestrian Movement. *Environment and Planning B: Planning and Design* 20:29–66.

Holley, George R.

1999 Late Prehistoric Towns in the Southeast. In *Great Towns and Regional Polities in the Prehistoric American Southwest and Southeast*, edited by Jill E. Neitzel, pp. 22–38. Amerind Foundation New World Study Series 3. University of New Mexico Press, Albuquerque.

Hudson, Charles M.

1990 *The Juan Pardo Expeditions: Exploration of the*

Carolinas and Tennessee, 1566–1568. Smithsonian Institution Press, Washington, D.C.
1997 *Knights of Spain, Warriors of the Sun: Hernando de Soto and the South's Ancient Chiefdoms*. University of Georgia Press, Athens.
Jantz, Richard L., Paul Marr, and Claire A. Jantz
2010 Body Proportions in Recent Native Americans: Colonization History versus Ecogeographic Patterns. In *Human Variation in the Americas: The Integration of Archaeology and Biological Anthropology*, edited by Benjamin M. Auerbach, pp. 292–310. Occasional Paper 38. Center for Archaeological Investigations, Southern Illinois University, Carbondale.
Jodry, Margaret A.
2005 Envisioning Water Transport Technology in Late Pleistocene America. In *Paleoamerican Origins: Beyond Clovis*, edited by Robsen Bonnichsen, Bradley T. Lepper, Dennis Stanford, and Michael R. Waters, pp. 29–42. Center for the Study of the First Americans. Texas A&M University Press, College Station.
Kantner, John
1997 Ancient Roads, Modern Mapping: Evaluating Chaco Anasazi Roadways Using GIS Technology. *Expedition* 39(3):49–62.
2004 *Ancient Puebloan Southwest*. Cambridge University Press, Cambridge.
Kelly, Robert L.
1995 *The Foraging Spectrum: Diversity in Hunter-Gatherer Lifeways*. Smithsonian Institution Press, Washington, D.C.
2003 Colonization of New Land by Hunter-Gatherers: Expectations and Implications Based on Ethnographic Data. In *Colonization of Unfamiliar Landscapes: The Archaeology of Adaptation*, edited by M. Rockman and J. Steele, pp. 44–58. Routledge, New York.
King, Adam
2003 *Etowah: The Political History of a Chiefdom Capital*. University of Alabama Press, Tuscaloosa.
King, Adam, Chester P. Walker, F. Kent Reilly III, Robert V. Sharp, and Duncan P. McKinnon
2011 Remote Sensing from Etowah's Mound A: Architecture and the Re-creation of Mississippian Tradition. *American Antiquity* 76:355–371.
Knight, Vernon J., Jr., and Vincas P. Steponaitis (editors)
1998 *Archaeology of the Moundville Chiefdom*. Smithsonian Institution Press, Washington, D.C.
Krist, F. J., Jr., and D. G. Brown
1994 GIS Modeling of Paleo-Indian Period Caribou Migrations and Viewsheds in Northeastern Lower Michigan. *Photogrammetric Engineering and Remote Sensing* 65(9):1129–1137.
Lekson, Stephen H.
1999 *The Chaco Meridian: Centers of Political Power in the Ancient Southwest*. AltaMira Press, Walnut Creek, California.
2008 *A History of the Ancient Southwest*. School for Advanced Research Press, Santa Fe, New Mexico.
Limp, W. F.
1990 Digital DeSoto: An Automated Analysis on Alternative Routes. Paper presented at the Second DeSoto Conference, University of Arkansas Museum, Fayetteville.
Meeks, Scott C., and David G. Anderson
2006 The Vacant Quarter Hypothesis: Changing Late Prehistoric Political Landscapes in the Tennessee River Valley from a Paleoecological Perspective. Paper presented at the XIX ANQUA Biennial Meeting, Bozeman, Montana.
2007 Large Scale Late Prehistoric Abandonments in the Southeastern United States: Examining Mississippian Demographic Shifts from an Ecological Perspective. Paper presented at the 72nd Annual Meeting of the Society for American Archaeology, Austin, Texas.
Meltzer, David J.
2002 What Do You Do When No One's Been There Before? Thoughts on the Exploration and Colonization of New Lands. In *The First Americans: The Pleistocene Colonization of the New World*, edited by Nina G. Jablonski, pp. 27–58. Memoirs No. 27, Wattis Symposium Series in Anthropology. California Academy of Sciences, San Francisco.
Moore, John, and Michael Moseley
2001 How Many Frogs Does It Take to Leap around the Americas? Comments on Anderson and Gillam. *American Antiquity* 66:526–529.
Myer, William E.
1928 Indian Trails of the Southeast. *Bureau of American Ethnology Annual Report* 42:727–857.
Nolan, Kevin C., and Robert A. Cook
2010 An Evolutionary Model of Social Change in the Middle Ohio Valley: Was Social Complexity Impossible during the Late Woodland but Mandatory during the Late Prehistoric? *Journal of Anthropological Archaeology* 29: 62–79.
Pandolf, K. B., B. Givoni, R. F. Goldman
1977 Predicting Energy Expenditure with Loads while Standing or Walking Very Slowly. *Journal of Applied Physiology* 43(4):577–581.

Pauketat, Timothy R.
2007 *Chiefdoms and Other Archaeological Delusions.* AltaMira Press, Lanham, Maryland.

Preziosi, Donald
1979 *The Semiotics of the Built Environment: An Introduction to Architectonic Analysis.* Indiana University Press, Bloomington.

Rademaker, Kurt
2006 Geoarchaeological Investigations of the Wayñuna Site and the Alca Obsidian Source, Peru. Unpublished Master's thesis, University of Maine, Orono.

Richards-Rissetto, Heather
2010 Exploring Social Interaction at the Ancient Maya City of Copán, Honduras: A Multi-scalar Geographic Information Systems (GIS) Analysis of Access and Visibility. Unpublished Ph.D. dissertation, University of New Mexico, Albuquerque.

Rissetto, John D.
2009 Late Pleistocene Hunter-Gatherer Lithic Exploitation and Mobility Patterns in Eastern Cantabria, Spain. Unpublished Ph.D. dissertation, University of New Mexico, Albuquerque.

Santee, W. R., W. F. Allison, L. A. Blanchard, and M. G. Small
2001 A Proposed Model for Load Carriage on Sloped Terrain. *Aviation, Space and Environmental Medicine* 72(6):562–566.

Sauer, Carl O.
1944 A Geographical Sketch of Early Man in America. *Geographical Review* 34:543–554.

Smith, Bruce D.
1978 *Mississippian Settlement Patterns.* Academic Press, New York.

Snead, James E.
2002 Ancestral Pueblo Trails and the Cultural Landscape of the Pajarito Plateau. *Antiquity* 76(293):756–765.
2008 *Ancestral Landscapes of the Pueblo World.* University of Arizona Press, Tucson.

Steele, James, and Marcy Rockman
2003 "Where Do We Go from Here?" Modeling the Decision-Making Process during Exploratory Dispersal. In *Colonization of Unfamiliar Landscapes: The Archaeology of Adaptation*, edited by Marcy Rockman and James Steele, pp. 130–143. Routledge, New York.

Suárez, Rafael, and J. Christopher Gillam
2008 The Paleoindian Database of Uruguay: Collections Survey and GIS Data Development. *Current Research in the Pleistocene* 25:200–202.

Surface-Evans, Sarah
2009a This Way or That Way: Hypothetical Least-Cost Models of Regional Interaction among Shell Mound Archaic Settlements in the Central Ohio Valley. Paper presented at the 74th Annual Meeting of the Society for American Archaeology, Atlanta.
2009b Hunter-Gatherer Cultural Landscapes: A Case Study for a GIS-Based Reconstruction of the Shell Mound Archaic in the Falls of the Ohio Region of Indiana and Kentucky. Unpublished Ph.D. dissertation, Michigan State University, East Lansing.
2010 Inferring Agency from Landscape-Scale Modeling of Archaic Hunter-Gatherers. Paper presented at the 75th Annual Meeting of the Society for American Archaeology, St. Louis, Missouri.

Surovell, Todd
2009 *Toward a Behavioral Ecology of Lithic Technology: Cases from Paleoindian Archaeology.* University of Arizona Press, Tucson.

Tanner, Helen Hornbeck
1989 The Land and Water Communication Systems of the Southeastern Indians. In *Powhatan's Mantle: Indians in the Colonial Southeast*, edited by Peter H. Wood, Gregory A. Waselkov, and M. Thomas Hatley, pp. 6–20. University of Nebraska Press, Lincoln.

Uchiyama, Junzo, Peter Jordan, and J. Christopher Gillam
2009 Understanding Neolithization of East Asian Cultural Landscapes. Paper presented at the 74th Annual Meeting of the Society for American Archaeology, Atlanta.

Van Dyke, Ruth M.
1999 Space Syntax Analysis at the Chacoan Outlier of Guadalupe. *American Antiquity* 64(3):461–473.

Vita-Finzi, C., and E. S. Higgs
1970 Prehistoric Economy in the Mount Carmel Area of Palestine: Site Catchment Analysis. *Proceedings of the Prehistoric Society* 36:1–37.

Waldron, John, and Elliot M. Abrams
1999 Adena Burial Mounds and Inter-hamlet Visibility: A GIS Approach. *Midcontinental Journal of Archaeology* 24(1):97–111.

Ware, J. A., and G. J. Gumerman
1977 Remote Sensing Methodology and the Chaco Canyon Prehistoric Road System. In *Aerial Remote Sensing Techniques in Archeology*, edited by T. R. Lyons and Robert K. Hitchcock, pp. 135–168. Reports of the Chaco Center Vol. 2. Chaco Center, Albuquerque, New Mexico.

White, Devin
2007 Transportation, Integration, Facilitation:

Prehistoric Trails Network of the Western Paraguería. Unpublished Ph.D. dissertation, University of Colorado, Boulder.

Whitley, Thomas G., and L. M. Hicks
2003 A Geographic Information Systems Approach to Understanding Potential Prehistoric and Historic Travel Corridors. *Southeastern Archaeology* 22(1):77–91.

Winterhalder, Bruce
1986 Diet Choice, Risk, and Food Sharing in a Stochastic Environment. *Journal of Anthropological Archaeology* 5:369–392.

Yoffee, Norman
2005 *Myths of the Archaic State: Evolution of the Earliest Cities, States, and Civilizations*. Cambridge University Press, Cambridge.

Contributors

David G. Anderson
Department of Anthropology
University of Tennessee

Sean M. Bergin
School of Human Evolution and Social Change
Arizona State University

Scott Branting
The Oriental Institute and Department
of Near Eastern Languages and Civilizations
University of Chicago

Gordon R. M. Bromley
Lamont-Doherty Earth Observatory
Columbia University

Robert A. Cook
Department of Anthropology
The Ohio State University

Erin J. Hudson
Cibola National Forest

John Kantner
School for Advanced Research

Phillip O. Leckman
Statistical Research, Inc.

Patrick Livingood
Department of Anthropology
University of Oklahoma

Kevin C. Nolan
Applied Archaeology Laboratories
Department of Anthropology
Ball State University

Shaun M. Phillips
Statistical Research, Inc.

Kurt Rademaker
Climate Change Institute and
Department of Anthropology
University of Maine

David A. Reid
Department of Anthropology
University of Maine

Heather Richards-Rissetto
HUMlab, Umeå University, Sweden, and
Department of Anthropology
University of New Mexico

John D. Rissetto
Department of Anthropology
University of New Mexico

Sarah L. Surface-Evans
Department of Sociology, Anthropology,
and Social Work
Central Michigan University

Isaac I. Ullah
School of Human Evolution and Social Change
Arizona State University

Devin A. White
Integrity Applications Incorporated and
Crow Canyon Archaeological Center

Index